CHINESE
FOREIGN POLICY

CHINESE
FOREIGN POLICY

PRAGMATISM AND STRATEGIC BEHAVIOR

FOREWORD BY JONATHAN POLLACK

SUISHENG ZHAO, EDITOR

An East Gate Book

M.E.Sharpe
Armonk, New York
London, England

An East Gate Book

Copyright © 2004 by Suisheng Zhao

Library of Congress Cataloging-in-Publication Data

Chinese foreign policy : pragmatism and strategic behavior / Edited by Suisheng Zhao.
   p. cm.
"An East gate book."
ISBN 0-7656-1284-4 (cloth: alk. paper) — ISBN 0-7656-1285-2 (pbk.: alk. paper)
   1. China—Foreign relations—1976– I. Zhao, Suisheng.

DS779.27 .C532 2003
327.51—dc21   2003010362

Printed in the United States of America

∞

BM (c) 10 9 8 7 6 5 4 3 2 1
BM (p) 10 9 8 7 6 5 4 3

To my mother, brother, and sisters

Contents

List of Tables and Figures

Tables

Figures

List of Abbreviations

ABM	Anti-Ballistic Missile
APR	Asian Pacific Region
ARF	ASEAN Regional Forum
ASEAN	Association of Southeast Asian Nations
CAEA	China Atomic Energy Agency
CBM	Confidence building measures
CCP	Chinese Communist Party
CMC	Central Military Commission
COSTIND	Commission on Science, Technology, and Industry for National Defense
CPV	Chinese People's Volunteers
CTBT	Comprehensive Test Ban Treaty
CWC	Chemical Weapons Convention
EU	European Union
IAEA	International Atomic Energy Agency
HKSAR	Hong Kong Special Administration Region
KMT	Kuomintang or Nationalist Party
KPA	Korean People's Army
MFA	Ministry of Foreign Affairs
MFN	Most Favored Nation
MOFTEC	Ministry of Foreign Trade and Economic Cooperation
MTCR	Missile Technology Control Regime
NATO	North Atlantic Treaty Organization
NEFDA	Northeast Frontier Defense Army
NIEs	Newly Industrialized Economies
NMD	National Missile Defense
NPC	National People's Congress
NPT	Nonproliferation of Nuclear Weapons Treaty
NSG	Nuclear Suppliers Group
PLA	People's Liberation Army
RFE	Russian Far East
SCO	Shanghai Cooperative Organization
SOEs	State-owned enterprises
DPRK	Democratic People's Republic of Korea
TMD	Theater Missile Defense
UNSCR	United Nations Security Council Resolution
WMD	Weapons of Mass Destruction
WTO	World Trade Organization

Foreword

This collection of essays attests to the increasing vigor and diversity of perspectives in the study of Chinese foreign policy. Drawing on the contributions by a distinguished group of scholars from the United States and Asia, Suisheng Zhao has skillfully elucidated some of the pivotal policy dilemmas confronting China as it assumes a much larger international position and role. As characterized by Professor Zhao, the debate about uniqueness versus commonality in the study of international politics seems ever less relevant in analyzing and describing Chinese external behavior. The abundance and complexity of China's policy challenges and the wealth of data sources on Beijing's foreign policies have definitively answered this question. The study of China may be rooted in particular circumstances and constraints, but the factors and forces shaping its behavior are neither unique nor obscure.

Professor Zhao argues that the analysis of Chinese foreign policy mirrors the larger debate between realism and liberalism among international relations scholars. This debate reflects policy as much as theory. In Professor Zhao's view, pragmatic policy considerations increasingly dominate China's external behavior, with less attention to either ideological or value-based approaches. He also favors an international systems approach, as distinct from an emphasis on the domestic sources of foreign policy. Yet Professor Zhao does not seek to impose an overly rigid framework. Many of the contributions to this volume highlight that there is no need for an "either-or" analytic choice. The relevance of various approaches reflects the issues under study, and the explanatory power afforded by different perspectives. This approach contributes to the richness and diversity of the individual chapters.

Among the numerous themes discussed in this volume, the manifestations and implications of China's enhanced power in a systemic and regional context seems especially pivotal. In essence, China's emergent international policies constitute a test case of competing approaches in the study of international relations. At the same time, however, many of the authors remain keenly attentive to the overlay of past history in Chinese policy deliberations. Even as the contributors show appropriate attention to how Chinese behavior is increasingly shaped by changes in power relationships, all recognize the need to place this awareness in its larger historical context.

Many of the essays in this volume first appeared in the *Journal of Contemporary China*. Though still a relatively new publication, it has emerged under Professor Zhao's editorship as an important and timely source of scholarship and policy debate. In particular, the journal has provided an outlet for younger scholars both in the United States and China to present new approaches and sources of data to broader professional scrutiny. This volume as a whole reveals that the study of Chinese foreign policy is both vibrant and diverse. As such, this collection constitutes a very welcome addition to a range of issues that will increasingly shape the study and practice of international politics and foreign policy.

Jonathan D. Pollack

Introduction

1

Chinese Foreign Policy

Pragmatism and Strategic Behavior

Suisheng Zhao

After more than a century of struggle with economic weakness and political turmoil, China entered the twenty-first century as a rising power thanks to the progress of market-oriented economic reform. While many Western businessmen welcomed the massive economic opportunities provided by China's rise, China's long-term great power potential has prompted politicians in some Western capitals and Asian countries to wonder whether an increasingly strong and assertive China would become a rational, peaceful, and pragmatic power or an irrational, bellicose, and expansionist state. Scholars and policy analysts have also taken different positions.[1] Some have been alarmed and argued that the rising economic and military power of China by its own accord makes China a threat to Asian and global security because it may upset the balance of power and spark realignments in East Asia as well as the world. In particular, the neo-conservativists in the Bush administration of the United States have warned of the prospect of China emerging as a great power to challenge American predominance in the post–Cold War world and seek extended sovereignty by launching aggressive warfare against its neighbors. Neoconservativists thus have raised the oldest question in diplomacy once again: "How [can] the international community . . . manage the ambitions of a rising power [China]?"[2] In contrast to this view, some other scholars have held that China is a conservative power and "in the foreseeable future it will seek to maintain the status quo."[3] China's increasing integration into the international system, evident in its growing memberships in international security regimes and economic organizations, such as the World Trade Organization (WTO), has created "constraints on its foreign conduct as well as incentives to adapt to the prevailing norms in contemporary international relations."[4] Although the September 11, 2001, terrorist attack on the United States and the subsequent war on terrorism has pushed the debate to the back burner, the issue remains and will certainly return to the forefront and confront policy-makers in the West and Asia-Pacific if the threat of terrorism somehow subsides. This policy debate coincides with the theoretical debate between realism and liberalism among international relations scholars. While the realist argument holds that a rising China will become assertive and expansionist because, as the People's Republic of China's (PRC) capabilities increase, its intentions will become less benign, the liberal argument believes that China's reform and growing economic interactions with the capitalist world will make it more open and democratic, which will help to promote international stability and security. While both international relations theories have provided valuable insights, neither of them alone is able to unravel the puzzle of whether a prosperous and powerful China will be a major force of stability or a threat to international peace.

With an eye on the debate, this book tries to explore how unique and particularistic or general and common in comparison with other states China, constrained by its changing status in the international system, is pursuing its national interests and to examine China's strategic behavior in the global stage, particularly in its relations with major powers and its Asian neighbòrs. It represents a modest effort to help understand if China has acted prudentially in response to the particular set of opportunities and constraints that its position in the international system has created since the founding of the PRC, particularly in recent decades.

Focusing on China's sovereign state behavior, this book takes mostly an international system-centered approach, which assumes that state behavior is guided by the logic of national interests, defined in terms of survival, security, power, and relative capacities. This approach is in contrast to domestic-centered approaches that look into ideological preferences and objectives of key decision-makers and their factional conflicts or bureaucratic cleavages in explaining Chinese foreign policy essentially as an extension of domestic politics. Since China began taking steps to reform and to open up to the outside world in the late 1970s, many scholars have made fruitful efforts to explore the role of a variety of domestic factors in China's foreign policy-making process. The rationale for the increased interest in domestic-centered approaches is provided partly by the fact that "as compared to Chinese foreign policy in the Maoist era, the domestic context of Chinese foreign policy today has become both more important and more complex,"[5] and partly by the obvious shortcomings of sovieregn state–centered approaches in which the policy-making process is treated as a "black box" and foreign policy is reduced to a predictable reaction based solely on the national interest or on *realpolitik* considerations. Nevertheless, the rationale for domestic-centered approaches by no means overrules the merits of sovereign state–centered approaches. Given the progress of reform and opening up toward the outside world in recent decades, China has increasingly become a part of a larger international environment that provides opportunities for, as well as constraints on, its policy options. Consistently pursuing the overarching goal of economic modernization, Chinese leaders have to be more and more sensitive to China's position in the changing international environment, in which Beijing has had only a limited role in shaping. As a result, these leaders have developed a pragmatic strategy to work with the major powers and China's Asian neighbors and to adopt some established international norms beneficial to its foreign policy objectives while rejecting others it deems in conflict with its national interests. Pragmatism by definition is "behavior disciplined by neither set of values nor established principles."[6] Pragmatist strategy is therefore ideologically agnostic, having nothing, or very little, to do with either communist ideology or liberal ideals. It is a firmly goal-fulfilling and national-interest-driven strategic behavior conditioned substantially by China's historical experiences and geostrategic environment.

China's pragmatic foreign policy behavior took shape in Mao's final years and has fully developed in the post-Mao era. In the early 1970s, Mao developed a non-Marxist international strategy based on a perceived hierarchical structure of three worlds. Cooperation with the developing countries of the Third World as well as a developed Japan and Western Europe, which constituted Mao's Second World, could be a force to counter the alleged hegemonism of the two superpowers that constituted the First World. After the United States extended diplomatic recognition to Beijing in 1979, post-Mao leaders worked very hard to shape a strategic triangle in which China played a crucial role by maneuvering between the two superpowers, the United States and the Soviet Union. After the end of the Cold War, Beijing's foreign policy-makers envisioned and worked hard to promote a world of multipolarization (*duojihua*) against the speculation of some Western commentators that a unipolar world characterized by United States hegemony had emerged from the ashes of the Cold War. For its own interests, China emphasized the desirability and likely emergence of a multipolar community of sovereign nations mutually respecting the principle of noninterference. To work toward the perceived trend of multipolarization and to insure a favorable international environment for its modernization, pragmatic Chinese leaders have tried to avoid confrontational relations with the United States and other Western powers and, in the meantime, pursued a policy of defusing tensions along its immediate borders. They have concluded that the failure of the Soviet Union was largely due to its strategy of confrontation with the United States in a competition for the position of world superpower that exhausted its economic and military capacity. Shen Jiru in his book, *China Does Not Want to Be Mr. No*, suggested that, as one of the weaker poles in the multipolar world, China should not become the second "Mr. No" after the former Soviet Union to confront the United States and exhaust itself. Instead, China should defend its national interest by conducting a shrewd diplomacy, which "requires rationality and calmness,"[7] because "both the nation's problems and most of the possible solutions were perceived as coming from outside."[8] Pragmatic strategy has thus gained power both from reacting to and absorbing from the outside world. Pragmatic strategic behavior is flexible in tactics, subtle in strategy, and avoids appearing confrontational, but it is uncompromising with foreign demands that involve China's vital interest or that trigger historical sensitivities.

Chinese Foreign Policy Behavior under Constraints

Taking a pragmatic posture has resulted from an awareness among Chinese leaders that they are under many international and domestic constraints. Setting economic modernization as their top national objective, pragmatic Chinese leaders have paid special attention to China's economic relations, particularly its trade relations, with other countries. In this case, Chinese leaders' expectations from international economic interactions in general and trade in particular set a major constraint on China's foreign policy behavior. This is the argument made in the chapter by Rex Li. Drawing on the theory of trade expectations, Li investigates the proposition of whether high interdependence between China and its trading partners is more likely to pacific or to belligerent Chinese behavior. If Chinese decision-makers' expectations for future trade are high, they will be less likely to use force to deal with unresolved disputes with neighboring countries. If, however, they have a negative view of their future trading environment, they will likely take measures, including military action, to remove any obstacles that might forestall their pursuit of great-power status. For the moment, China's expectations of future trade are by and large optimistic, but there is evidence of growing Chinese suspicion of a Western "conspiracy" to contain China. This may alter Beijing's future perceptions. To ensure that the rise of China will not cause regional and global instability, the outside world should seek to integrate China into the international community by pursuing policies that will have a positive influence on China's expected value of trade. In the meantime, some elements of the balance of power strategy need to be introduced in order to curtail China's expected value of war.

In addition to trading expectations, China's foreign policy behavior is constrained by many other factors, including its perceived power position, established principles, and policy options in the changing world. Since the end of the Cold War, the new generation of Chinese leaders has been confronting a more and more complicated world as the more or less predictable and manageable bipolar competition and ideological confrontations have been replaced by unpredictable and hardly manageable ethnic, religious, and nationalistic conflicts among states as well as nonstate and transnational actors. They have found that China's foreign policy is constantly tested by different sets of contradictions. The chapter by Wu Xinbo examines four sets of contradictions that have constrained China's foreign policy behavior: self-images of great power versus poor country, "open-door" incentive versus sovereignty concern, principle versus pragmatism, and bilateralism versus multilateralism. China views itself as a major power and wants to play a role accordingly in the world arena, but it still lacks an adequate material basis to do so. The open-door policy requires China to be fully integrated into international society, but the strong concern over sovereignty makes it difficult for Beijing to embrace some of the mainstream values. China believes in a set of principles in international affairs, but consideration of its national interest causes Beijing to make a pragmatic compromise from time to time. Beijing has long been accustomed to dealing with others in bilateral settings, but the post–Cold War era is witnessing a rise of multilaterism in international politics, which is creating more and more pressure on China's traditional diplomacy. These contradictions have given rise to a dichotomy in China's foreign policy behavior. While Beijing cherishes the aspiration of being a major power and expects to be treated as such, it is highly selective in assuming the responsibility of a major power. While China seeks to integrate its economy into the international system, it stands vigilant on the sovereignty issue and rejects any interference into its internal affairs. While Beijing reacts to many international issues with strong rhetoric, in practice it acts in a deliberate and restrained manner. While Beijing expresses enthusiasm about multilateralism, it still feels more comfortable with bilateralism.

Of course, these contradictions may become less important in constraining China's foreign policy behavior as China comes closer to its goal of modernization and becomes fully integrated into regional and global economic and security communities. With the growth of China's overall national power, the dual-identity syndrome of great power versus poor country will diminish, and China will be able to play a more visible role in international affairs. With the full integration of China into regional and global communities and hence enhanced confidence in securing its national security and territorial integrity, China will likely take a more flexible attitude toward the issue of sovereignty, growing more reflective of the new reality in a more interdependent world. As China grows into a mature power and becomes deeply involved in international affairs, Beijing will approach many issues more from a pragmatic basis and less from a principled standpoint. Finally, when China becomes more experienced and confident with multilateral activities, it will find less tension between bilateralism and multilateralism and feel comfortable to conduct both types of activities in its diplomacy. This development would help narrow the gap between China's accorded status and expected status and hence strengthen the pragmatic strategy in pursuing China's national interests.

The underlying force behind the pragmatic strategy is Chinese nationalism, which has remained a bedrock political belief shared by most Chinese people, including many of the Communist regime critics, after the rapid decay of Communist ideology in post-Mao China. However, some analysts in the West are concerned with possible negative impacts of rising nationalism upon Chinese foreign policy, because nationalism has often resulted in irrational behavior and fueled interstate warfare in modern world history. Indeed, if Chinese nationalism is unconstrained, it may become a source of international aggression; however, Chinese nationalism is a particularly historical product that has been more defensive than offensive in relations with other countries because Chinese political elites have embraced nationalism largely due to its instrumentality for China's regeneration and defense against foreign imperialism.

The last chapter in Part I, by Suisheng Zhao, shows that the development of Chinese nationalism in recent decades has been a function of its utility to the Communist state in response to both domestic and external challenges. Nationalism is a double-edged sword as it may be used by the regime to replace the discredited Communist ideology as a new base of legitimacy, but it may also cause a serious backlash and place the government in a tight spot causing trouble from both internal and external sources. Domestically, nationalism is both a means for legitimizing the Chinese Communist Party (CCP) rule and a means for the Chinese people to judge the performance of the Communist state. As Nicolas Kristof observed, "All this makes nationalism a particularly interesting force in China, given its potential not just for conferring legitimacy on the government but also for taking it away."[9] Internationally, Chinese leaders are constrained by China's relative weakness in its power capacity to obtain their nationalist objectives, and they certainly do not want to stir up nationalist demands for an assertive international position that they are not in a position to satisfy. Nationalism, thus, could become a dangerous Pandora's box. Without constraints, it could release tremendous forces for unexpected consequences. It is not hard for pragmatic leaders to realize that the Boxer Rebellion–style xenophobia that prevailed during the Cultural Revolution may do more harm than good to the Communist regime. Its complex effects thus set a limit on the utility of nationalism to pragmatic Communist leaders. Therefore, pragmatic leaders have been careful to prevent the nationalist sentiment of Chinese people from getting out of hand. While pragmatic Communist leaders have consciously cultivated nationalism as a new glue to unite the nation against the so-called "splitting" (fenhua) and "Westernization" (xihua) of China by Western countries, nationalism has not made China's foreign policy behavior particularly irrational or inflexible as strong nationalist rhetoric has often been followed by prudential policy actions in foreign affairs. This behavior pattern is particularly clear in China's dealing with the major powers as well as its important Asia-Pacific neighbors. In particular, pragmatic Chinese leaders have tried very hard to assure that its foreign policy remains based on pragmatic consideration of China's national interest rather than being dictated by nationalist rhetoric.

Ideology, Strategic Culture, and Pragmatism

In the early years of the PRC, Communist leaders tried to apply a highly articulated and systematic Communist ideology to the realm of foreign policy on the assumption that it could provide them with an accurate guide to the choices they faced in the international arena. Communist ideology is "a formal system of ideas which provided a perceptual prism" through which the Communist leaders view the world and which, they believe, explain the reality.[10] Perceiving international relations through the ideological concepts, Chinese Communist leaders believed in the inevitable victory of anti-imperialism, socialist revolution, and national liberation struggles. Acting upon the belief that "the East wind was prevailing over the West wind," Mao Zedong adopted aggressive foreign policies against the capitalist world in the 1950s and tried to export the Chinese model of socialism in the 1960s. These Communist ideology-driven foreign policies, however, caused more harm than good to the Communist regime. After Mao's death and the inception of Deng's market-oriented economic reform, the importance of ideology has dramatically declined in its importance and "lost salience to the motivation of pragmatic power politics."[11]

It is indeed widely held that, before the CCP's shift of its emphasis from worldwide Communist revolution to national economic modernization in the late 1970s, China's foreign policy was largely driven by Communist ideology, and, as a result, rational calculations of power and interests were often relegated to a secondary position. Since the inception of China's reform in 1978, the emphasis on economic modernization and opening-up to the outside world have propelled China's foreign policy toward pragmatism and realism. The rhetoric of Marxism-Leninism has tended to be overshadowed by the pragmatic calculation of national interests. While this understanding of Chinese foreign policy is generally true, the inconsistencies between ideological predicts and policy practices existed even in the Mao years, as Chinese leaders had to devise pragmatic policies to achieve some specific and mundane objectives on issues involving China's vital national interests.

Lau Siu-kai's chapter on China's policy toward Hong Kong under the British rule, which was an integral part of its foreign policy, examines a distinctive case that reveals the limited influence that ideological fervor had on foreign policy even in the early years of the PRC. Lau argues that the Hong Kong policy was primarily driven by utilitarian calculations of national interests and the interests of the Chinese Communist Party. Its primary goals were to secure a less-threatening external political environment and to make calculated use of Hong Kong for China's economic development. By letting China's arch-imperialist foe in modern Chinese history, Britain, continue the rule of Hong Kong in 1949 and thereafter, Communist leaders took a significant exception to its ideological lines. But the act served China's vital national interest as it represented a rational attempt by the CCP to avert war with the West over a piece of land which, unlike the Korean Peninsula, did not pose a military threat to China. By allowing Hong Kong to remain a British colony, China deliberately manipulated the differences between Britain and the United States and ameliorated the pain inflicted on China by the latter's policy of containment. While China's decision to maintain Hong Kong's status quo in 1949 might be only an expedient and short-term move in view of the current national security needs of the new regime, this move did bear long-term fruit. As the international situation since 1949 continued to be unfavorable to China, any attempt by China to recover Hong Kong would have jeopardized its national interests. In addition, while the economic value of Hong Kong to China was not obvious in 1949, the industrial takeoff of Hong Kong since the late 1950s had made Hong Kong of critical economic importance to China, which was reeling from the trade embargo imposed by the West and the termination of Soviet aid resulting from the Sino-

Soviet confrontation. As a result, China gradually came up with an explicit and stable policy toward Hong Kong under the principle of "long-term consideration, full utilization," the thrust of which was instrumentalism and pragmatism.

China's decision to enter the Korean War in 1950 is another case involving China's vital national interest; hence, pragmatic calculation of national interest played a crucial role. Andrew Scobell's chapter documents that the Chinese troops' cross of the Yalu River into North Korea was the culmination of a deliberative decision-making process. China's leadership pondered the matter of whether to enter the Korean conflict for months before finally acting. The process was torturous, as there was widespread reluctance as well as significant opposition within the Chinese elite to becoming embroiled in Korea. While Mao certainly dominated the decision-making process, other figures, including soldiers, played major roles. Mao made up his mind only after months of deliberation and discussion with Chinese civilian and military leaders to conclude that China had no choice but to enter the war with the United States, which was going to invade China after wining the war in Korea. Diplomats, economic bureaucrats, and military commanders all pondered the question of whether or not it was in China's best interests to intervene in Korea. Their feelings of "proletarian solidarity" with the oppressed peoples of the world, particularly the strong ties forged with the Korean Communist troops who fought side by side with them during the Anti-Japanese War and the Chinese Civil War, were not sufficient at the outset of hostilities on the peninsula to dispose them toward intervention. They were finally convinced of the necessity of intervention and supported the intervention only after Mao made support of the war a patriotic crusade to defend the motherland and linked patriotism with the cause of communism. Scobell's study suggests that the attitudes of Chinese soldiers and statesmen conform roughly to the pattern found by Richard Betts in his study of the attitudes of American leaders during the Cold War. That is, military figures tended to be no more hawkish than civilian leaders and, in fact, often were more dovish. On issues of national security and foreign policy, Chinese soldiers and statesmen hold basic dispositions generally consistent with their counterparts in other countries.

This conclusion represents one answer to a controversial question, namely, how unique or common is Chinese foreign policy behavior in comparison with other states? In recent years, some scholars' efforts to apply the concept of strategic culture to explain China's strategic thinking and policy behavior have brought about a debate on the fundamental tradition and orientation of Chinese statecraft.[12] The central issue is the compatibility or incompatibility between Chinese traditional strategic culture and Western thinking on international relations and the impact of traditional Chinese culture on modern Chinese strategic thinking. Conventional views hold that China has a unique cultural predisposition to seek nonviolent solutions to problems of statecraft and to be defensive-minded, favoring sturdy fortifications over expansionism and invasion. This assumption, however, is challenged by some recent studies, which argue that, like many other states, China possesses a stark realist strategic culture that views war as a central feature of interstate relations.[13] However, some Chinese scholars continue to insist that ancient Chinese traditions in philosophy and statecraft are unique or at least substantially different from Western traditions and that this strategic cultural tradition still has a major impact upon contemporary Chinese strategic thought and behavior.

The chapter by Zhang Junbo and Yao Yunzhu, two Chinese military scholars, represents this school of thought. Their comparative study of Chinese and Western strategic thinking from a philosophical perspective generalizes disparities of these two traditions in the following three paradigms: "justice" versus "interests," "human factors" versus "weapon factors," and "stratagem" versus "strength." In their view, the Chinese military tradition focuses on "man" by stressing ethical and moral dimensions in war, valuing the "human factor" more than the "weapon factor" and preferring victory won by wisdom to victory won by brutal force. The Western strategic culture, on the other hand, shows its emphasis on the "material" by centering on interests, the weapon factor, and strength in military calculations. As cultural legacies, these distinctive preferences still exert influences on modern strategic thinking and military decision-making in China. For example, Mao Zedong judged military conflicts by moral standards and held that if a state acquired *dao* (justice or moral strength), it would eventually win no matter how inferior it was in physical strength to its opponent. If a state did not acquire *dao*, it was doomed to failure no matter how powerful it might be. The strong desire to uphold justice, combined with the need to safeguard national interests, justified the risk of a direct confrontation with the world's strongest power, the United States, when Mao made the decisions to send the Chinese troops to Korea in the early 1950s and to provide military assistance to Vietnam in the late 1960s and early 1970s. Mao also stressed the human factor and believed that, as long as China demonstrated domestic harmony and a resolve to fight, no enemy, including the two superpowers, would risk a general war against China. Killing power of weapons was not decisive. Instead, China's huge human resources, its unbeatable national spirit, and its history of always denying final victory to any foreign invader were taken as counterthreats to the use of even nuclear weapons. These two Chinese scholars believe that Chinese military strategists today continue to depend very much on stratagem as a means of winning. While a part of this dependence can be accounted for by the fact that the People's Liberation Army (PLA) had no weapon advantages to rely on in most of its war experiences, much of it has to be explained by the dominance of traditional thinking. Westerners frequently complain about the ambiguity and secrecy surrounding the Chinese military doctrine, its deployment, its budget, its weapon research and development program, and its current order of battle, and attribute lack of transparency to the nature of China's social system. But Zhang and Yao hold that this is due to the overlook of the role played by deep-rooted philosophical thinking, in which ambiguity is not only a means to achieve an end but also an art to be explored with imagination. Concealment, deception, and secrecy, all salient ingredients of traditional military stratagem, have much more to do with China's traditional culture than with its current social system.

This school of thought, however, has been challenged by more and more Chinese scholars in recent years. Zhang and Yao's chapter acknowledges that in the ongoing debate about China's strategic culture, the major issue has shifted to how to define China's national interests and how to reconcile the long-term and short-term interests of China. Indeed, while some scholars insist moral factors play a continuous role in the foreign policy decision-making process, a strong undercurrent goes in the reverse direction to the pragmatic consideration of China's national interests. Yan Xuetong, a Berkeley-trained Chinese scholar based in Beijing, published a book, *Guanyu Zhongguo de Guojia Liyi Fenxi* (Analysis of China's National Interests) in 1996. Yan claims that his book is "designed to clarify the confusing concept of national interest, provide an analysis of China's national interest after the Cold War and propose some strategic suggestions for realizing national inter-est."[14] This serious scholarly work gained unusual popularity because Yan argues that China is facing a competitive international environment, and it is therefore crucial that China's leaders place emphasis on its economic, political, and security interests. Yan suggests that while China should avoid military conflict with other powers, particularly the United States, it should be assertive in defending China's national interest against any external erosion. Yan's argument has found strong supporters among Chinese strategic analysts and foreign policy-makers as they are calling for a more interests-oriented approach in crisis manage-ment in line with Deng Xiaoping's position that "national interests" should be the "over-riding consideration" in policy-making. As more and more Chinese analysts have downgraded ideological and moral factors in favor of national interest in international relations, an increasing number of Chinese strategists have also begun to rethink the inter-relating roles of the human factor and the weapon factor. The performance of high-tech weapons in the 1991 Gulf War served as an eye opener to the Chinese leaders. While holding that human beings play more important roles than weapons in wars of older times, Chinese strategical analysts have accepted that new military technology may have changed the role in modern wars. Increased efforts on the acquisition of new weapons and military application of new technologies are a testimony to this new thinking.

This development shows that the tide has turned in China's strategic thinking toward the direction of pragmatism, as the consideration of national interest rather than ideology or morality has become crucial in shaping China's foreign policy behavior. Indeed, China has acted prudentially, like most other countries, in light of the opportunities and con-straints provided in the international system in recent decades. The third chapter by Suisheng Zhao analyzes Beijing's perception of the change in the international system from the Cold War bipolar confrontation to a new post–Cold War multipolar world and China's consequential foreign policy adjustment in the early 1990s. To act in accordance with its own interests and to exert its influence on international affairs, Beijing worked hard to find and shape an international system that was in its favor or at least not to its disadvantage after the end of the Cold War. Losing its leverage as a balancer in the Cold War bipolar contest between the United States and the Soviet Union, Beijing had to face a unipolar reality in which the United States became the only superpower. While working hard to encourage a multipolarization that would provide more room for maneuvering, Beijing made pragmatic accommodations to the unipolar world through its foreign policy adjust-ment right after the Tiananmen incident, which resulted in sanctions being imposed by Western countries. These policy adjustments included at least the following three aspects. First, Beijing played down its pretense to being a global power and acted more as a re-gional power, focusing its policy objectives primarily on seeking better relations with, and greater influence on, neighboring Asia-Pacific countries. Second, Beijing's policy toward the United States and other Western powers became characterized by a combination of ever-ready concessions in spite of an official anti-Western sentiment and rhetorical tough-ness. Third, Beijing tried to be more and more cooperative in multilateral activities with-out and within the UN system, striving to establish an image as an independent and responsible power in the community of nation-states.

Particularly in the multilateral arena, China has taken a very pragmatic approach to accept many prevailing international norms and to gain access to some international regimes but has rejected some others in light of its calculation of their impact upon its national interests. Jing-dong Yuan's chapter on the evolution of China's nonproliferation policy is a case study of how China calculated its national interests to selectively endorse the international norms and regimes in arms control, disarmament, and nonproliferation. China clearly changed its position on nonproliferation from one of dismissal in the 1970s and 1980s to one of selected support to serve its national interests in the 1990s, when Beijing began undertaking serious efforts to integrate itself into the formal international nonproliferation regime through accession to key treaties and conventions and by active participation in multilateral negotiations, particularly in such forums as the Conference on Disarmament in Geneva. By adhering to international treaties and conventions, Beijing not only demonstrated its commitment to nonproliferation principles, but also placed itself, to some extent, under international legal constraints. There are many specific explanations about the evolution of Chinese nonproliferation policies. But the core of these explanations is based on the national interest calculation. China has gradually realized that proliferation of weapons of mass destruction (WMD) and delivery systems can affect its own security interests negatively. In this regard, the existing international nonproliferation regimes, such as the Nonproliferation of Nuclear Weapons Treaty (NPT), offers tangible benefits for China, not the least of which would be the prohibition of Japan, the Koreas, and Taiwan to acquire nuclear weapons. While the Comprehensive Test Ban Treaty (CTBT) imposes constraints on China's own nuclear weapons modernization programs, Beijing is willing to pay the price if such mechanisms would prevent countries such as India from joining the nuclear club. In addition, with its rising national comprehensive capacity, China does not want to be seen as an outcast or impediment to international nonproliferation efforts, which would damage its efforts to establish its image as a responsible power. China's policy change to a certain degree has also been influenced by its asymmetrical economic interdependence in its bilateral relationship with the United States. Beijing's need for advanced U.S. technologies obliges it to undertake the necessary policy adjustments required by Washington.

China's conformation to international norms and regimes, however, has been carefully calculated. While it has acceded to most international treaties and conventions that are broadly based with universal membership (e.g., NPT, Chemical Weapons Convention [CWC]) and has by and large complied with their norms and rules, it has been less than forthcoming and occasionally quite critical, with regard to the largely Western-initiated, supply-side, multilateral export-control regimes such as the Nuclear Suppliers Group (NSG), the Australia Group (AG), the Wassenaar Arrangement, and the Missile Technology Control Regime (MTCR). China is highly critical of these cartel-like regimes, because it considers them to be discriminatory, selective, imbalanced, unequal, and arbitrary. China has its own security concerns when considering the various nonproliferation regimes and their impact. This is particularly revealing in its positions on MTCR compliance and missile transfer issues. Ballistic missiles have increasingly become a critical element in China's defense modernization programs, and a "niche" or "comparative advantage" in its arms exports. Other than the drive for commercial gains and defense modernization programs, geostrategic considerations, particularly retaining its bargaining leverage with the United States on issues such as National Missile Defense (NMD) and Theatre Missile Defense (TMD), have been important factors behind China's missile transfer activities. China is increasingly concerned with the ultimate goal of the U.S. nonproliferation policy—what it views as the U.S. drive for absolute security. Consequently, Beijing wants to retain flexibility and bargaining leverage with Washington. For Beijing, development and deployment of ballistic missiles serve the purpose of deterring independence elements in Taiwan and discouraging potential U.S. intervention in a military operation in the Taiwan Strait. In this case, Yuan concluded, the pace and future direction of Chinese nonproliferation policies will be closely linked to Beijing's overall assessment of its security interests, threats, and policy priorities.

Strategic Relations with the Major Powers and Asia-Pacific Neighbors

Chinese scholars and officials often start their analysis of international relations from *liliang duibi* (balance of forces) in the world, a Chinese term similar to the conception of distribution of power in Western literature on international relations. The dynamics of international politics are understood as a change of power distribution across the world. As Wang Jisi indicated, "Without a study of *liliang duibi*, policy-makers in Beijing presumably would not be able to adjust foreign policy accordingly."[15] As the demise of the Cold War strategic triangle temporarily weakened China's immediate strategic leverage in the global balance of power, Beijing's foreign policy-makers have tried to resist the emergence of a unipolar world that is perceived not to be in China's favor. To find an alternative, Beijing's foreign policy analysts have suggested that "the world has undergone a transition to multipolarization since the end of the Cold War. A relative balance of power has resulted in an effective check on all global powers."[16] Chinese analysts have presented three possible types of multipolar systems. One is a three-polar world, in which the European community constitutes one pole, North America forms the second pole, and the Asia-Pacific region is rising as the third pole. The second is a five-polar structure, which consists of the United Sates, Germany, Russia, China, and Japan. The third one presents a pattern of "one superpower and four big powers" (*yi chao duo qiang*). The one superpower refers to the United States and the four big powers are the European Union, Japan, Russia, and China. Bilateral relations differ between these five powers. Some are diametrically opposed on the ideological level, some are in sharp political conflict, some have serious economic conflicts, some are political allies but full of economic contradictions, some have different social systems but complement each other economically, and some have normal political ties but are rather cool in their economic relations.[17]

The most frequently mentioned pattern in Chinese literature of strategic studies has been that of "one superpower and four big powers." Chinese strategists admit that although the Cold War sapped the vitality of the United States, it has remained the sole superpower because of its comprehensive national strength, whether in terms of its economy, scientific and technological strength, military might, or foreign influence. The United States, as the sole superpower, has had strong security interests and competed energetically for its dominant position in the world. In particular, the articulation of ambitious goals under the Bush administration has given a whole dimmension to the unipolar world order that the United States has sought to maintain. In this case, China must live with and even accommodate the United States' superpower status while trying to retain its independent power aspirations and building a united front with other nations to protect its national interests. This view also holds that the United States has failed to enact an effective strategy of controlling the globe unilaterally. The U.S. government not only is subject to strong domestic resistance to foreign military intervention due to the American public's concern with their domestic problems, but it also faces the reality that the traditional control of its allies has become more difficult as other big powers have adopted more independent policies. One major feature of global politics after the Cold War is that a growing number of big powers have become bold enough to say no to the sole superpower. In the words of Qian Qichen, "The development of the world multipolarization tendency has brought to an end of the era in which one or two superpowers tended to dominate the world."[18] To manage its relations with major powers while actively pushing for a multipolar world, China has designed a new network of partnerships since the mid-1990s. This network of partnerships practically covers all the major powers and regional organizations, including Russia, France, the United States, the United Kingdom, the Association of Southeast Asian Nations (ASEAN), the European Union (EU), and Japan.

The chapter by Joseph Y.S. Cheng and Zhang Wankun suggests that promotion of partnerships with major powers reflects an attempt on the part of China to redefine its position in new international strategic relations. It also reveals the importance attached to relations with major powers on the part of the Chinese leadership who looks upon China as a rising power and pursues legitimate interests on the world stage. Within the network of partnerships, Chinese leaders have hoped to secure a multipolar world in which the major powers would establish relationships based on equality, mutual trust and respect, and mutual checks and balances. Stable, balanced relations among the major powers would provide a better guarantee of regional and global balance and stability. This idea of partnership is in accord with China's strategic objectives of maintaining a peaceful international environment in which China may concentrate on economic development; improving China's relations with all major powers and regional organizations; pushing for multipolarity; facilitating progress in China's economic diplomacy to open up markets and attract investment; advancing technology and management expertise from diverse sources; and improving China's image in the international community. To achieve these strategic objectives, Chinese leaders have appealed to their counterparts to abandon the Cold War mentality and actively identify common interests, with the hope that differences and contradictions in social systems and values would not affect the healthy development of state-to-state relations. The use of phrases such as "facing the twenty-first century" and "straddling over the present and the next centuries" in the partnership relationships between China and other major powers signifies that Chinese leaders hope to maintain such relationships among major powers on a long-term and stable basis. This is the central theme in China's partnership network with all major powers and regional organizations.

Among all partnerships, China's strategic partnerships with Russia and the United States are the most significant as they represent not only a redefinition of the two important bilateral relationships in the post–Cold War era but also an extension and reshaping of the Cold War strategic triangle in the new international environment. The Sino-Russian strategic, cooperative partnership is the first partnership established between China and a major power and has become the most developed partnership with well-established operational mechanisms. While some scholars see the Sino-Russian strategic partnership as a serious threat to Western, and specifically to American global interests and an attempt to shore up the forces of authoritarianism via militaristic nationalism in the death of ideological legitimation, others dismiss the argument as misconceived or greatly exaggerated because the partnership is not an alliance, has no treaty commitments, and is bound by neither shared interests, strategic objectives, nor common adversaries as the two countries' national interests have little in common and are likely to diverge and even possibly conflict in the long run. In a response to this controversy, Lowell Dittmer in his chapter argues that the Sino-Russian strategic partnership, formed after decades of rancorous verbal and sometimes lethal dispute, is genuine, bespeaking a genuine desire on both sides to put the past behind them and to forge a more friendly and mutually profitable relationship. It is a product of the strategic adjustments on the part of the Chinese and Russian leaderships in response to the post–Cold War situation to push for global multipolarization and to limit the predominance of the United States in international affairs. Chinese leaders highly evaluate this strategic, cooperative partnership because it is the only relationship with possible leverage to pose a credible alternative to the lone superpower and an entry ticket back to what Jiang Zemin calls "great power strategy" (*da guo zhanlue*). The partnership is held together by dovetailing strategic and material interests and institutional complementarity. Certainly there are underlying problems and suspicions between the two countries. It is also limited by the interest each partner retains in closer relations with the center of international economic gravity in the West. But the partnership represents a stable and meaningful commitment to bilateral relations, whose content is left vague to allow for unpredictable vicissitudes in the far less-structured post–Cold War era. It also represents the attempt of two large and precarious multiethnic continental empires to form a mutual-help relationship that would be uniquely useful to them in the face of a relatively hostile international environment.

The partnership with the United States, however, has been more complicated and troublesome. The Sino-U.S. relationship is the most crucial and important one in all of China's foreign relations. But it is also the most frustrating foreign policy challenge for China's pragmatic leaders who have hoped to establish and maintain a "friendly and cooperative relationship" with the United States, the unwieldy superpower holding the key to China's future of economic modernization. This relationship has experienced significant ups and downs in recent decades. China was a friendly nonally and strategic partner of the United States in the triangular competition with the Soviet Union during the last years of the Cold War. After the disintegration of the Soviet Union, China was left as the only major Communist power in the world and, to an extent, replaced the Soviet Union as the "evil empire" in the eyes of some American politicians. As a result, China's relations with the United States have experienced a dramatic change since the end of the Cold War due to the disputes over the issues of human rights, intellectual property rights, trade deficits, weapons proliferation, and Taiwan. While the Clinton administration in its second term came to seek to build a constructive strategic partnership with China, China was redefined as a "strategic competitor" and a potential regional rival in the security arena and only a trading partner in the first year of the Bush administration. The United States thus became a source of both hope and frustration for China's pragmatic leaders. To pursue its national interests, pragmatic Chinese leaders did not make a sharp reaction to President George Bush's initial negative attitude toward China. Instead, they steadfastly worked to build a cooperative relationship with the United States. This pragmatic policy paid off as the Sino-U.S. relationship returned to a "strategic partnership" in the U.S.-led international coalition against terrorism by the second year of the Bush administration. This pattern of up-and-down has compelled many Chinese scholars to determine what is the most important source for building a cooperative and stable relationship with the United States: geopolitical interests or ideological and cultural traditions.

The chapter by Zi Zhongyun, a Chinese scholar, explores the impact of ideology on the Sino-U.S. relationship from a historical perspective. At the turn of the century when China and the United States encountered each other, their destinies took opposite trajectories. While the United States saw its dream of national greatness fulfilled with the continuous rise of its national power, China witnessed a sharp downfall from the peak of its self-perception as the great central kingdom. Since that time, all progressive intellectuals and reformers took inspiration from Western ideas of democracy and freedom, while struggling at the same time against oppression and exploitation by Western powers. One of the basic urges of Americans of all sorts in dealing with China throughout the years was to influence, educate, and change China to its liking. However, the development of China takes its own course beyond the control of the United States. Yet, under different circumstances, neither side could help placing hopes on the other. In particular, U.S. relations with the CCP have undergone twists and turns. During China's civil war period, the CCP leaders once hoped the United States would wield its influence toward dissuading the Kuomintang (KMT) from launching armed suppression of the CCP forces, but disappointed by U.S. policy they began to see the United States in a negative light, not only its foreign policy, but its domestic system and ideology as a whole. In the 1950s, with the intensification of the Cold War, ideological confrontations between the two countries proved more difficult to resolve. The breakthrough of Sino-U.S. relations in the early 1970s was based on strategic and geopolitical considerations, leaving aside ideological differences. The Sino-U.S. relationship was best in the 1980s but underwent a sharp turn after the events of 1989. Leaving aside other factors, the age-long cultural paradigm of unrealistic hopes placed on each other played a certain role. For many Americans, the 1989 events were a heavy blow to their expectation of China, which they had thought were becoming "more like us." On the Chinese side, while the government had tried to play down ideological factors in dealing with Western countries, the chronic raising of the "human rights" issue with China as the main target, both in U.S.-China relations and in international stances, reminded the Chinese leaders of the U.S. policy toward China during the Cold War consisting of "containment" plus "peaceful evolution." The attitude of Chinese intellectuals and young students toward the United States also underwent a great change from admiration of the United States to a certain degree of resentment of American pressure on China. The pursuit of a strong and prosperous China remains the common and deep-rooted national aspiration prevailing among Chinese of all ages and social strata, and hence they identified themselves more with the government in opposition to foreign pressure, whatever their views on other subjects.

While Zi's chapter concludes that ideological aspiration and strategic interest have interacted to influence the ups and downs in Sino-U.S. relations, the chapter by Liu Ji, a former Chinese government official, makes an explicit call for strategic interests to be the most important consideration in handling the relationship. He urges political leaders in both countries to make rational choices based on strategic interests rather than on ideological and moral grounds. Drawing lessons from the history of Sino-U.S. relations, Liu believes that the United States made two wrong choices and missed opportunities to establish a beneficial relationship with China by choosing to work with the KMT government instead of the CCP at the end of World War II and by refusing to recongize the newly founded PRC at the end of China's civil war. At this turning point in history at the dawn of the twenty-first century, ideological and cultural differences between China and the United States should not become the cause of conflict, because these two countries could establish a cooperative relationship supported by strategic interests. Among these interests are that China and the United States are remote neighbors and have no common borders besides the Pacific Ocean and, therefore, do not have any geographic elements for confrontation and cannot have any territorial disputes. China and the United States also complement each other economically. The twenty-first century will be a century of intense economic competition. While the United States wants to win the competition in the huge market of China, China will also need huge funds and technology from the United States to realize the overarching goal of modernization. Although Liu Ji, like all other Chinese scholars and officials, holds that the Taiwan issue is a huge obstacle to Sino-American cooperation and the unification of China is a matter beyond dispute and bargaining, he suggests that China and the United States first focus on their most fundamental strategic interests in the twenty-first century and thereby build up a friendly Sino-American partnership. Liu's view on strategic interests to a great extent explains China's steadfast effort to maintain a cooperative relationship with the United States in recent decades.

Managing China's relations with its Asian neighbors constitutes another strategic challenge for China's pragmatic leaders. China's rapid economic growth over the past decades, coupled with its increasing defense spending, has stimulated much interest as well as trepidation among policy-makers and analysts of its Asian-Pacific neighbors. As a rising power, will China contribute to regional prosperity and security or become a destablizing force in the Asia-Pacific region? Will a rising China be tempted to use military power to assert its territorial claims and achieve its national goals against its weaker neighbors and use its increased influence to attempt to shape the regional and international political systems in accordance only with Chinese interests? Will China be a revisionist force within the international arena and, therefore, pose a grave threat to China's neighboring countries, particularly those that have territorial disputes with China? These are unanswered questions that have caused concern within the Asian-Pacific region. Just like in the China threat debate, while some worry that a powerful China would act unilaterally, coercively, and exploitatively under the assumption that China's economic and security interests are inversely related to those of other states, others believe that "a powerful future China might determine that its interests are better served through cooperation, consultation, providing public goods, managing stability, and facilitating regional prosperity through peaceful and fair economic competition."[19]

The fourth chapter by Suisheng Zhao attempts to shed light on this debate by examining the changes in China's way of managing its relations with its Asia-Pacific neighbors in recent decades. China often calls its Asian neighbors "periphery countries" (*zhoubian guojia*). Although it was always aware of the importance of maintaining stable relations with these periphery countries for its national security, Beijing was not able to make an integrated policy to maintain good relations with neighboring countries before the 1980s. This situation began to change in the early 1980s when pragmatic Beijing leaders made a deliberated effort to devise an integrated regional policy, known as *zhoubian zhengce* (periphery policy) or *mulin zhengce* (good-neighboring policy). This policy change came when Chinese leaders tried to abandon ideology as the determining guide and to develop friendly relations with neighbors regardless of their ideological tendencies and political systems and their relations with either the Soviet Union or the United States. Beijing's new periphery policy is aimed at exploring the common ground with Asian countries in both economic and security arenas and conveying the image of a responsible power willing to contribute to stability and cooperation in the region. Economically, Beijing's good-neighboring policy, just like any other part of its foreign policy, is closely related with its overarching objective of economic modernization and with its desire for its share of the rapid economic growth in the region. In the traditional security arena, the periphery policy is to prevent their neighboring countries from becoming military security threats by settling border disputes "through consultations and negotiations." It is indeed a prudent policy that China has made the effort to find shared strategic interests with neighboring countries.

Implementing this policy, China has improved its relations with most of its periphery countries since the 1980s. In particular, China has made impressive progress to find a peaceful settlement in the most challenging land and maritime territorial disputes with its neighboring countries. Allen Carlson's chapter examines the Chinese approach to its major continental and maritime boundaries during the 1980s and 1990s. He finds that the Chinese position on border relations during these two decades were characterized by a set of striking continuities and subtle changes. While foreign policy elites in Beijing consistently worked to stabilize China's borders, they also deemphasized the use of confrontational claims and increasingly made use of diplomatic measures and international legal agreements to accomplish this goal. Such a development reveals that during this crucial period China took a set of relatively conservative, stability-seeking, diplomatic initiatives and representational practices. As a result of such efforts, for the first time in the twentieth century, China was able to create a relatively stable set of territorial boundaries with each of its continental neighbors. This transition greatly enhanced China's national security. Such a conservative set of territorial practices is indicative of the relatively status quo agenda that is currently guiding Chinese behavior with its Asian-Pacific neighbors. The persistence with which the Chinese leaders have enacted such practices stands in stark contrast to the dire warnings issued by many of those who tend to emphasize China's aggressive and revisionist agenda within Asia.

In spite of the improvement in China's relations with its Asia-Pacific neighbors, pragmatic Chinese leaders still have to overcome the concerns of its weaker Asian neighbors over the potential China threat. Ho Khai Leong's chapter looks at the relations between China and the ten states of the Association of Southeast Asian Nations (ASEAN) and indicates that although China now enjoys official diplomatic relations with all the ASEAN member states, all these states have retained a certain wariness about China. Despite reassurances and determination by both sides to increase interaction and cooperation, there remain areas of contention and contestation in China and ASEAN relations as China's market potential, military capability, and its enormous size have both excited and threatened the Southeast Asian states. After its long isolation, China has been making goodwill gestures to the Southeast Asian states and has been able to integrate itself successfully into the political life, economies, and security interests of the region. A network of multilevel dialogues, official, semi-official, and unofficial, between China and ASEAN and frequent exchange of visits by senior leaders of both sides have increased mutual trust, support, and understanding. The PRC obviously wants to reassure the ASEAN countries that its intention in the region is purely nonmilitary in nature and its principles of peaceful coexistence are alive and well. ASEAN, on the other hand, has reciprocated and maintained an approach of constructive engagement with this giant neighbor. These diplomatic efforts have contributed significantly to the stability and peaceful development in the region. However, Ho suggests that there is an element of ritualism that runs through the interactions. While the Southeast Asian states have been willing to engage an emerging China, they are also wary of the potential risks when dealing with it, as China is still in a dispute over the Paracel and Spratly island groups in the South China Seas with several ASEAN nations, including Vietnam, Brunei, Malaysia, and the Philippines. The South China Seas dispute will continue to be on top of the agenda for Sino-ASEAN relations in the foreseeable future. In addition, military and economic rivalries have intensified. As the globalization trend surges ahead with great momentum, China and ASEAN will be confronted with a more competitive situation in terms of direct foreign investments and high-tech capital investments. The increasing presence of Taiwan in Southeast Asia adds to the complexity of this economic rivalry. China has shown more flexibility and pragmatism in its approach to modernization and foreign relations, including its relations with ASEAN. However, the prospects for greater economic and political interaction will depend largely on how China continues to hold to a pragmatic foreign policy in the post-Deng and post-Jiang periods.

This book is divided into three sections. The first section analyzes some major constraints on Chinese foreign policy behavior. The second section examines the debate on the impact of ideology, strategic culture, and pragmatic consideration of national interests on Chinese foreign policy. The third section elaborates on China's pragmatic foreign policy behavior by investigating China's strategic relations with the major powers and its Asia-Pacific neigbors. Most chapters were published in the *Journal of Contemporary China* (*JCC*) during the past five years. As the editor of the *JCC*, I am very grateful for the authors who have made special efforts to update and revise their articles that are collected in this book. In addition, I would like to thank my assistant, Mary Bramlett Vars, for her editorial assistance. A special debt of appreication goes to Patricia Loo at M. E. Sharpe for her efficient efforts at the publication stage of the book.

Notes

1. For some views in the China threat debate, see Denny Roy, "The China Threat Issue: Major Arguments," *Asian Survey* 36, no. 8 (August 1996): 758–771; David Shambaugh, "Containment or Engagement of China: Calculating Beijing's Responses," *International Security* 21, no. 2 (Fall 1996): 180–209; Gideon Rachman, "Containing China," *Washington Quarterly* 19, no. 1 (Winter 1996): 129–140; Alastair Iain Johnston and Robert S. Ross, *Engaging China: The Management of an Emerging Power* (London: Routlege, 1999).

2. Nicolas D. Kristof, "The Real Chinese Threat," *New York Times Magazine*, August 27, 1995, p. 50.

3. Robert S. Ross, "Beijing as a Conservative Power," *Foreign Affairs* 76, no. 2 (March/April 1997): 34.

4. Weixing Hu, Gerald Chan, and Daojiong Zha, "Understanding China's Behavior in

World Politics: An Introduction," in Weixing Hu, Gerald Chan, and Daojiong Zha, eds., *China's International Relations in the 21st Century: Dynamics of Paradigm Shifts* (Lanham, MD: University Press of America, 2000), p. 2.

5. Joseph Fewsmith and Stanley Rosen, "The Domestic Context of Chinese Foreign Policy: Does 'Public Opinion' Matter?" in David M. Lampton, ed., *The Making of Chinese Foreign and Security Policy in the Era of Reform* (Stanford, CA: Stanford University Press, 2001), p. 151.

6. Lucian Pye, "After the Collapse of Communism: The Challenge of Chinese Nationalism and Pragmatism," in Eberhard Sandschneider, ed., *The Study of Modern China* (New York: St. Martin's Press, 1999), p. 38.

7. Shen Jiru, *Zhongguo Budang Bu Xiansheng: Dangdai Zhongguo de Guoji Zhanlue Wenti* (China Does Not Want to Be Mr. No: Problems of International Strategy for Today's China) (Beijing: Jinri Zhongguo Chubanshe, 1998), p. 62.

8. Andrew J. Nathan and Robert S. Ross, *The Great Wall and the Empty Fortress: China's Search for Security* (New York: W.W. Norton, 1997), pp. 32–33.

9. Nicolas D. Kristof, "Guess Who's a Chinese Nationalist Now?" *New York Times*, April, 22, 2001.

10. Steven I. Levine, "Perception and Ideology in Chinese Foreign Policy," in Thomas W. Robinson and David Shambaugh, eds., *Chinese Foreign Policy: Theory and Practice* (Oxford, UK: Clarendon Press, 1995), p. 30.

11. June Teufel Dreyer, *China's Political System: Modernization and Tradition*, 3rd ed. (New York: Longman, 2000), p. 305.

12. Robert S. Ross and Paul H.B. Godwin, "New Directions in Chinese Security Studies," in David Shambaugh, ed., *American Studies of Contemporary China* (Armonk, NY: M. E. Sharpe, 1993), p. 145. For some monographs that utilize a strategic culture or cultural approach in this regard, see Shu Guang Zhang, *Deterrence and Strategic Culture: Chinese-American Confrontations, 1948–1958* (Ithaca: Cornell University Press, 1992); Jonathan Adelman and Chih-Yu Shih, *Symbolic War: The Chinese Use of Force, 1840–1980* (Taipei: Institute for International Relations, 1993); Alastair Iain Johnston, *Cultural Realism: Strategic Culture and Grand Strategy in Chinese History* (Princeton, NJ: Princeton University Press, 1995).

13. See, for example, Johnston's *Cultural Realism*.

14. Yan Xuetong, *Guanyu Zhongguo de Guojia Liyi Fenxi* (Analysis of China's National Interests) (Tianjin: Tianjin Renmin Chubanshe, 1996), p. 1.

15. Wang Jisi, "International Relations Theory and Study of Chinese Foreign Policy: A Chinese Perspective," in Thomas W. Robinson and David Shambaugh, *Chinese Foreign Policy: Theory and Practice* (Oxford, UK: Clarendon Press, 1995), p. 489.

16. Feng Lidong, "An Interview with Yang Chengxu, Director of International Studies Institute," *Banyuetan* (Bimonthly Talk), January 10, 1995, p. 67.

17. Da Zhou, "Multipolar or One Superpower and Four Big Powers," *Shijie Zhishi* (World Affairs), January 16, 1995, pp. 4–5.

18. "Qian Qichen on International Situation and China's Diplomacy," *Xinhua*, June 16, 1995.

19. Denny Roy, *China's Foreign Relations* (Lanham, MD: Rowman & Littlefield, 1998), p. 2.

Part One

Understanding Chinese
Foreign Policy Behavior

2

Security Challenge of an Ascendant China

Great Power Emergence and International Stability

Rex Li

The reemergence of China as a great power is arguably the single most important development in the post–Cold War world. The rapid economic growth of the People's Republic of China (PRC) over the past decade, coupled with its high level of defense spending, has stimulated much interest as well as trepidation among policy-makers and analysts across the world. Although the continued augmentation of Chinese power is not predetermined, the profound effects of China's growing process cannot be underestimated.

Over the past decade, there has been a heated debate in the West over the potential challenge of an increasingly strong and assertive China to the Asia-Pacific region and to the world more generally. What are the international repercussions of the rise of the "sleeping giant"? Will a rich China contribute to regional and global prosperity or threaten Western economic interests? Will a rising China be tempted to use military power to assert its territorial claims and achieve its national goals, or will it be constrained by the potent forces of global economic interdependence? Will a more open and reform-oriented China gradually move toward political liberalization and democratization, thus enhancing stability and security in the Asia-Pacific? Will a more prosperous and powerful China be a peaceful, responsible, and constructive member of the international community that will respect and adhere to the rules of international organizations and regimes, or will it throw its weight around and challenge the norms of international society? There has certainly been no shortage of publications that address these important questions. However, while scholars and analysts may occasionally refer to some international relations theories or make some theoretical points in their discussion, little systematic effort has been made to place the entire debate on the China challenge within specific theoretical frameworks.[1]

This chapter attempts to provide a comprehensive analysis of the nature and implications of China's reemergence in the world from the realist and liberal perspectives. It argues that while both realism and liberalism have made useful contributions to the current debate, neither of them is adequate in offering satisfactory answers to the questions raised above. It is, therefore, necessary to combine the strengths of the two theories to reach a proper understanding of China's rise and its ramifications.[2] More specifically, I will apply Dale C. Copeland's theory of trade expectations[3] to the analysis of the factors that will shape the PRC's policy toward the outside world in the coming decades.

Given their different intellectual origins and theoretical orientations, realism and liberalism have provided divergent interpretations of the consequences of a rising China. Most realists believe that an economically powerful China will become more assertive and expansionist because of structural constraints. As the PRC's capabilities increase, they argue, its intentions will become less benign. Liberals, however, contend that China's reform and growing economic interactions with the capitalist world will make it more open and democratic and a promoter of international stability and security. Both realism and liberalism recognize the salience of high economic interdependence in determining China's future behavior, but they fail to offer a dynamic theory that will demonstrate precisely the conditions under which interdependence will have a positive or negative impact on Chinese security policy.

Drawing on Copeland's theory, I will show the conditions under which high interdependence between the PRC and its trading partners will lead to a pacific or belligerent Chinese security policy. If Chinese decision-makers' expectations of future trade are high, they will likely pursue policies that will enhance regional and global security. On the other hand, if they have a negative view of their future trading environment, they will likely take measures, including military action, to remove any obstacles that might forestall their pursuit of great-power status. Thus, the key to a smooth and successful integration of China into the international community is to maintain a high degree of economic interdependence within the country while promoting positive expectations of future trade among its leaders.

This chapter is organized around four main sections. The first two sections present the realist and liberal interpretations of the growth of Chinese power and its security implications, respectively. The third section analyzes the key factors shaping future Chinese perceptions and policies within the framework of the theory of trade expectations. On the basis of the research findings presented here, the final section considers how the international community should respond to the challenge of a rising China.

Realist Interpretations of the Rise of China: Power and Competition in an Anarchic World

The realist view of international relations is based on the assumptions that the world is essentially anarchic and that there is no central authority governing the behavior of individual states. To protect their national security and survival in such a self-help system, states must seek to acquire or maximize their power through economic and military means. It is, therefore, the structure of the international system that determines the behavior of states within it.

For realists, the emergence of China as a potential great power in the international system must be understood within the context of the end of bipolarity and the advent of a "unipolar world" following the disintegration of the Soviet Union. Soon after the revolutions in Eastern Europe in 1989, Charles Krauthammer argued in *Foreign Affairs* that the bipolar system would be replaced by one of "unipolarity." The center of world power, in his view, "is the unchallenged superpower, the United States, attended by its Western allies."[4] It would take decades to reach the stage of multipolarity. For the moment, the United States remains "the only country with the military, diplomatic, political and economic assets to be a decisive player in any conflict in whatever part of the world it chooses to involve itself."[5] The leading role played by the United States in the 1990–1991 Gulf crisis was cited as an example of U.S. military strength and political decisiveness in shaping the "new world order." Indeed, the 1990s can be characterized as a decade of unipolarity.[6]

However, realists believe that in a world of unipolarity other powers will rise to challenge the predominant position of the United States, and that the "unipolar world" may not last a long time. In an article entitled "The Unipolar Illusion," Christopher Layne uses neorealist theory[7] and historical evidence to explain great power emergence in the international system. The emergence of great powers, according to Layne, is "a structurally driven phenomenon."[8] The rise and fall of great powers are based on economic power that grows within states at different rates. There are winners and losers. In a unipolar system, states with successful economic expansion tend to become more ambitious and more capable of challenging the status quo, defending their increased overseas interests and commitments, and disrupting the dominance of the world's major power.[9] Any state that wishes to advance its status in an anarchic world must attempt to balance its position against the dominance of the hegemonic power. As states compete with each other, they tend to imitate their rivals' successful policies, moving toward "sameness." States with the capability of becoming major powers naturally seek to challenge the hegemon's preponderance and to pursue great-power status that in turn produces a shift from unipolarity to multipolarity, thus having a "structural impact" on the international system. Indeed, Layne notes that there was widespread concern in Europe, Asia, and the Third World about America's "unchallenged dominance in international politics" in the immediate post–Cold War era.[10]

The country that has the greatest suspicion of a unipolar system dominated by the United States is probably China,[11] which falls within the type of rising powers described by Layne. Over the past two decades, China has achieved remarkable economic success and accelerated its defense modernization. While the Chinese have imitated many of America's successful policies, it perceives the United States as a major rival in the international arena. Indeed, Chinese officials and analysts argued in the early 1990s that the "new world order" (*shijie xinzhixu*) advocated by George Bush was simply an American vision of a world order based on U.S. values and interests and designed to maintain American "hegemony" (*bachuanzhuyi*) in the post–Cold War world.[12] "In a unipolar world," says Layne, "others must worry about the hegemon's capabilities, not its intentions. The preeminent power's intentions may be benign today but may not be tomorrow."[13] In the eyes of Chinese leaders, however, post-1989 U.S. intentions are not even benign. They suspect that the United States attempts to undermine the legitimacy of their regime through various means following the Tiananmen events and the collapse of the Communist governments in Eastern Europe and the former Soviet Union.[14] Thus, China has been determined to challenge the predominant position of the United States in the international system and to defend what it perceives as its vital economic and security interests. It has been suggested that "China's primary foreign policy goal today is to weaken American influence relatively and absolutely, while steadfastly protecting its own corner."[15] "With certain new equipment and certain strategies," Thomas Christensen argues, "China can pose major problems for American security interests . . . without the slightest pretense of catching up with the United States by an overall measure of national military power or technology."[16]

Another structural change that encourages China to seek great-power status is the changing balance of power in the Asia-Pacific region since the end of the Cold War. With the breakup of the Soviet Union and a reduced U.S. military commitment to the region, realists predict, a "power vacuum" will emerge that will likely be filled by powerful regional players.[17] While Japan and India are often referred to as potential candidates who may wish to fill the vacuum in the event of an American withdrawal from the Asia-Pacific region, it is China that many realist analysts believe is the "hegemon on the horizon."[18] Why would China want to achieve a predominant position in the region? For realists, "a state's freedom to choose whether to seek great-power status is in reality tightly constrained by structural factors." More important, "eligible states that fail to attain great-power status are predictably punished."[19] China is seen as such a state that is foreordained to become a major power in the Asia-Pacific region. The size, population, and resources of the country, combined with the enormous potential of its economic and military strength, will empower China to achieve great-power status. As Kenneth Waltz argues, "For a country to choose not to become a great power is a structural anomaly. . . . Countries with great-power economies have become great powers, whether or not reluctantly."[20] If eligible states "do not acquire great-power capabilities, they may be exploited by the hegemon."[21] Assuming that the existing hegemon (the United States) will decline or pull out from the region, China would have to face the challenge of a potential hegemon, which could be Japan, its historical rival in East Asia. Chinese leaders know perfectly well that China was invaded and exploited by foreign powers during the nineteenth century precisely because it failed to attain great-power status while Japan and the European states did. After all, China suffered hugely from Japanese imperialism when it was economically and militarily weak.[22] To avoid another "century of shame and humiliation" (bainian chiru), China would need to achieve great-power status.

Of course eligible states will not become great powers unless their policy-makers make a "unit-level decision" to respond to the structural factors that drive them toward that direction.[23] But all the available evidence suggests, realists argue, that the policy-makers in Beijing have chosen to take advantage of the structural change in the international system. In fact, ever since economic reform and the open-door policy were introduced by Deng Xiaoping in 1978, Chinese leaders have been actively developing China's "comprehensive national strength" (zhonghe guoli) in order to be in a position to compete with other great powers politically, economically, and militarily.[24] The end of bipolarity has undoubtedly provided China with an excellent opportunity to elevate its status in the hierarchy of the international structure and to fulfill its great-power aspirations. Seen from the realist perspective, the rise of China is primarily a consequence of the fundamental change in the structure of the international system. Certainly, China's economic expansion since the 1980s has made it easier for its leaders to challenge America's unipolar position in the post-Soviet world, but the growth of Chinese power is a structurally driven phenomenon that will in turn contribute to the structural transition from unipolarity to multipolarity.

As the emergence of the PRC as a great power in the twenty-first century seems highly probable, realists warn that a rising China will present the international society with an immense challenge that will not be easy to manage. Specifically, an economically and militarily powerful China may pose a long-term threat to the stability and security of the Asia-Pacific region, and the world in general. As the late Gerald Segal puts it, "If, in 2020, the world faces a united, authoritarian China with the world's largest GDP, perhaps the world's largest defense budget . . . it will be too late to do much about it."[25] Western analysts point to the rapid development of Chinese military capabilities that has been made possible by China's sustained economic growth in recent years.[26] "International politics, like all politics," Hans Morgenthau writes, "is a struggle for power. Whatever the ultimate aims . . . power is always the immediate aim."[27] Realists tend to see power in terms of a zero-sum game where one actor's gain is another's loss. As John Mearsheimer notes: "States are principally concerned about their relative power position in the system; hence they look for opportunities to take advantages of each other. If anything, they prefer to see adversaries decline, and thus will do whatever they can to speed up the process and maximize the distance of the fall."[28] Therefore, the growth of Chinese power would mean the relative decline of the power of other countries. If and when China achieves its great-power status, realists fear, it will throw its weight around and will not play by the rules of the international community.

From a realist standpoint, a great power's behavior is determined not so much by its intentions but by its capabilities. As a state's economy expands, it will use its newfound power to extend its spheres of influence and to defend its economic interests whenever and wherever these interests are challenged. This pattern of great-power emergence has recurred many times in history: Britain, France, Germany, Japan, the former Soviet Union, and the United States, all went through similar paths.[29] "Big countries are more likely to be difficult to live with," Gideon Rachman notes, "if they have a strong sense of cultural superiority or historical grievances about their treatment by the rest of the world."[30] As an emerging power with a great civilization and a history of being humiliated by foreign countries, China will be likely to behave in the same way as other rising powers did in the past.[31] Thus, Denny Roy predicts that "China's growth from a weak, developing state to a stronger, more prosperous state should result in a more assertive foreign policy" and that "an economically stronger China will begin to act like a major power: bolder, more demanding, and less inclined to cooperate with the other major powers in the region."[32] Indeed, to the proponents of "civilizational *realpolitik*" like Samuel Huntington, China represents one of the hostile non-Western civilizations that will challenge the security interests of the West in the post–Cold War era.[33]

Clearly, realists are pessimistic about the international repercussions of great-power emergence. According to power transition theory, a rising power will seek to challenge the status of the leading power in the international hierarchy, which could result in a war between them.[34] As Robert Gilpin observes, a non–status quo power will attempt to change the international system by extending "its territorial control, its political influence, and/or its domination of the international economy" until "the marginal costs of further expansion are equal to or greater than the marginal benefits of expansion."[35] China is widely known as a dissatisfied and non–status quo power seeking to "right the wrongs" of its humiliating history and alter the existing rules of the international system that are thought to be created and dictated by the West.[36] Indeed, Beijing's irredentist claims in East and Southeast Asia have been cited by realists as the most convincing evidence for their argument that a more open and prosperous China will not necessarily be a pacific China. While it is inconceivable that China will attempt to reclaim its historical preeminence in Asia by conquering other countries, the PRC's neighbors are increasingly worried that it may use or threaten to use force to deal with unresolved territorial disputes.

Chinese declarations on and actions in the South China Sea since the 1980s have exacerbated this type of fear. Beijing has become much more assertive in handling the competing claims to the Spratly Islands.[37] For example, in an armed clash with Vietnam in March 1988, China took six islands in the Spratly area. During the conflict, it was reported that three Vietnamese transport ships were sunk and seventy-two seamen killed.[38] In February 1992, "The Law of the People's Republic of China on Its Territorial Waters and Contiguous Areas" was passed by China's National People's Congress to support Chinese claims to the South China Sea Islands (and the Diaoyu/Senkaku Islands in the East China Sea) and to authorize military eviction of "trespassers." In the same year the China National Off-Shore Oil Corporation signed a contract with the U.S. company Crestone Energy Corporation to explore oil and gas in the Vanguard Bank area. China promised to provide naval backing for the operation should there be any crisis in the region. In February 1995, China occupied Philippine-claimed Mischief Reef, which for the first time led to a direct confrontation between Beijing and an Association of Southeast Asian Nations (ASEAN) member state over the disputed islands in the South China Sea.[39]

The trepidation of China's Asian neighbors heightened substantially in 1995–1996 when Beijing held a series of large-scale military exercises and missile tests in the Taiwan Strait. The PRC's belligerent response to the unofficial visit of Lee Teng-hui, the former Taiwanese president, to the United States and Taiwan's first direct presidential election gave rise to much uneasiness among East Asian countries. To prevent the war game from turning into a military confrontation between Beijing and Taipei, the United States swiftly dispatched two aircraft carriers to the area. Although Lee managed to win the presidential election amid the crisis, Taiwan's economic progress was disrupted and the stability of the entire Asia-Pacific region was seriously threatened.[40] While the Chinese government has insisted that the cross-strait crisis was purely an "internal affair," realists are quick to point out that the event has indicated unambiguously China's willingness to achieve its national goals through military means. If Chinese leaders are prepared to use force against their compatriots, they ask, who can guarantee that they will not resort to the use of force in the future to intimidate countries that have disputes with China? It is argued that the PRC was sending a clear signal to its neighbors that "it now feels strong enough to play Asia's power game by its own rules and win."[41] "Over the past several years," Kurt Campbell and Derek Mitchell believe, "the training regimen, doctrine, writings, weapons procurement, and rhetoric of the People's Liberation Army have all turned to focus on a Taiwan attack scenario."[42]

Realists are particularly critical of the liberal assumption that economic interdependence reduces the possibility of military conflict. Both Taiwan and ASEAN states are important trading partners of China, yet their close economic relationships have not prevented the PRC from acting assertively in the South China Sea and across the Taiwan Strait. Gerald Segal argues that "China's behavior . . . suggests that China does not feel that the fruits of economic interdependence are at risk when it pursues its irredentist agenda or seeks greater international status, or else that these are short-term prices worth paying for a greater good."[43] Economic interdependence, realists contend, can in fact increase the likelihood of armed confrontation among trading nations as they seek to gain or maintain their access to vital resources and materials that are essential to the pursuit of wealth and power in an anarchic world.[44] Indeed, China's activities in the South China Sea are driven in part by the consideration that "the rich natural resources of the Sea are crucial to the survival and prosperity of an overpopulated mainland with ever declining resources."[45] Over the past two decades, China's demand for energy has increased rapidly as a result of greater energy consumption in the country.[46] In November 1993, China became a net oil importer for the first time in more than a quarter-century, and it is likely to become more dependent upon oil imports in the future.[47] According to a recent report, by 2020, China would have to import as much as 60 percent of its oil needs.[48] Without new reserves and with an 8 percent annual growth rate, China's current reserves could be exhausted within two decades.[49] In order to sustain its economic growth and achieve great-power status in the twenty-first century, the Chinese government would have to defend existing oil supplies and to find new energy reserves. Given the tremendous importance of the South China Sea to China's national development, realists point out, Beijing will press its claim to the area even though it may involve military confrontation with ASEAN states.

Similarly, Taiwan's dynamic economic interaction with China does not guarantee permanent peace across the Taiwan Strait because any attempt to separate the island from the mainland will be seen by China as both a threat to its security and an obstacle to the achievement of its great-power status. In the eyes of Chinese leaders, the nationalist cause of building a united, prosperous, and powerful China will not be complete without reclaiming the sovereignty of Taiwan. Thus, economic interests could be forfeited, if necessary, for the sake of military security and power maximization.

For realists, the challenge of a more stalwart and confident China to international security seems formidable, but there is no commonly acceptable strategy to deal with the China challenge. Those who believe that the transition from unipolarity to multipolarity is inevitable and that the rise of China is a structurally driven phenomenon tend to advise against attempts to suppress the emergence of new great powers. Any attempts by the United States to prevent emerging powers from rising will be futile or counterproductive. Instead, America should develop a strategy to safeguard its interests during the difficult process of structural transition. Christopher Layne argues that "it makes no sense to alienate needlessly states (such as China) that could be strategically useful to the United States in a multipolar world." The United States, he goes on to say, should concentrate its energy on "redressing the internal causes of relative decline" that "would be perceived by others as less threatening than a strategy of preponderance."[50] Some realist scholars believe that "despite the prevailing global unipolarity, contemporary East Asia is bipolar" and will be dominated by a maritime power and a land power—the United States and China.[51] The "lesser great powers" like Russia and Japan, Robert Ross contends, would not be in a position to challenge the bipolarity in the region due to geographic constraints. The United States and China will be strategic rivals competing for power and influence, but geography and stable bipolarity will contribute to regional peace and order.[52]

Others realists, however, are less certain that it is in the interests of the world to see the reemergence of China as a great power. Specifically, they believe that a stronger China will likely be a regional hegemon in the Asia-Pacific region[53] and a long-term adversary of the United States.[54] More seriously, the growth of Chinese power could provoke a remilitarized Japan and further arms buildup in the region that is "ripe for rivalry."[55] Realists maintain that the strength of nationalism as a driving force in international politics has not been weakened by growing economic interdependence. States only pursue cooperation with each other if it helps enhance their national interests or advance their status in the international system. "The fundamental goal of states in any relationship," as Joseph Grieco puts it, "is to prevent others from achieving advances in their relative capabilities."[56] Thus, "international institutions are unable to mitigate anarchy's constraining effects on inter-state cooperation."[57]

Consequently, various strategies of "containment" have been advanced. By far the most extreme form of containment is advocated by Charles Krauthammer who argues that the United States must do everything within its power to contain an aggressive and expansion-ist China, including developing and strengthening U.S. security relations with China's neighboring countries and supporting Chinese dissidents to destabilize the Communist regime in Beijing.[58] A slightly less-confrontational strategy is proposed by Denny Roy who believes that Asian and Western countries should try to "slow the growth of China's military and economic power."[59] This goal, he suggests, can be achieved by three means: (1) organization of a coordinated effort to transfer less capital and technology to China; (2) promotion of greater regional autonomy within China, thereby undermining Beijing's ability to marshal resources for aggressive actions; and (3) induction of the United States to main-tain its military forces in the region. Realizing that to impede China's economic develop-ment is probably not a feasible policy option, Roy later puts forward a strategy of "enmeshment"[60] that is more or less the same as Gerald Segal's proposal of the "constrain-ment of China." Segal argues that Chinese behavior cannot be moderated by economic interdependence alone and that the balance of power strategy is indispensable. China can be "constrained," he maintains, if Asian and Western countries are willing to act together to defend their interests "by means of incentives for good behavior, deterrence of bad behavior, and punishment when deterrence fails."[61]

A somewhat similar strategy is recommended by Gideon Rachman who argues for a combination of "economic engagement and strategic containment." The West should en-courage economic and trade interactions with China and allow China to join such organi-zations as the World Trade Organization (WTO) provided that no special concessions are given to the Chinese government.[62] In the meantime, Western countries and China's neigh-bors must unite in dissuading Beijing from using force to settle disputes in East Asia. They should make it absolutely clear that any Chinese military actions will provoke a strong and concerted response from the outside world that will severely damage China's economic interests. The key to this strategy, according to Rachman, is to sustain and expand the U.S. military presence in the Asia-Pacific region.

Liberal Interpretations of the Rise of China: Economic Modernization, Democratization, and Peace

Unlike the realists who tend to stress the importance of structural constraints to state ac-tions, liberals believe that the behavior of a state is determined largely by domestic factors such as culture, ideology, and political structure. In the liberal view, a government that is democratically elected is less likely to go to war against the will of its own people, and a state that is more interested in economic development and trade is unlikely to invade its trading partners. Thus, democracy and economic interdependence help mitigate the effects of anarchy and promote peace and international cooperation.

Liberals see China's gradual reemergence as an influential player in the international system as primarily a consequence of its successful economic reform and open-door policy over the past two decades. A China that is committed to reform and trade should be welcomed by the international community, for economic change will gradually transform the country into a more open and democratic one that will in turn be a stabilizing force in Asia-Pacific and global security. In the words of Kenneth Lieberthal, "a reform-minded and modernizing China will continue to advance toward a market-driven system guided by law rather than by corrupt families and will better meet the material needs of its citizens, eventually creating a middle class with a moderating influence."[63] As China becomes more prosperous, it is argued, its emerging middle class will demand more political freedom and a greater degree of participation in the decision-making process. Indeed, economic decentralization and increasing competition for economic benefits among different groups, organizations, and regions have resulted in the rise of interest-group politics in the PRC. Moreover, rapid technological change combined with growing economic and cultural interactions between China and the external world have made it difficult for the regime to maintain tight social and political control. There now exist numerous semi-official and unofficial social organizations and publications, leading to a scholarly debate on the possible emergence of a civil society in China.[64] Thus, economic development is seen as "a vital part of and a crucial condition for the realization of democratization."[65]

Liberals have often referred to the transformation of Taiwan, South Korea, and other East Asian countries where economic modernization was followed by political liberalization and democratization.[66] Indeed, the process of political change, according to many scholars, is already underway in China. Before the Tiananmen events in 1989 the Chinese authorities tolerated, if not encouraged, moderate moves toward liberalization. Despite occasional campaigns against Western "spiritual pollution," professionals and intellectuals became more influential in both articulating their views and providing policy advice to the government. There were also suggestions of legal and political reforms, including making the National People's Congress, China's parliament, more effective in scrutinizing the work of the government. However, those who had advocated political pluralism faced a serious setback and even persecution after June 1989 when Chinese leaders decided to "keep one foot on the economic accelerator and the other foot on the political brake."[67] Nevertheless, liberals point out, the pre-Tiananmen economic and political change was so profound that "China's citizens had not retreated to the older pattern of submission"[68] and that "the old tools of Communist indoctrination are no longer effective."[69] Since the late 1980s, they maintain, the trend toward greater liberalization has accelerated as a result of further economic progress, and "China is undergoing greater political change than is generally understood in the West."[70] William Overholt observes, "The media report a range of views rather than just the party line. Individuals' political speech is relatively free, short of calling for the organized overthrow of the leadership. People change jobs and move around much more freely. Citizens can sue the state." More significantly, competitive elections at the local level have been introduced that allow Chinese villagers to choose their local officials.[71] All these, according to Overholt, "are not token changes."[72] They will eventually lead to more substantial reform of institutions of a higher level such as the National People's Congress, and in a longer term, a fundamental change in the political system.[73]

The assumption that a democratic China will be a peaceful China is based on the theory of democratic peace that is central to the liberal perspective on international relations.[74] Liberals believe that democracies are restrained by constitutional mechanisms from fighting wars, as an unjustified war will not be supported by the people who have to bear the burdens of armed conflict. Moreover, democracies do not fight democracies, it is argued, because of their shared democratic ideals and moral values, greater transparency in communication, and more peaceful approaches to the resolution of conflict. Finally, within the "pacific union," war can be prevented by trade interaction and economic interdependence among democratic nations.[75] While the relationship between democracy and peace will no doubt stimulate further academic debates,[76] liberals are convinced that democratic regimes are more peaceful than other types of regimes. The proposition that democracies do not fight each other is so persuasive that even some realists subscribe to it. As Stephen Krasner, a prominent realist scholar, says, "Although the validity of the claim that democracies have not fought has been challenged on the basis of classification . . . and the small number of available cases . . . , the fact remains that the absence of war among democracies is an exceptionally powerful finding in a field where such unambiguous evidence is hard to come by."[77] Based on the democratic peace theory, therefore, liberals argue that a China that is moving toward political liberalization and democratization, albeit at a very slow pace, is less likely to use force to resolve territorial disputes with neighboring countries and to pursue its great power ambition by military means.[78] More important, a China that is increasingly linked to the world economy and interdependent with its trading partners is less likely to take aggressive actions that will be detrimental to its own economic interests.

Indeed, the argument that the likelihood of war can be reduced by international trade and economic interdependence has long been advanced[79] and continues to influence the thinking of many liberal scholars and policy-makers. Liberals believe that "China's growing economic power and its increasing number of links with global markets will correspondingly boost the costs of aggressive or noncooperative behavior" and that "proliferating economic links will enhance other states' abilities to influence China to play a more constructive role in the world community."[80] As Vincent Cable and Peter Ferdinand point out, "the nature of current Chinese development has involved building up strong economic linkages with its neighbors through trade and investment flows . . . which would make military confrontation all the more costly."[81] Chinese leaders, says Michael Yahuda, have recognized that economic interdependence plays a vital part in sustaining China's economic growth, maintaining its social stability, and legitimizing the rule of the Chinese Communist Party.[82] Some liberals reject the realist assumption that China "has 'adapted' its outward behavior but has not 'learned' a new way of thinking."[83] Christopher Findlay and Andrew Watson argue that "China's interaction with the world economy has created a level of trade interdependency that has transformed both China's international role and the way in which the rest of the world relates to China."[84] Similarly, Stuart Harris notes that "cognitive learning has taken place in the economic field" in China, which "supports liberal rather than realist views of the world, including the possibility of cooperative international behavior not only among 'liberal' societies, but also among liberal and non-liberal (in the Western sense) societies."[85] Furthermore, Thomas Robinson argues that interdependence in one sphere helps facilitate interdependence in other spheres. "Once the door is open in one arena," he suggests, "it is easier to open the doors in others and to keep them open."[86] This view is shared by Samuel Kim who observes that "judging by the phenomenal growth of Chinese IGO and INGO membership, positive participatory experience, and a number of policy adjustments and shifts over such global issues as arms control and disarmament, UN peacekeeping, North-South relations, human rights (until Tiananmen), and science and technology, there has indeed occurred some Chinese global learning."[87] Given China's multidimensional contacts with the outside world, liberals believe, it is gradually involved in the global process of "complex interdependence" and restrained by its participation in international institutions and regimes.[88] "What matters most," as Thomas Moore and Dixia Yang argue, "is China's behavior in coping with interdependence, not whether this behavior reflects adaptation or learning as such."[89]

For liberals, the reemergence of China as a great power will represent a huge challenge to international society. But "the challenges which China's rise presents," they argue, "are far more welcome than the problems that would have existed if China had remained unambiguously among the ranks of the poor and underdeveloped."[90] As Cable and Ferdinand put it, "a more prosperous China will be a better neighbor than a China stricken by poverty and economic crisis."[91] Commenting on the challenge of a rising China to the United States, Barber Conable and David Lampton conclude that "it is far better that America face the problems of success in China than those of failure."[92]

For liberal scholars, the future of China's economic and political development will have profound repercussions on other countries in a world of global interdependence. Economically, China's reform and open-door policy have provided the outside world with a variety of business and investment opportunities, and many Asian and Western companies and consumers have benefited enormously from the economic change in China in terms of market growth and cheaper products. "Most of the economic fears that China arouses," according to The Economist, "ignore all the benefits of China's growth" and "elementary economics."[93] As the PRC is a trading partner of numerous nations and has extensive economic links with the wider world, an economic failure in the country will seriously undermine the stability and prosperity of the Asia-Pacific region that could have far-reaching consequences for world trade and the global financial market. In the meantime, a poor and unstable China will become a huge burden for the international community. If the world finds it difficult to tackle poverty and famine in the Third World, liberals argue, it will be impossible to deal with similar problems in a country that has a population of 1.3 billion.[94] Related to this issue is the possibility of mass migration of Chinese citizens to other countries following a major economic disaster, social upheaval, and political turbulence. The world simply cannot cope with a refugee problem of such a massive scale.[95] Moreover, an economically weak and politically insecure Chinese regime will likely be more nationalistic and less cooperative. To gain popular support and preserve political control of the country, the regime may seek to recover some "lost territories" through military means, thus heightening tension in the Asia-Pacific and destabilizing the region. Should the central leadership fail to maintain the political cohesion of China, internal strife or even civil war may ensue that, some scholars fear, could invite external intervention and lead to great power struggle for control over the fragmented kingdom.[96] Under those circumstances, Bryce Harland predicts, "China could once again become an object of international competition and stimulate rivalries that would make regional cooperation an idle dream."[97]

The liberal response to the rise of China is therefore not to contain it but to try to integrate the nation the international society. "For economic, environmental and security reasons alike," it is argued, "a major priority is now to bring China into the centre of global, as well as regional, governance . . . into an institutional arrangement which recognizes that, within a short space of time, China is beginning to matter as much for the rest of the world as Japan, the EU or the US."[98] The problem is that it has never been easy for the international system to accommodate newly emerging powers. Historically, the arrival of great powers has tended to cause instability or military conflict, but it is precisely because of unhappy historical precedents, liberals contend, that the outside world must seek to draw China into the international community peacefully. Kishore Mahbubani warns that the lesson that other great powers failed to suppress the rise of Germany after World War I should be remembered. "New powers must be accommodated," he says, "not contained. Adjustments must be made . . . adjustments need not take place on the battlefield."[99] A useful adjustment that the world can make, according to liberals, is to incorporate China fully into a liberal trading order. More specifically, they argue for China's accession to the World Trade Organization, which would promote further Chinese reform as well as global economic stability.[100] A similar argument is put forward by *The Economist*: "Far better to bring China inside the world trading order, at a time when it is showing a genuine commitment to profound economic change, than to leave it out in the cold, nursing grievances."[101] Liberals believe that in trade and other areas, such as environmental cooperation, the international community should make a greater effort to promote dialogue and collaboration with the PRC with the ultimate aim of encouraging it to act more constructively and responsibly.[102]

From the liberal perspective, both China and its trading partners have common interests in maintaining stability and prosperity in the post–Cold War world, and they should seek to maximize their absolute gains through international cooperation.[103] In this view, the realist preoccupation with relative gains would only perpetuate mutual suspicion about which state will gain more from cooperation and how it might use its enhanced capabilities to dominate other states.[104] The suggestion of a "China threat" to the world is seen as a reflection of realist fear of relative gains by China, but it is inaccurate in terms of Beijing's intentions as well as its capabilities. Nicholas Kristof notes that "what China is doing is in most cases perfectly natural, and even its territorial and military aspirations are reasonable." No Chinese leaders, Communist or otherwise, can afford to give up claims over Hong Kong, Taiwan, the South China Sea, and the Diaoyu Islands.[105] Any country in China's position will seek to improve the quality of its weapons and strengthen its military capabilities. However, China is not seeking to overturn the entire balance of power in the Asia-Pacific; it is merely assuming the role of a great power in the region.[106] Given the importance of a peaceful global environment to Chinese economic modernization, China would be unwilling to destabilize the existing regional and international order. Indeed, liberals point out that since the end of the Cold War China has developed cooperative relations with most of its Asian neighbors, promoted regional economic cooperation, participated in bilateral and multilateral security dialogues in Asia, signed the Nuclear Nonproliferation Treaty, and acted on most occasions as a responsible member of international organizations.[107] "China is a revisionist power," says Robert Ross, "but for the foreseeable future it will seek to maintain the status quo."[108]

As far as China's military capability is concerned, liberal scholars argue, it is exaggerated by the "China threat" school. Despite China's determination to achieve military modernization, its defense budget is relatively small compared with that of Japan and other great powers;[109] its weapons remain obsolete and limited, and its power projection capability is poor.[110] Most defense specialists agree that the People's Liberation Army will not possess the capability to project and sustain force outside Chinese borders for at least twenty years.[111] David Shambaugh has persuasively argued that China does not have the capability nor the conscious desire to exercise hegemony over East Asia in the early twenty-first century.[112] Proposals of containing or constraining China are therefore seen as unrealistic and irresponsible that will only provoke Chinese hostility, heighten regional tension, and undermine international cooperation. Lieberthal warns that containment would "engender precisely the type of Chinese behavior that will prove upsetting to the regional and international systems" and that it would "divide Asia, strengthen narrow nationalisms, and reduce prosperity, security, and the prospects for peace throughout the region."[113] Thus, "containment would be the wrong policy, at the wrong time, against the wrong country"[114] that would help create a self-fulfilling prophecy. "The issues facing both the world and China," liberals contend, "are not how to 'contain' China, but how to manage the consequences of China's emergence as a major trading nation and how to ensure that trade interdependency evolves smoothly and beneficially."[115]

Western scholars and analysts have proposed various strategies to engage China, the soft version of which includes "comprehensive engagement"[116] and "constructive engagement"[117] that rely primarily on cooperative measures to integrate China into the international community. Others have put forward more assertive approaches such as "conditional engagement"[118] and "coercive engagement"[119] that place more emphasis on punitive measures should Beijing fail to follow the rules and regulations of the international institutions and regimes. Still others suggest that the outside world should try to "incorporate" the CCP leaders "by creating vested interests for them in the current international system."[120] Despite the differences in their strategies, scholars of the "engagement school" believe that engagement "offers the greatest leverage to influence the domestic evolution of Chinese society in a more liberal and open direction" and that "there is a likely correlation between China's domestic liberalization and its ability and readiness to comply with international rules and norms."[121] Thus, engagement is defined by Alastair Johnston and Robert Ross as "the use of non-coercive methods to ameliorate the non-status-quo elements of a rising major power's behavior" in the hope that "this growing power is used in ways that are consistent with peaceful change in regional and global order."[122] In reality, however, the strategies of engagement adopted by individual governments differ widely depending on their historical and current relations with China.[123]

Will China Be a Threat? The Trade Expectations Perspectives

Both realism and liberalism have offered valuable theoretical perspectives on the rise of China and its implications. As each of them has its strengths and weaknesses in explaining the nature of the China challenge, it would be fruitful to combine the insights of the two theories.

First, the reemergence of China as a great power is generally interpreted by realists as a consequence of the structural change in the post–Cold War international system. But the liberal view that domestic political and economic change plays a significant role in China's rise must also be taken into account. After all, Chinese leaders made a conscious "unit-level decision" to achieve great-power status well before the late 1980s, although the changing international structure has provided a favorable environment for China to fulfill its ambitions.[124]

Second, based on historical precedents, realists believe that great-power emergence is destabilizing because rising powers tend to pursue expansionist policies to promote or protect their economic interests. This, it is argued, is determined by states' capabilities rather than their intentions. As a rising power, the PRC will follow the footsteps of its predecessors, which would antagonize other great powers and provoke armed combat in Asia. Such a deterministic view of the repercussions of China's rise is based on the assumption that history will repeat itself. But what happened in the past may or may not recur in the future. It also ignores domestic constraints on foreign policy such as decision-makers' perceptions and political structures. In this respect, the liberal theory of democratic peace seems far more convincing. All the available evidence seems to suggest that economic modernization does lead to political liberalization, however slowly, and that democracies do not fight democracies. If China manages to sustain its economic growth, maintain its social cohesion and national unity, and become a fully democratic country, it will likely be a peaceful and cooperative member of the international community, but it will probably take several decades for China to reach that stage, if it ever does. For the foreseeable future, what would be the main factors shaping China's calculation of the utility of the use of force should there be major barriers to the achievement of its national goals?

Third, the realist apprehension of the potential threat of a prosperous and strong China is largely a reflection of a zero-sum conception of power that one actor's gain will be another's loss. While it is true that each state is seeking to enhance its capabilities in an essentially anarchic world, it does not follow that the growth of one country's strength necessarily means the decline of that of others. This realist interpretation of power is too static and pessimistic, as it reinforces existing suspicion and fear among states and precludes the possibility that all actors involved in international cooperation can have shared benefits. The liberal emphasis on absolute gains that encourages China and the outside world to work with each other appears to be more conducive to the furtherance of prosperity and security. However, given the realist tendency in dominant Chinese security perceptions,[125] would China take a military action to resolve conflicts even though it is economically interdependent with other countries? If so, under what conditions would China choose such an option?

Finally, realists and liberals disagree fundamentally on the adequacy of economic interdependence in managing the effects of an emerging China in the international system. Liberal scholars believe that growing economic and trade interactions between China and the outside world will aid Chinese leaders to appreciate the value of pursuing a peaceful foreign policy. Realists, however, contend that interdependence alone will not restrain the behavior of a rising power. On the contrary, interdependence will increase the probability of conflict between the PRC and its neighbors, as China becomes more dependent on external resources such as oil and grain imports. Given the volatility of Chinese foreign policy behaviors in recent years, there is no conclusive evidence to support either of the two arguments, but it does seem that economic interdependence can lead to peace or war, depending on the circumstances.[126] One must, therefore, ask these crucial questions: Under what conditions will a strong but (inter)dependent China be a pacific China that is willing to resolve differences with other countries in a peaceful manner; and under what conditions will it take threatening and belligerent actions in pursuit of its national interests?

The answers to the above questions, I argue, would depend on China's expectations of future trade. My argument is based on Dale Copeland's theory of trade expectations.[127] Although realism and liberalism have made useful contributions to the theoretical debate on the causes of war, neither of them is adequate in our understanding of the dynamic relation between economic interdependence and war. This is clearly demonstrated by their inability to provide satisfactory explanations for the outbreaks of the two world wars.[128] Liberals argue that interdependent states are unlikely to go to war because of the benefits of trade. Yet the European powers did fight with each other during World War I, even though there had been a high level of trade among them before the war. Realists seem to be correct in predicting that a high degree of interdependence can lead to war due to the potential costs of economic vulnerability. However, Germany and Japan were much more dependent on outside resources in the 1920s than in the late 1930s when they initiated World War II. Realist theory, therefore, fails to prove the correlation between high dependence and war.

Realism and liberalism predicate their predictions of a state actor's decision to initiate war on "a snapshot of the level of interdependence at a single point in time." Copeland's theory, however, offers a new variable (i.e., the expectations of future trade) that "incorporates in the theoretical logic an actor's sense of the future trends and possibilities." Thus, the theory shows "the conditions under which high interdependence will lead to peace or war."[129] Peace can be maintained, Copeland argues, for as long as "states expect that trade levels will be high into the foreseeable future. . . . If highly interdependent states expect that trade will be severely restricted," he goes on to say, "realists are likely to be right: the most highly dependent states will be the ones most likely to initiate war, for fear of losing the economic wealth that support their long-term security."[130] Using historical evidence Copeland analyzes Germany's decisions to fight World War I and World War II. In both cases, the findings reveal the same pattern of behavior: German leaders decided to go to war because of negative expectations for future trade and high dependence on other countries. "For any given expected value of war," Copeland argues, "we can predict that the lower the expectations of future trade, the lower the expected value of trade, and therefore the more likely it is that war will be chosen." It does not matter whether the present trade level is high or low. If expectations for future trade are low, the expected value of trade will be negative, as illustrated by Germany's decisions to take the military option prior to the two world wars. On the other hand, the expected value of trade can be positive, if expectations for future trade are high, even though present trade level is low or zero. For instance, the relationships between the Soviet Union and the United States and Western countries were relatively amicable during 1971–1973 and after 1985 because of the Soviet leaders' positive expectations for future trade.[131]

Following Copeland, I argue that whether a powerful China would pose a threat to regional and global security would depend largely on Chinese decision-makers' expectations for future trade. If Chinese leaders feel that current trade with the outside world would continue to expand, which would help enhance the wealth and power of the Chinese nation, there would be little incentive for them to resort to the use of force to settle unresolved disputes. If, however, Chinese leaders are convinced that their trade prospects would deteriorate substantially in the near future, which would undermine their ability to maintain the long-term prosperity and security of China, they would be likely to take the military option to avert vulnerability and decline. In other words, the PRC's future behavior would be determined not only by the level of interdependence with its trading partners but by its expectations of the future trading environment.

Since the late 1970s, China has gradually emerged as a major trading nation in the world, and its economic and trade relations with most countries have broadened considerably. Indeed, China has been actively involved in global economic activities, and is fully integrated into the Asia-Pacific economy. The PRC is now a member of all the major international and regional economic organizations, including the World Trade Organization, the World Bank, the International Monetary Fund, the Asian Development Bank, and the Asia Pacific Economic Cooperation (APEC).[132] From 1980 to 1997, the Chinese government approved 162 foreign financial institutions to develop business in China.[133] Over four hundred of the world's top five hundred multinational corporations have now invested in the country.[134] As a result, there has been a huge growth in China's foreign trade over the past two decades. From 1978 to 2002, China's exports grew from US$9.8 billion to US$325.6 billion, and its imports grew from US$10.9 billion to US$295.2 billion. Between 1983 and 2002, actual foreign direct investment in China increased from US$916 million to US$52.8 billion.[135] In terms of total trade volume, China has become the seventh largest trading nation in the world.[136] China has also benefited from its involvement in a regional division of labor and economic cooperation in East Asia. It is integrated into a number of subregional economic groupings or "growth triangles" such as the Hong Kong–Guangdong–Shenzhen triangle and the Northeast China–Korea–Japan triangle. In addition, China is closely involved in the development of two new subregional groupings: the Yellow Sea Economic Zone that includes Liaoning and Shandong provinces, Japan, and South Korea; and the Tumen River project that seeks to promote economic cooperation among China, Japan, North Korea, South Korea, Mongolia, and Russia.[137] At the ASEAN Plus Three summit in November 2002 at Phnom Penh, Chinese Premier Zhu Rongji and ASEAN leaders decided to establish a China-ASEAN Free Trade area by 2010 that could become the world's third largest trading bloc.[138]

No doubt, China's integration into the world economy has brought about much benefit to the country, but it has also increased Chinese vulnerability in a world of growing interdependence. Indeed, foreign direct investment has become the single most important source of foreign capital for the PRC.[139] It is estimated that "foreign investment may now account for one-quarter of all Chinese exports."[140] In 1996, the total value of foreign-funded firms' import and export trade reached US$137.1 billion, accounting for 47 percent of the national total of foreign trade.[141] According to a *Beijing Review* report, 18 million people, about 10 percent of China's nonfarming population, are employed by foreign-funded firms. The investment by these firms covers a whole range of areas that are vital to Chinese economic modernization, including infrastructure, energy, communication, and high-tech projects.[142] The Chinese government has also relied heavily on foreign investment to develop the central and western regions of China that are still very poor. In 1996, for example, a total amount of US$1.34 billion of foreign government loans was utilized for 69 projects in these underdeveloped regions. In addition, 125 key projects in the PRC are supported by foreign government loans that include the construction of metropolitan underground railways, power plants, airports, telephone networks, and other large-scale development plans.[143] By June 2002, the World Bank Group had lent a total of US$33.9 billion to China, which supported 239 projects in all major sectors of the Chinese economy and in a wide range of regions across the country.[144]

In the past decade China has increased its foreign borrowings substantially. Its total external debt is believed to have risen from US$24,000 million in 1987 to US$116,280 million in 1996.[145] Besides, many of China's reform projects, such as enterprise restructuring, infrastructure improvement, financial reform, poverty reduction, human development, and environmental protection, are currently supported by the World Bank.[146] Of all the major sectors of the Chinese economy, energy is probably the most critical one in terms of sustaining the PRC's modernization program. In this sector the role of foreign capital is becoming more significant. For example, a joint venture has been established at the Pingshao coal mine, and the construction of a power station in Guangxi Zhuang is financed entirely by foreign investment. In the areas of petroleum and natural gas, a greater effort has also been made to attract foreign capital. By 1997, China had signed 126 contracts with 65 foreign oil companies.[147] Moreover, the progress of Chinese reform is dependent on the availability of advanced foreign technology and equipment. The contract value of Chinese technology imports amounted to US$159.23 million in 1997. Indeed, imported technologies play an important part in major Chinese industries ranging from energy, electronics, and computer software to telecommunications, information, and other high-tech industries.[148] Clearly, Chinese leaders are aware that the success of China's economic modernization rests ultimately with its access to the global market and with inflows of external funding. If, for political or security reasons, the world were to reduce the level of economic interactions with or apply trade sanctions against China, it would have a devastating effect on Chinese economic development.

For the moment, China's expectations of future trade with both its Asian neighbors and Western nations are by and large positive. In a speech to an academic symposium in Beijing, Chen Jian, a senior Chinese official, said that "the international situation has moved at a speed faster than expected in a direction favorable to China. . . . The ongoing reform and opening up policies and the economic development in China . . . are based on the judgment that world peace can be maintained and a new world war will not erupt for the near future."[149] Similarly, Wu Yi, former minister of Foreign Trade and Economic Cooperation, has noted, "We are immersed in the irreversible general trend toward worldwide economic integration . . . economic cooperation with various countries makes it easier than any time in the past to reach a common view, and can be carried out in a wider area and at a higher starting point. This in turn portends that possibility for successful cooperation is much greater in the future."[150] This type of optimistic assessment of the future trading environment is echoed by many Chinese leaders, officials, and scholars.[151] Despite the recent financial turmoil in East and Southeast Asia, they believe that the economic dynamism in the Asia-Pacific region will continue into the twenty-first century and that China will benefit from further economic growth and cooperation in the region. For example, citing the view of a Chicago professor and Nobel Prize winner, a Chinese commentator maintains that "the prospects of most rapidly growing economic entities of East Asia are still bright. . . . Even if the economy of these countries stops growing in the coming five years," it is argued, "their average speed of economic increase in the next 25 years will surpass that of the world."[152] Undoubtedly, China's accession to the WTO will entail an immense economic and possibly political challenge to the Beijing government,[153] but the Hu Jintao leadership appears as sanguine about China's future trading environment as its predecessor. In any case, Chinese leaders know that the potential market and business opportunities that the PRC can offer to the outside world are so attractive that no country wants to miss out. It is, therefore, unlikely that any countries would want to sever trade relations with China in the near future.

However, there is no guarantee that Chinese leaders' positive anticipation of future trade will remain unchanged. In fact, there has been growing concern among Chinese elites over the past few years that some external forces do not wish to see a strong and prosperous China and that they are trying to prevent the PRC from fulfilling its great-power potential.[154] Beijing is especially sensitive to the suggestion of a "China threat,"[155] fearing that it will arouse anti-Chinese sentiments in Asian and Western societies and jeopardize the relationship between China and its trading partners. Criticisms of Chinese policies, such as China's Tibet policy, human rights record, and arms sales to Third World countries, have been perceived as Western attempts to weaken China's position both internally and in the international community. In particular, America's China policy has been viewed by PRC elites with skepticism.[156] Despite former U.S. President Bill Clinton's reference to China as a "strategic partner," his policy was characterized by most Chinese analysts as one of engagement plus containment (bian jiechu, bian ezhi).[157] Many of the measures taken by the Clinton administration, the Chinese believe, reflected in varying degree the U.S. intention of guarding against (fangfan), constraining (zhiyue),[158] or containing (ezhi) China. For example, Chinese leaders suspected that the 1996 U.S.-Japan Joint Declaration and the 1997 revised Guidelines for U.S.-Japan Defense Cooperation were targeted primarily against China. They were particularly concerned about the possible inclusion of the Taiwan Strait in the ambiguous "surrounding areas" covered by U.S.-Japan defense cooperation,[159] which reflected their perturbation of a joint effort of Washington and Tokyo to "constrain" China.[160]

Not surprisingly, America's continued support for Taipei is seen as a means of obstructing the PRC from achieving reunification with Taiwan. Beijing's suspicion of U.S. intentions heightened when China was depicted as America's "strategic competitor" by some foreign policy advisors of the George W. Bush administration.[161] In April 2001, President Bush said in public that the United States would do "whatever it took to help Taiwan defend itself." In the meantime, he approved the sale of a massive arms package to Taipei that would enhance Taiwan's capability to break potential Chinese blockades. Despite the need to secure Beijing's support for its international campaign against terrorism, Washington has not abandoned its commitments to Taiwan. If anything, it has developed closer defense ties with the Taiwanese military and allowed senior Taiwanese leaders and officials to visit the United States. A leaked Pentagon report has allegedly suggested that nuclear weapons could be used against China in the event of a conflict across the Taiwan Strait.[162] It is clear that on a variety of strategic, political, and economic issues, the perceptions of Chinese and American policy-makers differ profoundly.[163] While the events of September 11 and the "war on terror" may have provided a new opportunity for U.S.-China cooperation, the expansion of America's antiterrorist networks in Central, South, Southeast, and Northeast Asia has exacerbated Chinese fear of a strategic encirclement of

China. Chinese leaders and elites are convinced that the Bush administration is seeking to maintain America's unipolar position in the global system through the development of a National Missile Defense system and a Theater Missile Defense system in Asia as well as other unilateral actions.

While the Chinese economy is relatively strong and China has abundant human and natural resources, the country would not be able to sustain its economic growth, let alone achieve a great-power status, if it were to be isolated from the global economy. This explains why the Western debate on the "China threat" (*Zhongguo weixie*) has caused tremendous concern in China and has been followed closely by Chinese analysts.[164] They respond to Western misgivings of the rise of China by arguing that "a peaceful environment could be conducive to the emergence of a great power," and that "the rising power would value peace more than any other countries do." The assumption that the rise of great powers in the international system is always associated with the causes of war, they contend, is based primarily on traditional Western political thinking that is being challenged by "contemporary practice."[165]

Despite Chinese anxiety of containment, Beijing's expectations of future trade remain high at the moment. As Chinese leaders' expected value of trade is positive, they see no benefit in taking the military option to deal with unresolved issues. Consequently, the Chinese have been involved in numerous official and unofficial security dialogues, including the ASEAN Regional Forum and many "track two" meetings. Their attitudes toward multilateral security cooperation have also become more positive.[166] Moreover, China has adopted a more flexible approach to its territorial disputes such as the conflict in the South China Sea. While Chinese leaders' position on the Spratly and Paracel islands has not changed, they are willing to shelve the issue of sovereignty for the time being and to negotiate joint development of the resources in the area with other claimants.[167] At the November 2002 ASEAN summit in Phnom Penh, China signed an important accord with ten Southeast Asian governments aimed at avoiding armed conflict over contested areas of the South China Sea.[168] Furthermore, China has been remarkably restrained in handling recent disputes with Japan over the sovereignty of the Diaoyu/Senkaku Islands in the East China Sea. Despite popular demand for taking tougher actions against Japan, the Chinese government has chosen to deal with the issue through diplomatic avenues.[169] On the Taiwan issue, China did not respond to Chen Shui-bian's victory in the 2000 presidential election in a belligerent manner. One can also discern a relatively muted reaction from Chinese leaders to the DPP (Democratic Progressive Party) government's bolder attempts to promote a Taiwanese identity. Growing economic integration across the Taiwan Strait has no doubt served as a disincentive for Beijing to use force against Taiwan. It is quite obvious that China has optimistic expectations of future trade and does not want to pursue policies that will dramatically alter the current trading environment. Indeed, the Chinese government has been reluctant to utilize the devaluation of the Renminbi as a means of stimulating its exports, as such a measure could further destabilize the fragile economies of Southeast Asia that would jeopardize China's future trading environment. While recognizing the negative effects of the Asian financial crisis on Chinese economic development, China sees it as an opportunity of strengthening its political and economic influence in East Asia and demonstrating its credibility as a responsible member of the international community.[170] For the moment, the governments of the United States, the European Union,

Japan, and ASEAN states have all made it clear that they wish to engage China both bilaterally and multilaterally. If, however, these countries were to change their policy to one of containment in the future, leading to a negative trading environment, Beijing might respond by taking military action to secure what it perceives as its vital interests. This could result in a situation similar to that in Germany after 1897 when it felt that other great powers were seeking to undermine German economic and security interests. Indeed, German leaders' trade expectations before World War I were so low that they came to the conclusion that "only a major war would provide the economic dominance of Europe needed for long-term German survival."[171]

Whether China will choose the military option in the future is also dependent on the Chinese assessment of the expected value of war in relation to the expected value of trade. In other words, China will have to think about its capabilities vis-à-vis the capabilities of other states in deciding to go to war.[172] According to Copeland, a state's expected value of war will be positive if its economic and military power is more superior than that of its adversaries. If a larger state can easily conquer a smaller and weaker one and absorb its economy, the expected value of war will also be positive. On the other hand, if the large state is facing an adversary of similar size or strength, it will have a lower or negative expected value of war because of the high costs involved and low possibility of victory. The most important consideration, however, is the state's estimate of the expected value of war relative to the expected value of trade. "If the expected value for trade is lower than the expected value for invasion," Copeland argues, "war becomes the rational choice, and this is so even if the expected value of invasion is itself negative: war becomes the lesser of two evils."[173]

If Chinese decision-makers were to reach a conclusion that the outside world is determined to impede China's economic progress and suppress its reemergence as a great power, their expected value of war will become greater than the expected value of trade. To guarantee that it can exploit the vast deposits of valuable resources in the South and East China Sea, control important shipping lanes in the areas, and gain strategic advantages over its adversaries, China might contemplate taking the military option. It might decide to use force to take over Taiwan, occupy the Spratly Islands, and reclaim the Diaoyu/Senkaku Islands from Japan. But in estimating the expected value of war, the PRC will have to consider the balance of power between itself and other states in East Asia. If it were to fight with Taiwan or an individual ASEAN state alone, its expected value of war might be positive. If, however, it had to face a more powerful country or group of countries, the expected value of war would be negative. Barring the intervention of Russia, it would be impossible for China to fight a war if it were to confront a military alliance consisting of the United States, Japan, Taiwan, and ASEAN states. It is generally agreed that China's war-fighting machines are well behind those of America, Japan, and some other Asian states, and that it would take some time for Beijing to develop its power-projection capability. In any case, other countries would continue to improve the quality of their weapons. Nevertheless, should the PRC be in such an isolated and desperate position, Chinese leaders might still choose the option of war because it would be seen as "the lesser of two evils."

When analyzing a state's trade expectations one must also take into account the effects of diplomacy and bargaining, as Copeland suggests. A state can make some economic, political, and military concessions to induce its trading partners to relax trade restrictions, thus raising its expectations for future trade. If the price for a higher level of trade is seen to be reasonable, the state would be willing to pay it. However, if the price is unacceptable because it would undermine the state's "internal stability or its external power position," there would be very little that the state could do to improve its trade expectations.[174]

In the case of China, it has made some economic and political concessions to induce the outside world to trade with and invest in the country. On most issues, Chinese leaders find the price of higher trade level reasonable and are willing to make compromises. The concept of "one country, two systems," for example, was basically formulated to assure the Western world that China's priority was economic development. In order to retain the confidence of foreign investors in Hong Kong, Chinese leaders have promised that the territory's capitalist system will remain unchanged for at least 50 years from 1997.[175] China's flexibility on the South China Sea disputes also reflects its desire to maintain harmonious relations with ASEAN countries that are propitious for China's trading environment. Similarly, to demonstrate its respect for the rules and norms in international trade, the Beijing government has taken strong measures to prohibit the violation of intellectual property in Chinese provinces, albeit without much success. Another example of China's willingness to compromise for economic and trade benefits is its agreement to sign the Nuclear Nonproliferation Treaty and the Comprehensive Test Ban Treaty. For the same reason, China did not vote against the U.S.-sponsored United Nations Security Council Resolution 1441 that forced Saddam Hussein to allow UN inspectors to conduct vigorous weapons inspections in Iraq.

On other issues, however, Chinese leaders feel that the price of a higher trade level is unacceptable as it will undermine the PRC's domestic stability, national unity, territorial integrity, regime survival, or security interests. For example, while China appears to be determined to reform its state-owned enterprises (SOEs), especially since the Chinese Communist Party's 15th congress in September 1997 and the appointment of Zhu Rongji as the Chinese premier in March 1998, it has rejected Western pressure to close down all the nonprofit-making enterprises immediately, as the result of these measures would be a huge rise in unemployment that could cause much hardship for many Chinese workers in the absence of a sound social security system. This could, in turn, provoke widespread discontent among the urban dwellers that would seriously destabilize Chinese society and possibly engender a social upheaval in the country. Instead, China has adopted a more gradual approach that concentrates on "improving the performance of the large SOEs considered strategic for national development, while letting go of the very large number of smaller enterprises, many of which are simply not viable."[176] Meanwhile, the Chinese government has introduced a series of reforms that are linked to a modern enterprise system, including pension, welfare, health care, and housing reforms.[177]

Beijing is also loath to succumb to any foreign pressure to change its policy toward Tibet that is viewed as an "inalienable" part of Chinese territory. To grant more independence to Tibet, Chinese leaders fear, would encourage other regions to break away from the center, thus jeopardizing China's "national unity and territorial integrity." Similarly, China took a highly assertive and confrontational posture in 1995–1996 when it was convinced that Lee Teng-hui was pursuing a "two China policy" in the guise of "pragmatic diplomacy."[178] It is inconceivable that Beijing did not expect strong reaction from the outside world to its belligerent actions that heightened tension across the Taiwan Strait, but it rather risked the possibility of confronting U.S. military forces and destabilizing the entire Asia-Pacific region, which could be detrimental to Chinese economic interests. Equally important is Chinese leaders' reluctance to compromise on issues relating to regime survival. The Tiananmen tragedy in 1989 is a classic example showing how far the Communist leadership is prepared to go when it comes to quelling activities that would challenge its authority and legitimacy to govern the country. It is, therefore, not surprising that the Chinese government has treated its political dissidents harshly, despite constant criticisms and protests from the West. Thus, the price for a higher trade level that the outside world expects Chinese leaders to pay will play a substantial part in shaping China's expectations for future trade.

An Ascendant China: Challenge for the International Community

Whether the reemergence of China as a great power in the post–Cold War international system is caused by structural factors (as the realist argues) or by unit-level decisions (as the liberal suggests), the challenge that China presents to the rest of the world is formidable. The best way of abating the likelihood of military conflict between the great powers, as Copeland suggests, is to "alter leaders' perceptions of the future trading environment in which they operate."[179] China's current expectations of future trade are on the whole positive, but there are growing suspicions among Chinese leaders and intellectuals of external forces seeking to "contain" China. Such a fear could magnify at a time when nationalistic sentiment is rising in Chinese society[180] that might lead to low expectations of future trade. To ensure that China's rise will not cause regional and global instability, the outside world should pursue policies that would enhance Chinese decision-makers' confidence in their future trading environment. This will not be a simple task due to China's innate distrust of other great powers as a result of its unpleasant encounters with Japan and Western powers in the nineteenth century. Given the complexity of Chinese domestic politics and enormous ideological and institutional constraints, China may not always respond to external efforts positively.[181] However, if China's trading partners hope to integrate the country into the international community peacefully, they must do what they can to raise PRC leaders' expectations for future trade.

To begin with, the outside world should try to convince Chinese leaders that it has no intention of hindering China's economic development, impairing its national cohesion, and thwarting its attempts to achieve great-power status. China should also be reassured that its sovereignty and territorial integrity are recognized and respected, provided that Beijing does not take any provocative or aggressive actions against its neighbors when dealing with territorial disputes. Some observers believe that as Chinese leaders have not yet reached a consensus on the future role of China in the world, there are still opportunities for other major powers to exert positive influence on their intentions.[182] As Rodolfo C. Severino, Jr., a senior government official of the Philippines, puts it, "The course of events will depend as much on perception as on reality."[183] The international community, he suggests, can try to "encourage the perception among China's leaders that the world is a benign place and that China's security and economic growth rest on integration rather than on the unilateral exercise of power."[184] Thus, the outside world should refrain from taking any actions that might augment Chinese fear of containment. Instead, it should actively support policies that will help integrate China into the global economy and international regimes,[185] and continue to engage China on a wide variety of security issues through bilateral and multilateral meetings as well as various official and unofficial channels of security dialogue.

To raise its expectations for future trade China has been and will be willing to make economic and political concessions when negotiating contracts and trade agreements with its trading partners. Thus, the outside world will have some leverage to steer China into a certain direction, and it should take the opportunity to encourage further economic reform, openness, and trade liberalization in the country. As liberals rightly argue, economic liberalization will gradually lead to greater political liberalization and democratization in China that will in turn help preserve peace and stability in the Asia-Pacific region. However, the international community must be patient with the pace of change in China and more sensitive to Chinese perceptions. This is not to say that the outside world should accede to any Chinese demands or policies. On the contrary, it should be prepared to raise its concern over particular Chinese policies, debate with China on issues of fundamental disagreement, and stand firm on matters of principle. Nevertheless, they must recognize that the process of democratization in China will be a lengthy and thorny one, given the lack of democratic tradition in Chinese history. An evolutionary path toward democracy is preferable to a violent change of regime in China that will likely produce an unstable and ineffective government, which would be incapable of handling the crises and upheavals associated with rapid political transformation in such a vast country. A chaotic China could not possibly pursue a rational and coherent policy toward other countries. In this regard, the warning of some liberal scholars of the linkages between democratic transition and war should be heeded.[186]

Moreover, the outside world can use economic leverage to entice China to seek peaceful solutions to its territorial disputes with littoral states. On the South China Sea, the PRC's proposal of "shelving the dispute and working for joint development" is a positive step in the right direction. The Chinese government must be urged to pledge not to occupy any reefs around the Spratlys and intimidate other claimants by military force, and to work out practical measures with Southeast Asian nations to jointly explore the marine resources in the area.[187] The emphasis on maintaining the status quo in the South China Sea should also be applied to the Taiwan Strait. China must be dissuaded from taking any military actions that would threaten the security of Taiwan and warned of the consequences of such conducts. At the same time, Taipei should not be encouraged to seek legal independence, as it would only ignite a major crisis in the region that could have highly destabilizing effects. As Chinese leaders see the Taiwan issue as a matter of territorial unity, they will never tolerate an independent Taiwan. If that is the price for a higher level of trade, it is almost certain that China will not pay for it. Thus, an unacceptable price for higher trade level will reduce China's expectations of future trade and increase its expected value of war.

If Chinese leaders feel that outside powers have no desire to undermine Chinese security interests and that China will continue to benefit from global economic interdependence, there is little reason for them to disrupt regional and international stability that is essential to Chinese prosperity. In other words, if China's expected value of trade is greater than its expected value of war, it should not initiate any military conflict. For a variety of reasons, however, misperception may at times arise that could affect Chinese decision-makers' estimate of expected value of trade. For example, the PRC's genial relations with Japan or ASEAN states could be swayed by trade frictions or territorial disputes in the East and South China Sea. Under certain circumstances, they might be tempted to use force to protect their economic interests or attain their political goals. An underestimation of U.S. resolve to defend Taiwan may encourage Chinese leaders to deal with the Taiwan issue through military means. Another potentiality is that China might opt for an assertive or adventurous foreign policy as a result of domestic pressures such as serious economic crises, social and political turmoil, rising nationalistic sentiments, leadership succession, and pressure from some quarters of the military.[188]

Apart from raising China's expected value of trade, the outside world should also lower its expected value of war by introducing some elements of the balance-of-power strategy. Thus, a robust American military presence in the Asia-Pacific region, combined with bilateral security ties between the United States and other Asian countries, would be indispensable to the reduction of China's expected value of war. The United States, Japan, and ASEAN states, however, must not form a military alliance in Asia unless there is unambiguous evidence that China is threatening regional security.[189] Otherwise, it would aggravate Chinese fear of encirclement and push Beijing toward the direction of further military expansion. Chinese leaders will select the military option if they conclude that the PRC's national security and survival are endangered by outside forces, even though their expected value of war is negative.

Accommodating emerging great powers has never been easy for the international community, but one must not assume that rising powers will necessarily be expansionist and aggressive. "Given the strategic head start the United States and its allies enjoy," Robert Ross argues, "Washington has the luxury of observing Chinese modernization before adopting a more assertive posture."[190] Similarly, Jusuf Wanandi, a leading Indonesian policy analyst, says that "there is plenty of time to see whether tough and perhaps dangerous policies need to be put into practice in order to cope with what might then be a difficult China."[191] For the time being, the goal of the outside world should be to cultivate a propitious economic and security environment that will help augment China's expected value of trade and curtail its expected value of war. As Joseph S. Nye, Jr., has rightly pointed out: "To discard the chances of a more benign future through a misguided belief in the inevitability of conflict would be a tragic mistake."[192]

Conclusion

This chapter has provided a systematic analysis of the Western debate on China's emerging role in the international system and its potential challenge to the world from the theoretical perspectives of realism and liberalism. While both theories have offered some illuminating insights, neither of them alone is able to unravel the puzzle as to whether a strong and prosperous China will be a force of stability or a threat to international peace.

Based on the theory of trade expectations, I have argued that whether China will be a peaceful and responsible member of the international community depends not only on the level of its economic interdependence with the outside world but on Chinese leaders' expectations of future trade. As realists would contend, the trade factor would not have much impact on China's consideration to choose between war or peace if its level of dependence were to be low, but the reality is that the Chinese economy is now interconnected and interdependent with the economies of most of its trading partners. Thus, Chinese decision-makers' expectations of future trade become pertinent to our analysis. If they expect that trade level will be high into the foreseeable future, they will be unlikely to take military actions that will damage their trade interests. In this sense, the liberal argument that inter-dependence reduces the probability of war would be valid. However, if Chinese decision-makers' expectations of future trade is low, then interdependence will increase the likelihood of war due to the high costs of severed trade relations, as realists would predict.

The evidence presented in the chapter suggests that China's expectations for future trade are relatively high at the moment. There are, however, indications that this positive assessment of the future trading environment could change because of growing apprehension among Chinese leaders and elites that some external forces are seeking to prevent China from realizing its great-power potential. If Beijing's expectations of future trade were to become negative, its expected value of trade would decline accordingly, and it would be likely to choose the military option to achieve its aims, if necessary. Thus, the chapter has concluded that to ensure that the rise of China will not threaten regional and global stability, the international community should strive to pursue policies that will raise Chinese decision-makers' expected value of trade and abate their expected value of war. This will not be at all easy, but the alternative to it could be disastrous for China and the world alike.

Notes

1. See Denny Roy, "The 'China Threat' Issue: Major Arguments," *Asian Survey* 36, no. 8 (August 1996): 767–770; Avery Goldstein, "Great Expectations: Interpreting China's Arrival," *International Security* 23, no. 3 (Winter 1997/98): 62–71; Herbert Yee and Ian Storey, "Introduction," in Herbert Yee and Ian Storey, eds., *The China Threat: Perceptions, Myths and Reality* (London: RoutledgeCurzon, 2002), pp. 6–10.

2. Charles Kegley has called for a "melding" of realism and liberalism, suggesting that "the global system's evolving character encourages considering how a reconstructed theory that integrates the most relevant features of both theoretical traditions might be built." See Charles W. Kegley, Jr., "The Neoliberal Challenge to Realist Theories of World Politics: An Introduction," in Charles W. Kegley, Jr., ed., *Controversies in International Relations Theory: Realism and the Neoliberal Challenge* (New York: St. Martin's Press, 1995), p. 17. Similarly, Robert Keohane believes that we should try to "break down artificial barriers between acade-micians' doctrines" and to "synthesize elements of realism and liberalism" to create new theories. See Robert O. Keohane, "Institutional Theory and the Realist Challenge after the Cold War," in David A. Baldwin, ed., *Neorealism and Neoliberalism: The Contemporary Debate* (New York: Columbia University Press, 1993), p. 293.

3. Dale C. Copeland, "Economic Interdependence and War: A Theory of Trade Expecta-tions," *International Security* 20, no. 4 (Spring 1996): 5–41. Drawing on both realist and lib-eral insights, Copeland's theory is an excellent example of "melding" or "synthesizing" the two international relations approaches suggested by Kegley and Keohane, respectively. See note 2.

4. Charles Krauthammer, "The Unipolar Moment," *Foreign Affairs* 70, no. 1 (1990/91): 23.

5. Ibid., p. 24.

48   CHAPTER 2

6. Michael Mastanduno, "Preserving the Unipolar Moment: Realist Theories and U.S. Grand Strategy after the Cold War," *International Security* 21, no. 4 (Spring 1997): 49–88; William C. Wohlforth, "The Stability of a Unipolar World," *International Security* 24, no. 1 (Summer 1999): 5–41.

7. Kenneth N. Waltz, *Theory of International Politics* (Reading, MA: Addison-Wesley, 1979).

8. Christopher Layne, "The Unipolar Illusion: Why New Great Powers Will Rise," *International Security* 17, no. 4 (Spring 1993): 9.

9. Robert Gilpin, *War and Change in International Politics* (New York: Cambridge University Press, 1981); Paul Kennedy, *The Rise and Fall of the Great Powers: Economic Change and Military Conflict from 1500 to 2000* (London: Fontana, 1988).

10. Layne, "The Unipolar Illusion," pp. 35–37.

11. See the analysis in Rex Li, "Unipolar Aspirations in a Multipolar Reality: China's perceptions of US Ambitions and Capabilities in the Post–Cold War World," *Pacifica Review* 11, no. 2 (June 1999): 115–149.

12. China's views of George Bush, Sr.'s vision of a "new world order" were widely expressed in Chinese newspapers, journals, and books in the early 1990s. See, for example, Du Gong and Ni Liyu, eds, *Zhuanhuanzhong de shijie geju* (The World Structure in Transition), (Beijing: Shijie zhishi chubanshe, 1992), pp. 299–308.

13. Layne, "The Unipolar Illusion," pp. 13–14.

14. John W. Garver, "China and the New World Order," in William A. Joseph, ed., *China Briefing, 1992* (Boulder, CO: Westview Press, 1993), pp. 55–57; Bonnie S. Glaser, "China's Security Perceptions: Interests and Ambitions," *Asian Survey* 33, no. 3 (March 1993): 259–260; Rex Li, "China and Asia-Pacific Security in the Post–Cold War Era," *Security Dialogue* 26, no. 3 (September 1995): 332, 334.

15. David Shambaugh, "Containment or Engagement of China? Calculating Beijing's Responses," *International Security* 21, no. 2 (Fall 1996): 187.

16. Thomas J. Christensen, "Posing Problems without Catching Up: China's Rise and Challenges for U.S. Security Policy," *International Security* 25, no. 4 (Spring 2001): 7.

17. See, for example, William T. Tow, "Post–Cold War Security in East Asia," *The Pacific Review* 4, no. 2 (1991): 97; Charles McGregor, "Southeast Asia's New Security Challenges," *The Pacific Review* 6, no. 3 (1993): 272; J. Mohan Malik, "Conflict Patterns and Security Environment in the Asia Pacific Region—The Post–Cold War Era," in Kevin Clements, ed., *Peace and Security in the Asia Pacific Region: Post–Cold War Problems and Prospects* (Tokyo: United Nations University Press, 1993), pp. 33, 38; Barry Buzan and Gerald Segal, "Rethinking East Asian Security," *Survival* 36, no. 2 (Summer, 1994): 8.

18. Denny Roy, "Hegemon on the Horizon? China's Threat to East Asian Security," *International Security* 19, no. 1 (Summer 1994): 149–168; Denny Roy, "Assessing the Asia-Pacific 'Power Vacuum,'" *Survival* 37, no. 3 (Autumn 1995): 50–55.

19. Layne, "The Unipolar Illusion," p. 9.

20. Kenneth N. Waltz, "The Emerging Structure of International Politics," *International Security* 18, no. 2 (Fall 1993): 66.

21. Layne, "The Unipolar Illusion," p. 12.

22. Jonathan D. Spence, *The Search for Modern China* (London: Hutchinson, 1990), chaps. 7–11.

23. Layne, "The Unipolar Illusion," p. 9.

24. For an excellent analysis of the Chinese conception of "comprehensive national strength," see Gerald Chan, *Chinese Perspectives on International Relations: A Framework for Analysis* (London, UK: Macmillan, 1999), pp. 28–33.

25. Gerald Segal, "Tying China into the International System," *Survival* 37, no. 2 (Summer 1995): 70.

26. Chong-Pin Lin, "Chinese Military Modernization: Perceptions, Progress, and Prospects," *Security Studies* 3, no. 4 (Summer 1994): 718–753.

27. Hans J. Morgenthau, *Politics among Nations: The Struggle for Power and Peace*, rev. 5th ed. (New York: Alfred A. Knopf, 1978), p. 29.

28. John J. Mearsheimer, "Back to the Future: Instability in Europe after the Cold War," *International Security* 15, no. 1 (Summer 1990): 53.

29. Samuel P. Huntington, "America's Changing Strategic Interests," *Survival* 33, no. 1 (January/February 1991): 12; Aaron L. Friedberg, "Ripe for Rivalry: Prospects for Peace in a Multipolar Asia," *International Security* 18, no. 3 (Winter 1993/94): 16; Denny Roy, "The 'China Threat' Issue," p. 762.

30. Gideon Rachman, "Containing China," *The Washington Quarterly* 19, no. 1 (Winter 1995): 132.

31. For comparisons of Wilhelmine Germany and today's China, see Nicholas D. Kristof, "The Rise of China," *Foreign Affairs* 72, no. 5 (November/December 1993): 71–72; Arthur Waldron, "Deterring China," *Commentary* 100, no. 4 (October 1995): 18.

32. Roy, "Hegemon on the Horizon?" pp. 159–160.

33. Samuel P. Huntington, "The Clash of Civilizations?" *Foreign Affairs* 72, no. 3 (Summer 1993): 22–49; Samuel P. Huntington, "The West: Unique, Not Universal," *Foreign Affairs* 75, no. 6 (November/December 1996): 28–46. See also his book *The Clash of Civilizations and the Remaking of World Order* (New York: Simon and Schuster, 1996).

34. A.F.K. Organski, *World Politics* (New York: Alfred A. Knopf, 1958); A.F.K. Organski and Jacek Kugler, *The War Ledger* (Chicago: University of Chicago Press, 1980).

35. Gilpin, *War and Change in World Politics*, pp. 106–107. For a detailed elaboration of this argument based on historical evidence, see chapter 3, "Growth and Expansion."

36. Buzan and Segal, "Rethinking East Asian security," p. 6; Roy, "Hegemon on the Horizon?" p. 161; Shambaugh, "Containment or Engagement of China?" pp. 186–187.

37. For analyses of China's policy toward the South China Sea disputes in the 1980s and the early 1990s, see John W. Garver, "China's Push through the South China Sea: The Interaction of Bureaucratic and National Interests," *The China Quarterly* 132 (December 1992): 999–1028; Chen Jie, "China's Spratly Policy," *Asian Survey* 33, no. 10 (October 1994): 893–903; Michael Leifer, "Chinese Economic Reform and Security Policy: The South China Sea Connection," *Survival* 37, no. 2 (Summer 1995): 44–59.

38. *Far Eastern Economic Review*, August 13, 1992, p. 15.

39. *Far Eastern Economic Review*, February 23, 1995, pp. 14–16; June 1, 1995, pp. 20–21

40. For a variety of interpretations of the origins, development, and implications of the 1995–1996 Taiwan Strait crisis, see the special issues/sections in *The China Journal* 36 (July 1996); *Security Dialogue* 27, no. 4 (December 1996); *The China Quarterly* 148 (December 1996), and *Journal of Contemporary China* 6, no. 15 (July 1997). See also Suisheng Zhao, ed., *Across the Taiwan Strait: Mainland China, Taiwan the 1995–96 Crisis* (London: Routledge, 1999); Chang Pao-min, "The Dynamics of Taiwan's Democratization and Crisis in the Taiwan Strait," *Contemporary Southeast Asia* 18, no. 1 (June 1996): 1–16; Chen Qimao, "The Taiwan Strait Crisis: Its Crux and Solutions," *Asian Survey* 36, no. 11 (November 1996): 1055–1066; Weixing Hu, "China's Taiwan Policy and East Asian Security," *Journal of Contemporary Asia* 27, no. 3 (1997): 374–391; Edward Friedman, "Chinese Nationalism, Taiwan Autonomy and the Prospects of a Larger War," *Journal of Contemporary China* 6, no. 14 (March 1997): 5–32; Robert S. Ross, "The 1995–96 Taiwan Strait Confrontation: Coercion, Credibility, and the Use of Force," *International Security* 25, no. 2 (Fall 2000): 87–123; Allen S. Whiting, "China's Use of Force, 1950–96, and Taiwan," *International Security* 26, no. 2 (Fall 2001): 103–131.

41. "Stay Back, China," *The Economist*, March 16, 1996, p. 15.

42. Kurt M. Campbell and Derek J. Mitchell, "Crisis in the Taiwan Strait," *Foreign Affairs* 80, no. 4 (July/August 2001): 17.

43. Gerald Segal, "East Asia and the 'Constrainment of China,'" *International Security* 20, no. 4 (Spring 1996): 133.

44. Waltz, *Theory of International Politics*, p. 106.

45. Chen, "China's Spratly Policy," p. 896.

46. Alan Troner and Sarah J. Miller, *Energy and the New China: Target of Opportunities* (New York: Petroleum & Energy Intelligence Weekly, 1995), pp. 49–53.

47. Ken E. Calder, "Asia's Empty Tank," *Foreign Affairs* 75, no. 2 (March/April 1996): 56; Mamdouh G. Salameh, "China, Oil and the Risk of Regional Conflict," *Survival* 37, no. 4, (Winter 1995–1996): 135.

48. "Out of Puff: A Survey of China," *The Economist*, June 15, 2002, p. 16.

49. Salameh, "China, Oil and the Risk of Regional Conflict," p. 141.

50. Layne, "The Unipolar Illusion," pp. 45–46.

51. Robert S. Ross, "The Geography of the Peace: East Asia in the Twenty-First Century," *International Security* 23, no. 4 (Spring 1999): 82.

52. Ibid., pp. 81–118.

53. Theoretically, those who subscribe to the theory of hegemonic stability might see an "open" and "liberal" China as a benign hegemon that could help maintain a stable, liberal economic order in East Asia. However, no realist scholars have argued that a China governed by the CCP will be sufficiently committed to the market economy in the near future to assume such a role. For an analysis of hegemonic stability theory, see Robert Gilpin, *The Political Economy of International Relations* (Princeton, NJ: Princeton University Press, 1987), pp. 72–80, 85–92.

54. Richard Bernstein and Ross H. Munro, "The Coming Conflict with America," *Foreign Affairs* 76, no. 2 (March/April 1997): 18–32. See also their book, *The Coming Conflict with China* (New York: Alfred A. Knopf, 1997).

55. Friedberg, "Ripe for Rivalry," pp. 27–32.

56. Joseph M. Grieco, "Anarchy and the Limits of Cooperation: A Realist Critique of the Newest Liberal Institutionalism," in Kegley, ed., *Controversies in International Relations Theory*, p. 161.

57. Ibid., p. 151. See also John J. Mearsheimer, "The False Promise of International Institutions," *International Security* 19, no. 3 (Winter 1994/95): 5–49.

58. Charles Krauthammer, "Why We Must Contain China," *Time*, July 31, 1995, p. 72.

59. Denny Roy, "Consequences of China's Economic Growth for Asia-Pacific Security," *Security Dialogue* 24, no. 2 (June 1993): 190.

60. Roy, "Hegemon on the Horizon?" pp. 165–168; Roy, "The 'China Threat' Issue," pp. 770–771.

61. Segal, "East Asia and the 'Constrainment' of China," p. 134.

62. Rachman, "Containing China," pp. 129–139. A shorter version of this article first appeared in *The Economist*, July 29, 1995, pp. 13–14. For a similar view, see Greg Mastel, "A New U.S. Trade Policy toward China," *The Washington Quarterly* 19, no. 1 (Winter 1996): 189–207; Mastel, "Beijing at Bay," *Foreign Policy* 104 (Fall 1996): 27–34.

63. Kenneth Lieberthal, "A New China Strategy," *Foreign Affairs* 74, no. 6 (November/December 1995): 36.

64. Gordon White, Jude Howell, and Shang Xiaoyuan, *In Search of Civil Society: Market Reform and Social Change in Contemporary China* (Oxford, UK: Clarendon Press, 1996); Baogang He, *The Democratic Implications of Civil Society in China* (London, UK: Macmillan, 1997).

65. Chen Jian, "Will China's Development Threaten Asia-Pacific Security?" *Security Dialogue* 24, no. 2 (June 1993): 195.

66. See, for example, Barber B. Conable, Jr., and David M. Lampton, "China: The Coming Power," *Foreign Affairs* 71, no. 5 (Winter 1992/93): 146; Thomas W. Robinson, "Interdependence in China's Foreign Relations," in Samuel S. Kim, ed., *China and the World: Chinese Foreign Relations in the Post–Cold War Era* (Boulder, CO: Westview Press), p. 197; Yoichi Funabashi, Michel Oksenberg, and Heinrich Weiss, *An Emerging China in a World of Interdependence* (New York: The Trilateral Commission, 1994), p. 65.

67. Conable and Lampton, "China: The Coming Power," p. 140.

68. Ibid., p. 141.

69. Donald S. Zagoria, "Clinton's Asia Policy," *Current History* 92, no. 578 (December 1993): 404.

70. Ibid.

71. Tianjian Shi, "Economic Development and Village Elections in Rural China," *Journal of Contemporary* China 8, no. 22 (November 1999): 425–442.

72. William H. Overholt, "China after Deng," *Foreign Affairs* 75, no. 3 (May/June 1996): 68, 71.

73. Ibid., pp. 71, 75–76. See also Minxin Pei, "Is China Democratizing?" *Foreign Affairs* 77, no. 1 (January/February 1998): 68–82.

74. Michael Doyle, "Kant, Liberal Legacies, and Foreign Affairs, Part I," *Philosophy and Public Affairs* 12, no. 3 (Summer 1983): 205–235; Michael Doyle, "Liberalism and World Politics," *American Political Science Review* 80, no. 4 (December 1986): 1151–1169; Francis Fukuyama, *The End of History and the Last Man* (London: Hamish Hamilton, 1992); Bruce Russett, *Grasping the Democratic Peace: Principles for a Post–Cold War World* (Princeton, NJ: Princeton University Press, 1993); Spencer Weart, "Peace among Democratic and Oligarchic Republics," *Journal of Peace Research* 32, no. 3 (1994): 299–316; R. J. Rummel, "Democracies Are Less Warlike than Other Regimes," *European Journal of International Relations* 1, no. 4 (December 1995): 457–479.

75. Georg Sorensen, "Kant and Process of Democratization: Consequences for Neorealist Thought," *Journal of Peace Research* 29, no. 4 (1992): 398–399.

76. See, for example, the special issues of *Journal of Peace Research* 29, no. 4 (1992); *International Security* 19, no. 2 (Fall 1994); *European Journal of International Relations* 1, no. 4 (December 1995). See also the correspondence section in *International Security* 19, no. 4 (Spring 1995): 164–184.

77. Stephen D. Krasner, "International Political Economy: Abiding Discord," *Review of International Political Economy* 1, no. 1 (Spring 1994): 17.

78. While accepting the thesis that democracies do not fight democracies, some liberal scholars argue that the process of the transition from authoritarian to democratic systems can be rather destabilizing and may increase the possibility of war. See, for example, Edward D. Mansfield and Jack Snyder, "Democratization and the Danger of War," *International Security* 20, no. 1 (Summer 1995): 5–38.

79. See, for example, Richard Cobden, *The Political Writings of Richard Cobden* (London: T. Fischer Unwin, 1903); Norman Angell, *The Great Illusion* (London: Heinemann, 1935); Richard Rosecrance, *The Rise of the Trading State: Commerce and Conquest in the Modern World* (New York: Basic Books, 1986).

80. Audrey Kurth Cronin and Patrick M. Cronin, "The Realistic Engagement of China," *The Washington Quarterly* 19, no. 1 (Winter 1996): 145–146.

81. Vincent Cable and Peter Ferdinand, "China as an Economic Giant: Threat or Opportunity?" *International Affairs* 70, no. 2 (April 1994): 259.

82. Michael Yahuda, "How Much Has China Learned about Interdependence?" in David S.G. Goodman and Gerald Segal, eds., *China Rising: Nationalism and Interdependence* (London: Routledge, 1997), p. 22.

83. Christopher Findlay and Andrew Watson, "Economic Growth and Trade Dependency in China," in Goodman and Segal, eds., *China Rising*, p. 107.

84. Ibid.

85. Stuart Harris, "China's Role in the WTO and APEC," in Goodman and Segal, eds., *China Rising*, p. 151.

86. Robinson, "Interdependence in China's Foreign Relations," p. 198.

87. Samuel S. Kim, "China's International Organizational Behaviour," in Thomas W. Robinson and David Shambaugh, eds., *Chinese Foreign Policy: Theory and Practice* (Oxford, UK: Clarendon Press, 1994), p. 433.

88. Robert Keohane and Joseph S. Nye, *Power and Interdependence: World Politics in Transition* (Boston: Little, Brown, 1977); Stephen D. Krasner, ed., *International Regimes* (Ithaca, NY: Cornell University Press, 1983).

89. Thomas G. Moore and Dixia Yang, "Empowered and Restrained: Chinese Foreign Policy in the Age of Economic Interdependence," in David M. Lampton, ed., *The Making of Chinese Foreign and Security Policy in the Era of Reform, 1978–2000* (Stanford, CA: Stanford University Press, 2001), p. 228.

90. Funabashi, Oksenberg, and Weiss, *An Emerging China in a World of Interdependence*, p. 2.

91. Cable and Ferdinand, "China as an Economic Giant," p. 259.

92. Conable and Lampton, "China: The Coming Power," p. 149.

93. "Eating Your Lunch?" *The Economist*, February 15, 2003, p. 12.

94. For a recent study of the global implications of China's growing demand of grain import, see Lester Brown, *Who Will Feed China? Wake Up Call for a Small Planet* (Washington, DC: Worldwatch Institute, 1995).

95. Akihiko Tanaka, "China: Dominant, Chaotic or Interdependent?" in Trevor Taylor and Seizaburo Sato, eds., *Future Sources of Global Conflict* (London: Royal Institute of International Affairs, 1995), pp. 56–57; Lieberthal, "A New China Strategy," pp. 36–37; Cable and Ferdinand, "China as an Economic Giant," p. 259.

96. Lieberthal, "A New China Strategy," p. 36.

97. Bryce Harland, "For a Strong China," *Foreign Policy* 94 (Spring 1994): 51.

98. Cable and Ferdinand, "China as an Economic Giant," p. 261.

99. Kishore Mahbubani, "An Asia-Pacific consensus," *Foreign Affairs* 76, no. 5 (September/October 1997): 158.

100. Robert S. Ross, "Enter the Dragon," *Foreign Policy* 104 (Fall 1996): 18–25.

101. "China Opens Up," *The Economist*, November 20, 1999, p. 18.

102. Shaun Breslin, "The China Challenge? Development, Environment and National Security," *Security Dialogue* 28, no. 4 (December 1997): 497–508.

103. Arthur A. Stein, "Coordination and Collaboration Regimes in an Anarchic World," *International Organization* 36 (Spring 1982): 318.

104. Waltz, *Theory of International Politics*, p. 105.

105. Kristof, "The Rise of China," pp. 68–69.

106. Ibid., p. 72.

107. Robert S. Ross, "China and the Stability of East Asia," in Robert S. Ross, ed., *East Asia in Transition: Toward a New Regional Order* (Armonk, NY: M.E. Sharpe, 1995), pp. 115–117; Gary Klintworth, "Greater China and Regional Security," in Gary Klintworth, ed., *Asia-Pacific Security: Less Uncertainty, New Opportunities?* (Melbourne, Australia: Longman, 1996), p. 36.

108. Robert S. Ross, "Beijing as a Conservative Power," *Foreign Affairs* 76, no. 2 (March/April 1997): 34.

109. Gary Klintworth and Des Ball, "China's Arms Buildup and Regional Security," in Stuart Harris and Gary Klintworth, eds., *China as a Great Power: Myths, Realities and Challenges in the Asia-Pacific Region* (Melbourne, Australia: Longman, 1995), pp. 263–265; Andrew J. Nathan and Robert S. Ross, *The Great Wall and the Empty Fortress: China's Search for Security* (New York: W.W. Norton, 1997), pp. 146–148.

110. Ross, "Beijing as a Conservative Power," pp. 35–38; Klintworth and Ball, "China's Arms Buildup and Regional Security," pp. 265–267; David Shambaugh, "China's Military: Real or Paper Tiger?" *The Washington Quarterly* 19, no. 2 (Spring 1996): 24–29.

111. David Shambaugh, "China's Military in Transition: Politics, Professionalism, Procurement and Power Projection," *The China Quarterly* 146 (June 1996): 295.

112. David Shambaugh, "Chinese Hegemony over East Asia by 2015?" *The Korean Journal of Defense Analysis* 9, no. 1 (Summer 1997): 7–28.

113. Lieberthal, "A New China Strategy," pp. 47, 37–38.

114. Cronin and Cronin, "The Realistic Engagement of China," p. 165.

115. Findlay and Watson, "Economic Growth and Trade Dependency in China," p. 107.

116. Lieberthal, "A New China Strategy," pp. 47, 37–38.

117. Cronin and Cronin, "The Realistic Engagement of China," p. 165.

118. James Shinn, ed., *Weaving the Net: Conditional Engagement with China* (New York: Council on Foreign Relations Press, 1996).

119. Michael J. Mazarr, "The Problems of a Rising Power: Sino-American Relations in the 21st Century," *The Korean Journal of Defense Analysis* 7, no. 2 (Winter 1995): 7–39.

120. Fei-Ling Wang, "To Incorporate China: A New Policy for a New Era," *The Washington Quarterly* 21, no. 1 (Winter 1998): 77.

121. Shambaugh, "Containment or Engagement of China?" p. 184. See also Bates Gill, "Limited Engagement," *Foreign Affairs* 78, no. 4 (July/August 1999): 66; Asia Project Policy Report, *Redressing the Balance: American Engagement with Asia* (New York: Council on Foreign Relations, 1996), pp. 18–20. For an informative and well-argued collection of essays reviewing various dimensions of Sino-American relations and recommending a policy of engagement of China, see Ezra F. Vogel, ed., *Living with China: U.S./China Relations in the 21st Century* (New York: W. W. Norton, 1997).

122. Alastair Iain Johnston and Robert S. Ross, eds., *Engaging China: The Management of an Emerging Power* (London: Routledge, 1999), p. xiv.

123. See chapters 2–8 in Johnston and Ross, eds., *Engaging China*. For a perceptive analysis of Japan's approach to engagement of China, see Reinhard Drifte, *Japan's Security Relations with China since 1989: From Balancing to Bandwagoning?* (London: RoutledgeCurzon, 2003).

124. The importance of analyzing the domestic sources of foreign policy within the context of the anarchic international system has been stressed by Fareed Zakaria; see his review essay "Realism and Domestic Politics," *International Security* 17, no. 1 (Summer 1992): 177–198.

125. Glaser, "China's Security Perceptions," pp. 253–254; Li, "China and Asia-Pacific Security in the Post–Cold War Era," pp. 332, 337; Thomas J. Christensen, "Chinese Realpolitik," *Foreign Affairs* 75, no. 5 (September/October 1996): 37–40. For an incisive analysis of China's realist security perceptions in historical perspective, see Alastair Iain Johnson, *Cultural Realism: Strategic Culture and Grand Strategy in Chinese History* (Princeton, NJ: Princeton University Press, 1995).

126. As Thomas Moore and Dixia Yang argue, economic interdependence "is not alone likely to transform either Chinese world-views or Chinese foreign policy." See Moore and Yang, "Empowered and Restrained," p. 229.

127. Copeland, "Economic Interdependence and War."

128. Ibid., p. 6.

129. Ibid., p. 17.

130. Ibid., p. 7.

131. Ibid., pp. 19–20.

132. William R. Feeney, "China and the Multilateral Economic Institutions," in Kim, ed., *China and the World*, pp. 226–251; Christopher Findlay, "China and the Regional Economy," in Harris and Klintworth, eds., *China as a Great Power*, pp. 284–305.

133. Kou Bian, "Foreign-Funded Banks Land in Chinese Market," *Beijing Review*, December 29, 1997–January 4, 1998, pp. 15–16.

134. Shi Guangsheng, Chinese Minister of Foreign Trade and Economic Cooperation, "Remarks at the Reception Hosted by EU-China Business Association and Belgium Chinese Economic and Commercial Council," December 9, 2002, http://english.moftec.gov.cn/article/200212/20021200056451_1.xml.

135. Nicholas R. Lardy, *China in the World Economy* (Washington, DC: Institute for International Economics, 1994), pp. 30, 63; PRC Ministry of Foreign Trade and Economic Cooperation, "The Information on Import and Export Statistics (December 2002)," http://english.moftec.gov.cn/

article/200301/20030100064844_1.xml; "Statistics about Utilization of Foreign Investment in 2002 (January–December) of China," http://english.moftec.gov.cn/article/200301/20030100063756_1.xml.

136. Shi Guangsheng, "Remarks at the Reception Hosted by EU-China Business Association and Belgium Chinese Economic and Commercial Council."

137. Peter J. Rimmer, "Integrating China into East Asia: Cross-Border Regions and Infrastructure Networks," in Harris and Klintworth, eds., *China as a Great Power*, pp. 306–327.

138. "China and ASEAN Agree to Create World's Biggest FTA," *Digital Chosunilbo* (English edition), November 5, 2002.

139. Lardy, *China in the World Economy*, pp. 63–72.

140. Christopher Howe, "The People's Republic of China: Economy," in *The Far East and Australasia 1998*, 29th ed. (London: Europa Publications Limited, 1997), p. 234.

141. *People's Republic of China Year Book 1997/98* (Beijing and Hong Kong: PRC Year Book Ltd. and N.C.N. Limited, 1998), p. 270.

142. "Over 314,533 Foreign Projects Approved," *Beijing Review*, October 5–11, 1988, p. 5.

143. *People's Republic of China Year Book 1997/98*, p. 273.

144. The World Bank Group, "Country Brief: People's Republic of China," November 2002.

145. Howe, "The People's Republic of China: Economy," p. 233.

146. *The World Bank and China*, the World Bank Group, http://www.worldbank.org/html/extdr/offrep/eap/cn2.htm.

147. Howe, "The People's Republic of China: Economy," pp. 232–233.

148. "China's Technology Imports in 1997," *Beijing Review*, October 19–25, 1998, p. 22.

149. "Chinese Scholars Review 1996," *Beijing Review*, January 27–February 2, 1997, p. 10.

150. Wu Yi, "Prospects for China's Foreign Economic and Trade Development," *Beijing Review*, February 17–March 2, 1997, p. 16.

151. See, for example, Shi Guangsheng, Chinese Minister of Foreign Trade and Economic Cooperation, "Five Positive Changes for China's Opening to Outside World," November 14, 2002, http://english.moftec.gov.cn/article/200211/20021100050256_1xml; Song Yuhua, "Dangqian Zhongguo mianlin de guoji jingji huanjing" (China's Current International Economic Environment), *Guoji wenti yanjiu* (Journal of International Studies) 3 (May 2002): 49–54.

152. Dai Xiaohua, "'East Asian Model': A Few Problems, but It Works," *Beijing Review*, March 23–29, 1998, p. 8.

153. See the articles in the special section on "China and the World Trade Organization: Winners or Losers?" in *Journal of Contemporary China* 11, no. 32 (August 2002).

154. Wang Zhenxi, "Huanhe yu duojihua shitou qiangjing, baquan yu lengzhan siwei yicun-yijiujiuqinian guoji xingshi zongshu" (Hegemonism and the Cold War Mentality Coexist with the Potent Forces of Détente and Multipolarity—A Review of the International Situation in 1997), *Guoji guanxi xueyuan xuebao* (Journal of the Institute of International Relations) 4 (December 1997): 6.

155. Wang Zhongren, "'China Threat' Theory Groundless," *Beijing Review*, July 14–20, 1997, pp. 7–8.

156. Zhou Qi, "Lengzhan hou de ZhongMei guanxi xianzhuang-gongtong liyi yu zhengzhi" (An Appraisal of Post–Cold War Sino-U.S. Relations—Common Interests and Disputes), *Meiguo yanjiu* (American Studies) 9, no. 4 (December 1995): 30–50.

157. See, for example, Liang Gencheng, "Bian jiechu, bian ezhi-Kelindun zhenfu de dui Hua zhengce pouxi" (Engaging while Containing—An In-Depth Analysis of the Clinton Administration's China Policy), *Meiguo yanjiu* (American Studies) 10, no. 2 (June 1996): 7–20; Wang Jisi, "'Ezhi' haishi 'jiaowang'?-ping lengzhan hou Meiguo dui Hua zhengce" ("Containment" or "Engagement?"—On U.S.-China Policy in the Post–Cold War Era), *Guoji wenti yanjiu* (Journal of International Studies) 1 (January 1996): 1–6; Wang Haihan, "Lun Kelindun zhengfu de dui Hua zhengce ji qi qianjing" (On the Clinton Administration's China Policy and Its Future Trend), ibid., 1 (January 1997): 3–9; Niu Jun, "Lun Kelindun zhengfu dier renqi dui

Hua zhengce de yanbian ji qi tedian" (On the Evolution and Characteristics of the Clinton Administration's China Policy during Its First Term), *Meiguo yanjiu* (American Studies) 12, no. 1 (March 1998): 7–28. See also Qingguo Jia, "Frustrations and Hopes: Chinese Perceptions of the Engagement Policy Debate in the United States," *Journal of Contemporary China* 10, no. 27 (May 2001): 321–330. It is difficult to translate the exact meaning of "engagement" into Chinese. The terms most commonly used by Chinese elites and specialists to describe Western policy of engagement include *jiechu* (contact), *jiaowang* (interaction), and *canyu* (involvement). The term "containment" is normally translated as *ezhi* or *weidu* in Chinese publications.

158. Chinese scholars use *zhiyue* and *qianzhi* interchangeably to describe U.S. attempts to "constrain" China.

159. Zhang Dalin, "Ping 'RiMei anquan baozhang lianhe xuanyan'" (On "U.S.-Japan Declaration on Security Alliance"), *Guoji wenti yanjiu* (Journal of International Studies) 4 (October 1996): 24–28; Li Genan, "RiMei anbao tizhi zai dingwei" (The Renewal of the U.S.-Japan Security Treaty), *Waiguo wenti yanjiu* (Research on Foreign Issues) 2 (1996): 1–3; Su Hao, "'MeiRi anbao guanxi de tiaozheng yu Yatai anquan wenti' xueshu yantaohui zongshu" (Summary of the Symposium on "the Adjustment in U.S.-Japan Security Relations and Asia-Pacific Security"), *Meiguo yanjiu* (American Studies) 12, no. 1 (March 1988): 143–147.

160. See Thomas J. Christensen, "China, the U.S.-Japan alliance, and the Security Dilemma in East Asia," *International Security* 23, no. 4 (Spring 1999): 49–80; Rex Li, "Partners or Rivals? Chinese Perceptions of Japan's Security Strategy in the Asia-Pacific Region," *The Journal of Strategic Studies* 22, no. 4 (December 1999): 1–25.

161. Lanxin Xiang, "Washington's Misguided China Policy," *Survival* 43, no. 3 (Autumn 2001): 7–23.

162. BBC News, "US 'Has Nuclear Hit List,'" March 9, 2002, http://news.bbc.co.uk/1/hi/world/americas/1864173.stm.

163. Rex Li, "US-China Relations: Accidents Can Happen," *The World Today* 56, no. 5 (May 2000): 17–20; Aaron L. Friedberg, "11 September and the Future of Sino-American Relations," *Survival* 44, no. 1 (Spring 2002): 36–40.

164. See, for example, Yuan Ming and Fan Shiming, "Lengzhan hou Meiguo dui Zhongguo anquan xingxiang de renshi" (China's Security Role in Post–Cold War American Perceptions), *Meiguo yanjiu* (American Studies) 9, no. 4 (December 1995): 7–29; Yan Xuetong, "Xifangren kan Zhongguo de jueqi" (Western Perspectives on the Rise of China), *Xiandai guoji guanxi* (Contemporary International Relations) 9 (September 1996): 36–45. See also Herbert Yee and Zhu Feng, "Chinese Perspectives of the China Threat: Myth or Reality?" in Herbert Yee and Ian Storey, eds., *The China Threat: Perceptions, Myths and Reality* (London: RoutledgeCurzon, 2002), pp. 21–42.

165. Yuan Ming, "Ershiyi shijichu Dongbeiya daguo guanxi" (Great Power Relations in Northeast Asia in the Early 21st Century), *Guoji wenti yanjiu* (Journal of International Studies) 4 (October 1996): 23.

166. Shi Yongming, "Yatai anquan huanjing yu diqu duobianzhuyi" (The Security Environment in the Asia-Pacific and Regional Multilateralism), *Guoji wenti yanjiu* (Journal of International Studies) 1 (January 1996): 41–47. See also Alastair Iain Johnston and Paul Evans, "China's Engagement with Multilateral Security Institutions," in Johnston and Ross, eds., *Engaging China*, pp. 235–272.

167. Lee Lai To, *China and the South China Sea Dialogues* (Westport, CT: Praeger, 1999). For a Chinese view on Beijing's policy toward the South China Sea disputes, see Ji Guoxing, "China versus South China Sea Security," *Security Dialogue* 29, no. 1 (March 1998): 101–112. See also the rejoinder by Tim Huxley, "A Threat in the South China Sea?" *Security Dialogue* 29, no. 1 (March 1998): 113–118.

168. "ASEAN, China Sign Landmark Accord," *The Guardian*, November 4, 2002, http://www.guardian.co.uk/worldlatest/story/0,1280,-2143239,00.html.

169. Erica Strecker Downs and Phillip C. Saunders, "Legitimacy and the Limit of Nationalism: China and the Diaoyu Islands," *International Security* 23, no. 3 (Winter 1998/99): 114–146; Phil Deans, "Contending Nationalisms and the Diaoyutai/Senkaku Dispute," *Security Dialogue* 31, no. 1 (March 2000): 119–131.

170. Duan Hong, "Shixi Yazhou jingji weiji dui Dongya anquan de yingxiang" (A Preliminary Analysis of the Impact of the Asian Economic Crisis on East Asian Security), *Guoji wenti yanjiu* (Journal of International Studies) 4 (October 1998): 43–44.

171. Copeland, "Economic Interdependence and War," p. 33.

172. This line of argument is again based on Copeland's theory, ibid., pp. 20–21.

173. Ibid., p. 21.

174. Ibid., p. 22.

175. For analyses of the economic and financial interdependency between China and Hong Kong, see, for example, Hsin-chi Kuan, "Does Hong Kong Have a Future?" *Security Dialogue* 28, no. 2 (June 1997): 233–236; Niu Tiehang, "Stock Market Integration in Hong Kong and China," *Journal of Contemporary China* 6, no. 16 (November 1997): 487–512.

176. *The World Bank and China*, the World Bank Group, http://www.worldbank.org/html/extdr/offrep/eap/cn2.htm.

177. Ibid.

178. Cao Zhizhou, "Luelun Li Denghui dalu zhengce de 'taidu' benzhi" (On the Essence of Taiwanese Independence in Lee Teng-hui's Mainland Policy), *Taiwan yanjiu* (Taiwan Studies) 3 (September 1996): 13–18.

179. Copeland, "Economic Interdependence and War," p. 40.

180. A penetrating analysis of the growth of nationalistic sentiment in China since the early 1990s can be found in Suisheng Zhao, *In Search of a Right Place? Chinese Nationalism in the Post–Cold War World*, USC Seminar Series No. 12 (Hong Kong: Hong Kong Institute of Asia-Pacific Studies, Chinese University of Hong Kong, 1997). See also his article "Chinese Intellectuals' Quest for National Greatness and Nationalistic Writing in the 1990s," *The China Quarterly* 152 (December 1997): 725–745.

181. For a detailed examination of the main domestic variables that will influence China's responses to the policies of the outside world, see Shambaugh, "Containment or Engagement of China?"

182. Cronin and Cronin, "The Realistic Engagement of China," p. 164. For useful analyses of and debates on Chinese perceptions of the changing international environment, see Yong Deng and Fei-Ling Wang, *In the Eyes of the Dragon: China Views the World* (Lanham, MD: Rowman and Littlefield, 1999); Gerald Chan, *Chinese Perspectives on International Relations*, Part III; Michael Pillsbury, *China Debates the Future Security Environment* (Washington, DC: National Defense University Press, 2000); the special sections on "New Generation, New Voices: Debating China's International Future" in *Journal of Contemporary China* 10, no. 26 (February 2001) and 10, no. 27 (May 2001).

183. Rodolfo C. Severino, Jr., "Integrate and Persuade, but Gently," *Far Eastern Economic Review*, June 20, 1996, p. 32. The role of perception in world politics has long been recognized by scholars of international relations. For a classic study, see Robert Jervis, *Perception and Misperception in International Politics* (Princeton, NJ: Princeton University Press, 1976).

184. Severino, "Integrate and Persuade, but Gently."

185. Recent studies of China's interaction with international regimes indicate that China is willing to participate in international regimes so long as they do not obstruct the PRC's economic development and infringe on Chinese sovereignty. Nevertheless, based on an in-depth analysis of China's involvement in the ozone and climate change regimes, Elizabeth Economy has concluded that the international community has been able to exert "a set of relatively discrete influences that shaped China's foreign policy-making process and, to a lesser extent, the policy outcome." See Elizabeth Economy, "The Impact of International Regimes on Chi-

nese Foreign Policy-Making: Broadening Perspectives and Policies . . . but only to a Point" in Lampton, ed., *The Making of Chinese Foreign and Security Policy in the Era of Reform*, p. 253.

186. Mansfield and Snyder, "Democratization and the Danger of War."

187. Stein Tonnesson, "China and the South China Sea: A Peace Proposal," *Security Dialogue* 31, no. 3 (September 2000): 307–326.

188. It has been suggested, for example, that Jiang Zemin was forced to take bellicose actions to intimidate Taiwan in 1995–1996 so as to secure full support from the military in fortifying his position as Deng Xiaoping's successor. See Jianhai Bi, "The Role of the Military in the PRC Taiwan Policymaking: A Case Study of the Taiwan Strait Crisis of 1995–1996," *Journal of Contemporary China* 11, no. 32 (August 2002): 539–572.

189. Ralph A. Cossa has argued that "the US-Japan alliance is not anti-Chinese but pro-peace and pro-stability." See his article "Avoiding New Myths: US-Japan Security Relations," *Security Dialogue* 28, no. 2 (June 1997): 219–231.

190. Ross, "Beijing as a Conservative Power," p. 44.

191. Jusuf Wanandi, "ASEAN's China Strategy: Towards Deeper Engagement," *Survival* 38, no. 3 (Autumn 1996): 127.

192. Joseph S. Nye, Jr., "As China Rises, Must Others Bow?" *The Economist*, June 27–July 3, 1998, p. 25.

3

Four Contradictions Constraining China's Foreign Policy Behavior

Wu Xinbo

In the post–Cold War era, China's foreign policy behavior is mainly constrained by four contradictions: self-images of great power versus poor country, "open-door" incentive versus sovereignty concerns, principle versus pragmatism, and bilateralism versus multilateralism.

A Great Power and a Poor Country

Chinese leaders and the Chinese general public believe that China is a nation with a dual identity. On the one hand, China is considered a great nation for its long, unbroken history, its contribution to the progress of civilization, its vast territory and population, and its significant geographic location. China's greatness is also rooted in its permanent membership in the UN Security Council and in its nuclear capability. On the other hand, both political elites and ordinary people understand that today's China is still a poor country; its level of economic development and technological prowess lag far behind those of Western countries and some of its Asian neighbors.

As a great power, China would naturally like to possess more influence in international affairs. Beijing wishes that it was accepted as a major player in the world community and that its voice was listened to carefully not only in the Asia-Pacific region but also in other parts of the world. Like other major powers, China wants to shape the current, not just respond to it. However, as a poor country, China has neither the material strength nor a genuine enough interest to play a role commensurate to its great-power self-image. Unlike the United States that, as a superpower, defines its interests in a global context and possesses adequate means to pursue those interests, China is still a country whose real interests lie mainly within its boundaries and, to a lesser extent, in the Asia-Pacific region, where developments may have a direct impact on the country's national interests. Furthermore, the resources that China can mobilize to promote interests beyond its boundaries are very limited. In terms of interests and resources, therefore, it is fair to say that China is a regional power with some limited global interests.

The dual-identity syndrome has created a dilemma in China's foreign policy behavior. As a great power, Beijing has to respond to any crisis in the world, no matter when and where it occurs. On the other hand, if a crisis takes place in a locale beyond the geographical scope of China's sphere of interest, Beijing's response would be confined to announcing a set of general principles without making any serious diplomatic efforts to solve the problem. In the case of the North Atlantic Treaty Organization (NATO) bombing of Yugoslavia, for instance, China took a two-point principled position throughout the crisis: Force should not be used, and the legal rights of all ethnic groups in Kosovo should be respected and preserved.[1] However, Beijing did not put forward any concrete proposal about how the crisis should be managed, nor did it try to mediate between NATO and Yugoslavia. The explanation is that China realizes it is geopolitically irrelevant in the Balkans, and its influence in the region is negligible, especially when NATO chose to take action without first soliciting authorization from the United Nations. Although both Chinese leaders and the Chinese public were irritated by NATO's operations against Yugoslavia, Beijing had neither a strong national interest in that area to serve as an incentive nor the capacity to intervene.

The dual-identity syndrome has set up a pattern of action in China's response to developments outside the Asia-Pacific region that involves principles, not workable proposals, and words, not deeds. This has created some problems. Externally, China sometimes appears unwilling to take responsibility commensurate to its P-5 status,* and it is viewed by outsiders as a parochial power. Internally, China's failure to exert more visible influence on world affairs has constantly frustrated the general public and made them feel that China is far from being accepted as a great power.

To be sure, as China's material strength grows, so will its aspirations for playing a more significant international role. In the face of the perceived U.S. attempt to shape a unipolar world, China may seek to leave its fingerprints on more and more regional and international affairs. For instance, Beijing will stress the legitimate role of the UN Security Council in preserving world peace and security and will reinforce its consultation with Moscow on major international issues. However, a more active Chinese role in international affairs would require a long process of Chinese power accumulation.

"Open-door" Incentive and Sovereignty Concerns

The open-door policy initiated in 1979 has rendered China accessible to the markets, technology, and capital of developed countries and has greatly augmented China's comprehensive national capability. In return, it requires China to maintain good political and economic relations with developed countries, which means that China has to be responsive to their concerns and cooperative with them in international affairs. However, Beijing's behavior in this regard is constrained by a strong concern over its state sovereignty.

Generally speaking, there are three major factors that have shaped China's sovereignty concerns. The first is China's historical experience in modern times. During the "century of humiliation," China suffered from political, economic, and military aggression by Western powers and Japan, and this experience has caused the Chinese to cherish their sovereignty. The second is the gap between China and Western nations in state building. As a developing, socialist country China differs from Western countries in terms of political and legal systems and values, and this has made the country subject to attack from the Western world. As a result, China has to invoke the principle of the sacrosanctity of its sovereignty to fend off external intrusion into its internal affairs. Third, China is concerned over its territorial integ-

---

*One of the five permanent members of the UN Security Council.

rity with regard to Tibet, Xinjiang, and Taiwan. Beijing is very sensitive to any precedent that may legitimize these regions' separation from China or lead to foreign intervention.

The strong sovereignty concern has remarkably affected China's posture in international affairs. Again taking the Kosovo crisis as one example, although many countries believe that Milosevic's unwise policy in Kosovo was a major cause of the crisis, the Chinese media has never presented a comprehensive and balanced picture about the situation in Kosovo, nor did it criticize Milosevic's policy toward the ethnic Albanians. The reason is Chinese leaders believe Milosevic was trying to prevent the ethnic Albanians from gaining independence and hence sympathize with him in his attempt to preserve Yugoslavia's territorial integrity, even though privately they might disapprove of Belgrade's policy in Kosovo. Beijing strongly condemned NATO's bombing of Yugoslavia, not only because it was not authorized by the United Nations, but also, more important, Beijing worried that this would set a dangerous precedent in which a country or a group of countries could use force to intervene in another country's internal affairs. Even if NATO had tried to solicit UN authorization to use force against Yugoslavia, China would definitely have vetoed such a proposal. As a matter of fact, after NATO started the air strike against Yugoslavia, Chinese leaders and the Chinese public began to ask the same question: If they are bombing Yugoslavia today for the issue of Kosovo, will they bomb China someday if a crisis arises over Tibet, Xinjiang, or Taiwan? The Western media has been accusing China of failing to present a balanced coverage of the situation in Kosovo and of not understanding the humanitarian motives behind NATO's action. The problem is that China's own security concerns left Beijing with little room for maneuvering on this issue.

China's position on international peacekeeping operations is another relevant case. Since the end of the Cold War, the United Nations has been playing a more active role in peacekeeping activities. China began to participate in UN-sponsored peacekeeping operations in the late 1980s. However, bearing in mind that the United States fought the Korean War on behalf of the United Nations, Beijing is very cautious in endorsing and participating in peacekeeping efforts, stressing that peacekeeping operations should abide by the principles stipulated in the UN charter regarding state sovereignty and noninterference of internal affairs. Any involvement on behalf of the United Nations should also be conducive to preserving the sovereignty and territorial integrity of the country concerned and should first secure the consent of the party in concern.[2] While generally acknowledging that peacekeeping activities have become an effective means for the United Nations to preserve international peace and security, China does not want to warrant the organization unlimited rights in putting its hands in the internal affairs of other countries, as the Taiwan issue is always on the minds of the Chinese leaders.

Sovereignty concerns also keep China from openly commenting on other countries' internal affairs. Here North Korea is one example. As North Korea's economy seems to be approaching a dead end, the international society urges Pyongyang to follow China's suit in undertaking economic reform, and China is believed to be the only country that can persuade the Democratic People's Republic of Korea (DPRK) to do so. However, Beijing seems reluctant to take on this task of either giving unsolicited advice to Pyongyang or attaching the condition of serious economic reform to its aid to North Korea. Beijing's difficulty is that it has long held that one should respect every nation's right to choose its own political system and way of development commensurate to its national conditions.

China has maintained this position more emphatically since 1989 in an attempt to fend off Western nations' attacks on its own political system and human rights practices. Beijing also has long opposed attaching conditions to its foreign aid, as it believes this may threaten a nation's sovereignty and interfere with its internal affairs. The Chinese still have bitter memories of the Soviet Union's conditioned aid to China in the 1950s, which was an important cause for the Sino-Soviet split later that decade.

Generally speaking, China's sovereignty concerns have led it to embrace a restricted concept of sovereignty and noninterference into internal affairs. This not only limits China's flexibility in the international arena but also makes Beijing unable to respond convincingly to criticism from Western nations regarding its internal policies on political reform and human rights. As a result, China is viewed as embracing a nineteenth-century concept of sovereignty at the end of the twentieth century, a concept that is essentially outdated in an era of globalization.

On the other hand, driven by the need to integrate itself into the international economic system and to maintain sound political and economic relations with Western nations, China has been gradually adjusting its position on sovereignty issues. Such adjustments, nonetheless, differ by area. In the economic area, Beijing seems to have become reconciled to the idea of "limited sovereignty," that is, in order to benefit from economic interdependence, it has to compromise some of its sovereignty. In this regard, Beijing is learning strategically. On political and security fronts, however, China's adjustments appear to be tactical and superficial—Beijing is unwilling to allow any outside actor to be a legitimate influence on its political and security policies. Although China's growing contact with the international society has created even more pressure on its concept of sovereignty, evolution in its thinking on this issue depends mainly on its internal social-economic developments. Hence, the contradiction between "open-door" incentive and sovereignty will continue to constrain China's foreign policy behavior in the foreseeable future.

Principle and Pragmatism

China is famous for upholding the flag of principle in the world arena. While there are some real grounds for those principles, by and large they reflect the moral and idealistic elements in China's foreign policy thinking and draw mainly from three sources: the traditional Chinese thinking, which dreams of a world of universal harmony (*da tong shi jie*); the humiliating experience in its modern history that causes China to long for a fair and reasonable world order; and the legacy of Marxism-Leninism and Mao Zedong thought, which advocates for a world free of aggression and exploitation of capitalism, imperialism, and colonialism—a world free of power politics, bloc politics, and hegemonism. These principles include the following major points:

1. Five principles of peaceful coexistence.
2. Setting up a fair and reasonable political and economic world order.
3. No use of force or threat of the use of force in international relations.
4. All nations, big or small, strong or weak, rich or poor, are equal in international affairs.
5. China should always side with developing countries. It should never seek hegemony or superpower status.

Obviously, there is some overlapping within those principles; nonetheless, Beijing initiated them at various times (1950s, 1970s, and 1980s, for instance), stressing different aspects for different issues and at different phases. While China's seriousness about these principles is not questionable, in a real world, there are many problems that cannot be resolved solely by principle, and in China's foreign policy, the application of principle to a specific issue is not always an easy job. Under certain circumstances, Beijing has turned to pragmatism, which, as a compromise to principle, provides a more realistic approach to the issue and can best serve China's interests. One case in point is China's position during the Gulf War. When members of the UN Security Council met in November 1990 to discuss the proposal that the United Nations should authorize the use of all means (including the use of force) against Iraq, China faced a dilemma. On the one hand, it opposed Iraq's invasion of Kuwait, and Beijing made very clear this principled position. On the other hand, China had long advocated the principle that force should not be used to solve international disputes, which meant that in principle China would not approve the use of force to drive Iraq out of Kuwait.[3] Here the problem for China was not only the contradiction of two principles but also a conflict of principle and China's national interests. If China stood by the principle of no use of force in international disputes and vetoed the motion, it would cast a serious blow to already difficult Sino-U.S. relations after 1989. In the end, Beijing abstained from voting, which actually made it legitimate for U.S.-led allied forces to drive Iraqi troops out of Kuwait. In this case, China evaded choosing between two principles and embraced pragmatism (the need to put an end to Iraq's occupation of Kuwait as well as to improve Sino-U.S. relations).

Once there exists a conflict between principle and national interest, China will most often opt for a more pragmatic solution. China's frequent abstention in the UN Security Council is illuminating. If, for instance, China finds a proposal brought to the UN Security Council incompatible with its principles while actually not mattering much to its national interests, Beijing does not often veto the proposal; instead, it usually chooses to abstain from voting. By so doing, it avoids unnecessarily offending the United States or being the only country standing in the way. Beijing understands that if it abuses the veto power in the United Nations just out of concern for its principles, it may hurt China's own interests and have a negative impact on its international image. As China's permanent representative to the United Nations Qing Huasheng put it, on the issue of veto power, China should "take into account its own capacity and interests." In his opinion, China's abstaining from voting "does not mean to be cowardly, but a flexible response to a complicated situation."[4] However, if an issue has a direct or indirect bearing on China's key national interests—even if it has no conflict with China's principles—Beijing would definitely use its veto power. In a few cases that China vetoed motions brought to the UN Security Council, the Taiwan issue was the major concern. For instance, in the cases of UN peacekeeping operations in Guatemala and Macedonia, Beijing used its veto power to punish those two countries for their policies toward Taiwan.

If Beijing finds itself incapable of implementing a pronounced principle, it will try to strike a compromise between principle and pragmatism. China's position on the existing international order is a case in point. Since the mid-1980s, Beijing has advocated the establishment of a fair and reasonable international political and economic order and has time and again criticized the unjust nature of the existing world order. To set up a fair and reasonable world order has become one important guideline in China's foreign policy. However, Beijing seems to hold high this flag only in word, while it acts very pragmatically in deed. China understands it has no capacity to challenge the existing world order at the present time as well as in the foreseeable future. Also, Beijing does not view itself as a sheer loser in the current international structure. As a permanent member of the UN Security Council and one of the five recognized nuclear powers, China possesses some significant political and strategic weight in international affairs. The international environment is generally stable and secure, allowing China to concentrate on its domestic development, and the liberal international economic system and the relatively easy access to the markets, technology, and capital of the developed countries help boost the country's economic growth. Although the existing world order is not an ideal one, it is certainly much better than those that China experienced after the Opium War. For the above reasons, China chooses not to make unilateral attempts to alter the existing world order but seeks to make use of it while stressing the need to redress drawbacks of the present international political and economic system.

Sometimes the compromise between principle and pragmatism represents a recognition of the reality, and China's position on the U.S. military presence in East Asia is a good example. In principle, China does not approve of the stationing of troops on the soil of another sovereign country. But in reality, Beijing realizes that the U.S. military presence in South Korea helps stabilize the situation on the Korean peninsula, and U.S. troops deployed in Japan are useful in preventing it from becoming a full-fledged, destabilizing military power, even though Beijing is also concerned that the U.S. presence provides Washington with easy access to the Taiwan Strait. As a result, on a policy level, China does not challenge the U.S. military presence in East Asia; although, Beijing cannot publicly endorse it because of the constraint of the principle.

Both principles and pragmatism have their respective merits in China's foreign-policy practice. While principles paint China's foreign policy with a color of idealism and accord it a kind of moral power, pragmatism creates flexibility and allows China's foreign-policy behavior to maximize China's national interests. There exists, however, a constant tension between these two ends, as principle inevitably constrains flexibility and pragmatism undermines from time to time the relevance and credibility of principle. The bifurcation in China's foreign-policy behavior also puzzles outside watchers. If they take too seriously the rhetorical expression of principles, they may lose sight of the pragmatism driving China's foreign-policy behavior. If they pay too much attention to the pragmatism, however, they may underestimate the influence of idealism and morality. There may not exist a universal formula to understand the contest between principle and pragmatism in China's foreign-policy practice, and the actual weight of each factor differs from case to case. Yet, on the whole, it is fair to say that since the 1980s, as ideology and idealism fades away from China's foreign policy thinking, pragmatism has gained a larger sway vis-à-vis principles.

Bilateralism and Multilateralism

Like other countries, Beijing's participation in international affairs is pursued in both bilateral and multilateral contexts. Ideally, China should employ either bilateralism or multilateralism as necessary and appropriate, but in reality, Beijing favors bilateral rather than multilateral channels. One reason is that China lacks experience with multilateralism; as a matter of fact, China's involvement in multilateral activities is a development of the last twenty years. The other reason—a historical one—is that China harbors strong suspicions toward international mechanisms and believes they mainly serve the interests of the dominant powers.

Since the 1980s, China has come a long way in working with multilateral mechanisms. On the economic front, Beijing actively seeks to join regional and global institutions, organizations, and forums and views relations with them as basically beneficial to its economic modernization. In political areas, being increasingly aware of its responsibility as a permanent member of the UN Security Council as well as its major-power status, China has been trying to play a more active role in the United Nations and has shown more interest in participating in multilateral political activities, such as the Asia-European Summit meeting and the Northeast Asian–ASEAN dialogue. Even on the security front, Beijing has joined, willy-nilly, some regional multilateral efforts, such as the ASEAN Regional Forum (ARF) and the "four-way talks" on building new peace mechanisms on the Korean peninsula.

Under certain circumstances, however, China still has significant reservations about multilateralism. If China finds multilateral settings threatening to put it into a disadvantageous position, it will insist on pursuing bilateral channels. In the case of the ARF, for instance, Beijing endorses its role in promoting confidence-building but rejects the idea of bringing the South China Sea sovereignty dispute to its agenda. Beijing's opposition to the internationalization of the South China Sea issue is due to its concern that if the issue is debated in a multilateral forum, not only the ASEAN countries will form a united front against China but also other nonclaimant countries, such as the United States and Japan, may side with them. Therefore, China has been insisting that the South China Sea issue be addressed in bilateral settings. Another relevant case is the proposed trilateral dialogue between China, Japan, and the United States. In the post–Cold War era, relations between these three have become the underpinning of regional stability; many issues, bilateral and regional, would be best addressed in a Beijing-Tokyo-Washington trilateral context. However, when Japan raised this idea in 1997, China showed at best a tepid interest in it. Beijing is worried that after the redefinition of the U.S.-Japanese alliance, a trilateral dialogue may turn out to be a "two-to-one" game, with Tokyo and Washington aligning themselves on various issues, putting Beijing in a very disadvantageous position. Out of this concern, Beijing prefers dealing with Washington and Tokyo separately and bilaterally.

Beijing's other concern is that a multilateral mechanism may turn out to be the tool of other major powers. On the security front, for instance, since the redefinition of the U.S.-Japanese alliance, China has expressed more vocal opposition to bilateral alliances. However, this does not mean that China is ready to embrace multilateral security in the Asia-Pacific region. Beijing still suspects that given the United States' dominant power and influence in the region, Washington will be able to manipulate any regional security mechanism, and this certainly will not be in China's interests. Also, given some regional members' concerns over the increase in Chinese power and its behavior on issues such as the South China Sea disputes, the multilateral mechanism may serve as a constraint on China in some aspects. In an economic area, during the 1997 Asian financial crisis, Japan proposed to establish an Asian Monetary Fund (AMF) as a regional financial mechanism for dealing with future financial storms. China, partly out of the concern that Japan may use its huge financial power to dominate this organization, withheld its support, although from a regional perspective, the proposed AMF may have helped stabilize the Asian financial environment. As a matter of fact, given China's growing financial strength, its participation in such a mechanism would allow it to play a larger role in regional economic affairs.

Overall, a mixture of bilateralism and multilateralism will continue to exist in China's foreign policy behavior. While Beijing will stick to bilateralism as the major form of its interactions with most countries, its position on multilateralism will be selective and issue-specific, depending on how this may, in Beijing's calculation, affect China's interests. As a result, tension between bilateralism and multilateralism will inevitably limit China's maneuverability on the international stage.

Conclusion

In summary, the aforementioned four contradictions have given rise to a dichotomy in China's foreign policy behavior. While Beijing cherishes the aspiration of being a major power and expects to be treated as such, it is highly selective in assuming the responsibilities of a major power. While China seeks to integrate its economy into the international system, it stands vigilant on the sovereignty issue and rejects any interference into its internal affairs. While Beijing reacts to many international issues with strong rhetoric, in practice it acts in a deliberate and restrained manner. While Beijing expresses enthusiasm about multilateralism, it still feels more comfortable with bilateralism.

However, these contradictions will not stand as permanent constraints on China's foreign policy behavior. Instead, their influence will decline as a result of the evolution of a wide range of internal and external variables. As China achieves greater economic prosperity and builds up a more impressive overall national power, the dual-identity syndrome of great power versus poor country will diminish, and China will be able to play a more visible role in international affairs, a role commensurate with China's real major-power status. As China gets further integrated into international political, economic, and security regimes; as China's concerns over Tibet, Xinjiang, and Taiwan are assuaged; and as Beijing conducts significant political reforms, the concern over its territorial integrity and the interference of its internal affairs will diminish. As a result, China will likely adopt an updated concept of sovereignty, and its policies on those issues will be more flexible and reflective of the new reality in a more interdependent world. As China grows into a mature power and becomes deeply involved in international affairs, Beijing will approach many issues more from a pragmatic basis and less from a principled standpoint. Finally, as Beijing gains experience and confidence with multilateral activities, its preference for bilateralism and caution with multilateralism will be modified, allowing it to make better use of the world arena.

Notes

1. Shao Zongwei, "Nato Urged to Halt Air Attack," *China Daily*, March 26, 1999, p. 1.
2. *China Diplomatic Survey: 1995* (Beijing: World Affairs Press, 1995), p. 569.
3. *China Diplomatic Survey: 1991* (Beijing: World Affairs Press, 1991), pp. 389–390.
4. Zhou Dewu, "China Chins Up and Chest Out in the United Nations," *Global Times*, October 1, 1999, p. 24.

4

Chinese Nationalism and Pragmatic Foreign Policy Behavior

Suisheng Zhao

The rise of Chinese nationalism after the decay of communism in the late twentieth century has captured the attention of many Western observers. Although some scholars are cautious in exploring the limits of Chinese nationalism and in raising the question of whether Chinese nationalism is affirmative, assertive, or aggressive,[1] alarmists believe that Chinese nationalism is a course of international aggression.[2] This chapter attempts to explore if there is a direct link between Chinese nationalism and foreign policy behavior and if the rise of nationalism has changed China's pragmatic foreign policy behavior, making China particularly aggressive or inflexible.

The Resurgence of Chinese Nationalism

Nationalism, or patriotism (*aiguo zhuyi*) in the Chinese official vocabulary, has been on the rise as a powerful force in the last decade. It is not only openly promoted by the Communist state but also advocated by many Chinese intellectuals, liberal and conservative alike, and reflected in the sentiment of the general population (see Table 4.1).

As faith in communism declined among the Chinese people, the Chinese Communist Party (CCP) rediscovered the utility of nationalism. Shortly after the 1989 crackdown, the Communist state launched an extensive propaganda campaign of education in patriotism. The core of the patriotic education campaign was the so-called *guoqing jiaoyu* (education in national condition), which unambiguously held that China's *guoqing* (national condition) was unique and not ready for adopting a Western-style, liberal democracy. Instead, the current one-party rule would help maintain political stability, which was a precondition for rapid economic development.

The campaign emphasized Chinese tradition and history as the CCP tried to link Communist China with its non-Communist past and defined patriotism in terms that had everything to do with Chinese history and culture and almost nothing to do with imported Marxist dogma. While the Great Wall in northern China was celebrated as an armory of official patriotism,[3] the Humen Burning Opium site in Guangdong province reminded the Chinese people about the beginning of the "hundred years of suffering and humiliation" in the hands of foreign imperialism. The celebration of the Great Wall and many historical sites was accompanied by the revival of Confucianism and other Chinese traditional cultural activities.

The patriotic education campaign also emphasized national pride and territorial in-

Table 4.1

The Rise of Chinese Nationalism in the 1990s

At the State Level
A top-down effort to promote and make use of patriotism

- The Communist state presented itself as the defender of China's national interest against Western sanctions after the Tiananmen incident in 1989.
- A Communist state-led patriotic education campaign (1991) emphasized the unique *guoqing* (national essence) of China and promoted Chinese tradition and history, territorial integrity, and national unity.
- The Communist state fought for China's entry into GATT/WTO and maintenance of the MFN/PNTR status and bid for hosting the year 2000 and 2008 summer Olympic games.
- The Communist state launched military exercises to preserve China's national integrity against Taiwan independence prior to Taiwan's first direct presidential election (1995–1996).

At the Intellectual Level
An intellectual discourse advocating nationalism

- Many Chinese intellectuals shifted their course from seeking inspiration from the West in the 1980s to a deep suspicion of the West in the 1990s.
- A geopolitical-based intellectual discourse critical of the West emerged in response to Western writings, such as *The End of History, The Clash of Civilizations*, and *The Coming Conflict with China*.
- Chinese cultural nationalism developed and called for a *fanshi zhuanyi* (paradigm shift) to redefine intellectual discourse in terms of Chineseness.

At the Societal Level
A bottom-up populist sentiment against foreign pressures

- A series of "Say No" books (*The China That Can Say No, The China That Still Can Say No, How China Can Say No, Behind the Scene of Demonizing China*), and so on became the best-sellers in the mid-1990s.
- Popular support for the PLA military exercises in the Taiwan Strait in 1995–1996.
- Popular protest against a right-wing Japanese group, which erected a lighthouse on the Diaoyu (Senkaku) Islands over which China claims sovereignty (1996).
- "Mao Fever" (the mid-1990s).
- Large-scale student protests in front of the U.S. embassy in China against the NATO bombing of the Chinese embassy in Belgrade (1999).

---

tegrity. In the midst of Western-imposed sanctions after Tiananmen, the Chinese Communist regime made the accusation that "a small number of Western countries feared lest China should grow powerful, thereby exercised sanctions against her, contained her, and added great pressures on her to pursue Westernization and disintegration [*xihua he fenhua*]."[4] Patriotism thus was used to bolster CCP leadership in a country that was portrayed as besieged and embattled. Defending China's national interests, the Communist regime presented itself as the fighter for China's entry into the World Trade Organization (WTO), the maintenance of the Most Favored Nation (MFN) status, or the Permanent Normal Trade Relations (PNTR) in the United States, and the successful bid for the Olympic games in Beijing.

A third theme of patriotic education was national unity against ethnic separatist movements, which were among the myriad of social and political problems confronting the Chinese Communist leaders in the post–Cold War era. The central point was that "the Han nationality cannot do without ethnic minorities and vice versa, so that they will consciously safeguard national unity and the motherland's unification."[5] The national unity theme was particularly emphasized in ethnic minority concentrated areas, such as Tibet, Xinjiang, and Inner Mongolia, where so-called narrow-nationalism or separatism was targeted in the campaign. The Chinese nation was said to have a great power of national cohesiveness (ningjuli).

In the meantime, the mainstream of Chinese intellectual discourse experienced a drastic shift from enthusiastic worship of the West in the 1980s to "deromanticization" and deep suspicion of the West in the 1990s. Many Chinese intellectuals were particularly sensitive to the views expressed in major Western intellectual works that developed out of post–Cold War international politics. Among them were the three books pertinent to China's relations with Western countries: Francis Fukuyama's 1992 book, The End of History and the Last Man; Samuel P. Huntington's 1993 article, "The Clash of Civilizations" in Foreign Affairs, which was later expanded into a book with the same title; and The Coming Conflict with China, coauthored by Richard Bernstein and Ross H. Munro and published in 1997.[6] While some Chinese, liberal intellectuals, welcomed Fukuyama's argument in terms of the victory of liberalism, they were concerned that Western, liberal democracies would confront China's rising based on geopolitical considerations. Their concern was confirmed by Huntington's argument that geopolitical struggles in the post–Cold War world were not ideologically motivated but defined by different civilizations. They were thus convinced that a confrontation between different nation-states under the banner of nationalism was going to replace the opposition between communism and capitalism. Under these circumstances, some Chinese intellectuals argued that nationalism would be indispensable and a rational choice to advance China's national interests.[7]

In response to the alarming message that there was a Western conspiracy to contain China, some Chinese intellectuals began to write articles and books advocating nationalism. Although the emerging intellectual discourse on nationalism overlapped, to a certain extent, the patriotic education rhetoric of the Chinese government, its emergence was largely independent of official propaganda. Those who contributed to the intellectual discourse on nationalism were from various political backgrounds. As Wang Xiaodong (using the name Shi Zhong) observed, "Among those under the banner of nationalism, there is a full array of people: some of whom advocate authoritarianism, others who support expansionism; while some people believe in more state controls, and others uphold total freedom in the market economy. There are also those who propose a return to tradition and others opposing this restoration."[8] The important common denominator that brought together intellectuals of different political views was the concern over China's changing position in the post–Cold War world.

Nationalism was not the sole province of state propaganda and intellectual discourse. Populist sentiments were also part of the nationalist orchestra. It was expressed vividly by a series of instant best-sellers, known as "say no" books, published in the mid-1990s. The first of the popular "say no" books was *Zhongguo Keyi Shuo Bu* (The China That Can Say No), which aroused keen interest among many Chinese people and became an instant best-seller, selling more than 2 million copies in 1996. The book warned that the U.S.-led Western countries were organizing an "anti-China club." It called on China to say "no" to various unreasonable demands from the West and to the lack of nationalist consciousness (*quefa minzu yishi*) from the Chinese themselves. The authors claimed, "the nineteenth century was the century of humiliation for the Chinese. The twentieth century has been the century that the Chinese experienced all kinds of sufferings in the mankind . . . the twenty-first century will be the century for the Chinese to restore its glory."[9] After the publication of *The China That Can Say No*, a series of "say no" books appeared in China's bookstores, such as *Zhongguo Hai Shi Neng Shuo Bu* (The China That Still Can Say No), *Zhongguo Heyi Shuo Bu* (How China Can Say No), and *Renminbi Keyi Shuo Bu* (Chinese Currency Can Say No).[10] In addition to the books with straightforward "say-no" titles, many other books that depicted a confrontational relationship between China and the West were published in the mid-1990s and became integral parts of the say-no series.[11]

One interesting phenomenon in relation to the upsurge of nationalism at the popular level was the rise of "Mao Fever" in the 1990s. The Communist dictator, Mao Zedong, was praised in popular books as a "great patriot and national hero" because of his courage to stand firm against Western imperialism. Mao's thought was attributed to the rise of China, as indicated by the title of a book, *Mao Zedong Sixiang yu Zhongguo de jueqi* (Mao Zedong Thought and the Rise of China).[12] Books on Mao's life once again became popular among the Chinese youth.[13] Stuart R. Schram's book written in 1969, *The Political Thought of Mao Tse-Tung*, was translated into Chinese and published in Beijing with a new title *Mao Zedong* in 1987.[14] Although it did not sell well in the first four years after its publication, it suddenly became popular and sold over 240,000 copies in 1991.[15] Many Western visitors to China in the 1990s were impressed by the display of Mao's status and pictures in almost all the taxicabs. The buttons of Mao's portrait also became fashionable among many Chinese people. A Western scholar also indicated that the nostalgia for Mao was to a great extent a reflection of the populist sentiment of nationalism.[16]

Chinese Nationalist Perspectives

Nationalism is a modern concept, which combines the political notion of territorial self-determination, the cultural notion of national identity, and the moral notion of national self-defense in the anarchical world. Nationalism appeared with the emergence of the nation-state system in Europe and spread to the rest of the world after non-European countries were brought into the system.[17] The modern nation-state is a unique form of political organization, which was born as a result of the struggle between empire and nation and between tradition and modernity. Traditional empires were characterized as a mixture of universal principles (such as Christianity in the Roman Empire, Islam in the Ottoman Empire, and Confucianism in China) and particularistic features (such as ethnic composition, language, and ancient customs). The rise of modern nationalism basically meant the

Table 4.2

Perspectives of Chinese Nationalism

	Nativism	Antitraditionalism	Pragmatism
Sources of China's weakness	Imperialism and subversion of indigenous Chinese virtue	Chinese tradition and culture	Lack of modernization, particularly economic backwardness
Best approach to national revitalization	Return to Confucian tradition and self-reliance	Boundless adoption of certain foreign cultures and models of modernization	Whatever works, whether modern or traditional, foreign or domestic
Periods of dominance	• The Boxer Rebellion in the late 1890s • The Great Leap Forward in late 1950s • The Cultural Revolution in the late 1960s • The Cultural Fever in the 1980s	• The New Cultural Movement in 1916–1919 • The May Fourth Movement in 1919 • The Westernization in the 1980s	• The Self-strengthening Movement in 1864–1895 • The post-Mao reform in the 1980s

ascendancy of sentiments associated with the particularistic features of the nation over universal principles. The ascendancy of particularism led to the drive by nations to gain political independence and then to acquire and maintain equal status with other nations.[18]

China was an empire but not a nation-state before the nineteenth century as Chinese people were not imbued with an enduring sense of nationalism based on loyalties to the nation-state. According to Benjamin Schwartz, nationalism "represents a fundamental 'turn' in modern Chinese culture."[19] The catalyst for this turn was China's defeat by British troops in the 1840–1842 Opium War. This defeat paved the way for the eventual disintegration of imperial China and led the Chinese elite to reject the European concept of nationalism that would provide a new basis for China's defense and regeneration. In the twentieth century, all Chinese leaders—from Sun Yat-sen, Chiang Kai-shek, Mao Zedong to Deng Xiaoping and Jiang Zemin—have shared a deep bitterness resulting from China's humiliation and have determined to blot out that humiliation and restore China to its rightful place as a great power.

While pursuing the similar goal of national greatness or, in Chinese words, a *qiangguomeng* (the dream of a strong China), Chinese political elites have been divided on how to revive China and have developed three different nationalist perspectives: nativism, antitraditionalism, and pragmatism. Each perspective is rooted in a different assessment of the sources of national weakness and advocates a best approach to revitalize China (see Table 4.2).

The nativist perspective has been advocated largely by traditional elites and was at times supported by a vast population in China. It calls for a return to Confucian tradition and asserts that China's decline is primarily due to foreign transgressions. In the nativist view, national salvation must be attained through exclusive reliance upon indigenous virtue and ideas. Nativism is a powerful call because its stress on national independence is appealing to a nation that had, for about a century, been the victim of foreign imperialist expansion. The goal of building an independent modern nation-state that could resist any foreign threat has been infused into the discourse of Chinese nationalism since the late nineteenth century. In the early days of the People's Republic of China (PRC), Chinese Communist leaders brought with them traumatic memories of their country's inability to determine its own fate in the century of humiliation. To counter imperialist challenges, the Chinese Communist Party strove to make China completely independent of foreign influence by economic self-reliance and military strength. Mao's policy of *zili gengsheng* (self-reliance) was consonant with the potent spirit of nativism. Nativism was expressed strongly during the Cultural Revolution when Mao carried out a policy of autarchy by isolating China almost completely from the rest of the world. After the beginning of reforms in the 1980s, Mao's nativism lost momentum. In the early 1990s, there was a new surge of nativism among the elites who worried about the decline of the Communist ideology and rediscovered the value of the Chinese tradition, which the party had relentlessly attacked.

In contrast to nativism, antitraditionalism sees China's tradition as the source of its weakness and calls for the complete rejection of Chinese tradition and boundless adoption of Western culture. Antitraditionalism received its first forceful expression during the May Fourth Movement in 1919. One extreme but nevertheless instructive example of the May Fourth era is Qian Xuantong's letter to Chen Duxiu, who later became one of the founders of the CCP. In response to Chen's proposal to abolish Confucianism, Qian concurred in the letter that "it is now the only way to save China," and suggested that the Chinese language had to be replaced by Esperanto in order to save the country. One study found that Qian did not seem to have been concerned with the extent to which his program might be supported by China's masses, since in his view they were the problem.[20] From the May Fourth Movement through the Maoist years to Deng's reform era, there have been repeated attacks on China's cultural heritage. There was a new surge of antitraditionalism driven by liberal intellectuals in the early reform years of the 1980s when the party began to correct the miscarriages of justice and opened China to the outside world. The dramatic events revealed in this process shocked the Chinese people. Their national pride suffered a heavy blow from self-condemnation of their recent past and awareness of China's economic backwardness. Many intellectuals could not help asking why this had happened. While open criticism of the Communist system was forbidden, some intellectuals turned to criticism of Chinese culture in an attempt to discover causes for the failure in modernization. Many Chinese intellectuals blamed China's "feudal culture" for the country's absolutism, narrow-mindedness, and love of orthodoxy, and even called Chinese people the ugly Chinese.

Different from both nativism and antitraditionalism, pragmatism sees foreign economic exploitation and cultural infiltration as a source of China's weakness, but believes that the lack of modernization is the reason why China became an easy target for Western imperialism. Pragmatists would like to adopt whatever approach that may make China strong. This is illustrated by Deng Xiaoping's famous saying: "It doesn't matter if it is a black or

Table 4.3

International Orientations of Chinese Nationalism

	Nativism	Antitraditionalism	Pragmatism
Confrontation	• The Boxer Rebellion in 1897–1901 • The Cultural Revolution in 1966–1969		
Accommodation		• The Sino-Soviet alliance in the early 1950s • The total Westernization school of thought in the 1980s	
Adaptation			• The Self-strengthening Movement in 1864–1895 • The post-Mao reform period in the 1980s– 1990

white cat as long as it can catch rats." Early pragmatic thought can be traced to notable bureaucrats and scholars in the Self-strengthening Movement of 1864–1895. They were pragmatic nationalists in the sense that they wanted to make selective use of foreign methods to defeat foreign barbarians (*yi yi zhi yi*). In a similar way, contemporary pragmatists have identified economic modernization as the key to the revitalization of China. Pragmatic leaders have set economic growth as China's top priority because they know that the CCP's continued leadership vitally depends on its ability to improve the Chinese people's standard of living. The Communist regime has thus substituted performance legitimacy provided by surging economic development and nationalist legitimacy provided by the invocation of distinctive characteristics of Chinese culture for obsolete Marxism-Leninism.

International Orientations of Chinese Nationalism

Each of the three perspectives of Chinese nationalism has its unique foreign policy implications. Nativism infuses xenophobia into a confrontational policy toward foreign powers; antitraditionalism tries to adapt to the modern world by invocation of certain foreign models; and pragmatism lies between the two, asserting China's national interests by both reacting to and absorbing from the outside world (see Table 4.3).

Nativism is often related to confrontational antiforeignism, which is hypersensitive to perceived foreign insults and often makes for militant reactions. Nativism may thus turn into ultranationalism, which is characterized by the suspicion, dislike, or fear of other nations and is associated with feelings of national superiority and with superpatriotism, an intensity that makes ultranationalism, to a certain extent, similar to fundamentalism, which often converges on "those who offer most complete, inclusive and extravagant world views."[21]

The most extreme example of antiforeignism in connection with nativism in modern China may be seen in the strong xenophobia of the Boxer Rebellion in 1900 and again during the Cultural Revolution of the late 1960s. The hostility that developed toward foreigners and all things foreign during the Boxer Rebellion resulted in the burning of the British legation in Beijing and shops that sold foreign merchandise and books, and the killing of foreigners. When nativism prevailed during the Cultural Revolution, the dependents of Soviet diplomats, seeking to leave China, were forced to crawl on their knees to the airplane waiting for them. The British legation in Beijing was burned to the ground again. As Harry Harding points out, "The Cultural Revolution represented the revival of a submerged strain of thought, a strongly xenophobic Chinese nativism, similar in some respects to the antiforeignism of the Boxer Rebellion at the turn of the century."[22] At the height of the Cultural Revolution, China remained friendly only with the most militant Communist countries such as Albania, North Korea, and North Vietnam. Beijing encouraged Maoist insurgents to carry out armed struggles against the governments of almost every foreign country.

Nativism lost its momentum in the 1980s, but regained some ground in the 1990s when some scholars took Chinese culture as a symbol against Western cultural hegemony and cultural colonialism and suggested that Asian values are superior to their Western counterparts and should, therefore, be the basis for the values of the world in the twenty-first century. Some politically motivated nativists warned about "cultural pollution" and "peaceful evolution" conspiracies by the West. Nativists are particularly hostile to the United States because the United States represents an opposing value system of liberalism and individualism. Nativist sentiments are against Japan as well, due mainly to the humiliation and injustices suffered by the Chinese people at the hands of Japanese imperialists during World War II. In comparison, the hostility of nativism against the United States is due to the fact that the United States represents the Western value system while the animosity toward Japan does not have this value dimension. Thus, the United States is treated in a very different way in nativist sentiments. To a certain extent, nativism is anti-Americanism or anti-Westernism. Nativists have used confrontations with foreign powers to rouse emotional, nationalist reactions by tapping the deep-rooted feeling of Chinese cultural superiority and resentment at foreign efforts to belittle or humiliate China in an attempt to rally the Chinese people against any foreign infiltration.

In contrast to nativism, antitraditionalism calls for boundless adoption of foreign models of modernization, but its orientation is adaptation rather than accommodation to foreign powers. Adaptation is a human learning process in which old ways of coping with the outside world are changed into new ones. This change can create a readiness to respond to a given international situation in a characteristic and repetitive fashion. In the China case, disagreement about what foreign models China should adopt has resulted in adaptation efforts that did not follow the path delineated by any one foreign model. Course change and even maladaptation became common practice.

In the early period of the PRC, the foreign model that Mao brought to China was Marxism-Leninism. China was going to adopt the Soviet model of economic and political development and adapt to the Soviet-led Communist world. This orientation was evident in Mao's historical announcement on June 30, 1949, that China would "lean towards one side" in the struggle between imperialism and socialism. China must ally itself "with the Soviet Union, with every New Democratic country, and with the proletariat and broad masses in all other countries."[23] However, the adaptation to the Soviet-led world was a total failure. This maladaptation resulted in his passionate engagement in the polemics with the Soviet Union and repudiation of the Soviet brand of communism in the 1960s.

In a different fashion, liberal antitraditionalists in the 1980s called upon the Chinese people to rejuvenate the nation by assimilating nourishment from the West, adopting Western models of modernization, and adapting to the competitive capitalist world. This message was conveyed vividly in *Heshang* (River Elegy), a six-part television series broadcast in 1988. Turning key symbols of China's glorious past into symbols of its modern backwardness, the series used images of the West as symbols of the new civilization, which were summoning China. This antitraditionalism gave voice to a nationalism that was urgent but simplistic and idealistic because of its overcriticalness of Chinese tradition and overenthusiasm about Western culture. Many Chinese intellectuals soon realized that antitraditionalism could not make China strong and alive, because a viable dream should link the legacies of the past with the hopes for the future. After the initial "culture shock" in the early years of China's reform and opening to the outside world, Westernization prompted less and less enthusiasm among Chinese people in the 1990s. It was thus recorded as another maladaptation.

Pragmatic nationalism is national interest-driven. Its orientation in world affairs is assertive in defending and seeking China's national interests. As one advocate of Chinese pragmatism states, "Its main objective is to build a politically, economically, and culturally united nation-state when foreign and largely Western influences are seen as eroding the nation-state's very foundation."[24] For pragmatic nationalists, "both the nation's problems and most of the possible solutions were perceived as coming from outside."[25] Pragmatic nationalism thus gains its power both from reacting to and absorbing from the outside world.

Pragmatic nationalists are no less determined than both nativists and antitraditionalists to establish China as a powerful nation. They have constantly revealed their pride in China's national heritage and their dedication to making China a leading power in the world. However, unlike nativists who believe that Chinese cultural tradition may be of universal human value, pragmatic nationalists insist that the universal cultural claims, whether modern or traditional, Chinese or foreign, are subject to the promotion of China's national interest or the enhancement of national pride. They want to see China being a full-fledged participant in international affairs and occupying a prominent position in the world. Compared with antitraditionalists who look to foreign models for China's future, pragmatists are more critical toward any imported universal principles, including both Marxism and liberalism. They emphasize the gap between these Western models and the Chinese condition (*guoqing*). This has led some pragmatists to look toward an Asian path of modernization. The economic success of many East Asian countries seemed to have proved that modernization is not necessarily tantamount to Westernization and China can accumulate wealth without simply following any foreign models.

In this case, although pragmatic nationalists may be flexible in tactics, subtle in strategy, and even able to avoid appearing confrontational, they are uncompromising with foreign demands and may be arrogant in their singularity and dismissal of Western views and positions. This may make it difficult for some foreign countries to elicit cooperation on issues that trigger historical sensitivities, as pragmatic leaders are deeply committed to the preservation of national sovereignty, the reunification of China, and the attainment of national wealth and power. This assertion is evident in Beijing's Taiwan policy. On the one hand, Jiang Zemin made an eight-point proposal on January 30, 1995, suggesting that the two sides across the Taiwan Strait start negotiations "on officially ending the state of hostility between the two sides and accomplishing peaceful reunification step by step."[26] On the other hand, the People's Liberation Army (PLA) launched one after another wave of military exercises in the Taiwan Strait to stop the perceived Taiwan independence momentum and its collaboration with foreign forces only a half year after Jiang's peaceful venture. Through the use of force in an exemplary and demonstrative manner, pragmatic leaders showed their assertiveness in defending China's territorial integrity.

Characteristics of Chinese Nationalism

While nativism and antitraditionalism have continued to lurk in the background, pragmatic nationalism has become the dominant line of thinking among Chinese people and their leaders since the 1980s. A careful examination of pragmatic nationalism reveals its three characteristics: instrumentality, state-centricity, and reactivity.

Chinese nationalism has been characterized by its instrumentality to the Communist government in compensating for or, to a certain extent, replacing the all-too-evident weaknesses of Communist ideology since the 1980s. Pragmatic leaders have fashioned nationalism because it has the effect of removing differences within the country and replacing it with a common, hegemonic order of political values. Nationalism has been used to rally popular support behind a less popular Communist regime and its policies by creating a sense of communality among citizens. Pragmatic nationalism thus does not have a fixed, objectified, and eternally defined content. Instead, it has been continually remade to fit the needs of its creators and consumers. Narratives of pragmatic nationalism, including interpretations of nationalistic symbols, are "invented histories or traditions."[27]

The emergence of pragmatic nationalism in post-Mao China was in response to a legitimacy crisis of the Communist regime starting in the late 1970s when the regime was deeply troubled by what was popularly called *sanxin weiji* (three spiritual crises), namely, a crisis of faith in socialism, a crisis of confidence in the future of the country, and a crisis of trust in the party. After Communist ideology lost credibility, some intellectuals turned to Western, liberal ideas and called for Western-style democratic reform. *Sanxin weiji* thus resulted in the prodemocracy movement and the large-scale demonstrations in the spring of 1989. How to restore legitimacy and build a broad-based national support became the most serious challenge to the post-Tiananmen leadership. It was during this time that the instrumentality of nationalism was discovered. Deng Xiaoping and his successor, Jiang Zemin, began to wrap themselves in the banner of nationalism, which, they found, remained a most reliable claim to the Chinese people's loyalty and the only important value that was shared by both the regime and its critics. It was ironic that prodemocracy demonstrators in Tiananmen Square in 1989, while confronting the government, claimed that patriotism drove them to take to the streets. After the Tiananmen incident, pragmatic leaders began to emphasize the party's role as the paramount patriotic force and guardian of national pride. In the face of Western sanctions against China, they moved quickly to position themselves as the representatives of the Chinese nation and Chinese economic interests, including China's entry into the WTO (World Trade Organization), and maintenance of the low-tariff treatment on exports to the United States, known as Most Favored Nation (MFN) status. One dramatic example of its effort to identify the regime with Chinese national pride was the bid to host the year 2000 summer Olympic games in Beijing. Although China failed to get the games, Chinese popular resentment was directed at foreign countries and human rights groups who were blamed for the failed Olympic bid.

Nationalism is an effective instrument for the Communist regime. Prior to the Tiananmen tragedy, the passion of Chinese intellectuals had been focused on political liberalization and the most popular international political figure on Chinese university campuses was Gorbachev. Many college students took Gorbachev's perestroika as the ideal course of action for the CCP. In the 1990s, however, worry about disorder and disunity outweighed concern about political lethargy or oppression, let alone a demand for liberty and human rights. The economic difficulties of the former Soviet states provided an expedient pretext for the government to relegate political liberalization to the background. The need for political stability, a precondition for steady economic growth, became an overriding mission of the Chinese leadership. Under the banner of patriotism, pragmatic leaders made an all-out effort to rally support for the status quo.

Led by the state, pragmatic nationalism identifies the nation closely with the Communist state. Nationalist sentiment is officially expressed as *aigu*, which in the Chinese language means "loving the state," or *aiguozhuyi* (patriotism), which is love and support for China, always indistinguishable from the Communist state. As Michael Hunt observes, "By professing *aiguo*, Chinese usually expressed loyalty to and a desire to serve the state, either as it was or as it would be in its renovated form."[28] From this perspective, Chinese patriotism can be understood as a state-centric or state-led nationalism. The Communist state is portrayed as the embodiment of the nation's will and seeks for the loyalty and support of the people that are granted the nation itself. The Communist state tries to create a sense of nationhood among all its citizens by speaking in the nation's name and demanding that citizens subordinate their interests to those of the state. Freedom is sought not for individuals but for the nation-state. This means all power is given to the rulers of the Communist state. As asserted by *The People's Daily* editorial on National Day in 1996, "Patriotism is specific. . . . Patriotism requires us to love the socialist system and road chosen by all nationalities in China under the leadership of the Communist Party." By identifying the party with the nation, the regime makes criticism of the party an unpatriotic act.

Pragmatic nationalism is reactive. It intends to become strong in response to perceived foreign pressures. As Liu Ji said, "Chinese nationalism has a unique characteristic: in peaceful times, there are often 'internal struggles' (*wolidou*) within the nation . . . when confronted with foreign invasions, this nation is prone to respond with a narrow kind of nationalism."[29] Indeed, a reactive sentiment to foreign suppression is the starting point of Chinese nationalism. Unlike the formation of nationalism in Europe as an indigenous process driven by mercantilism and liberalism, nationalism in China was initially borrowed from the West by the Chinese elite to defend China against foreign invasions in the late nineteenth century. That is why Xiao Gongqin calls Chinese nationalism a *"yingji-zhiwei xing"* (reactive-defensive type) of nationalism. According to Xiao, Chinese nationalism has risen in response to "negative stimulus" (*buliang ciji*) from certain foreign forces. It is, therefore, reactive to specific issues and has little to do with abstract ideas, religious doctrines, or ideologies. "The intensity of the reaction is in proportion to the intensity of negative stimulus from abroad."[30]

The effort to identify the Communist regime with the Chinese nation was particularly effective when China faced challenges from hostile foreign countries. The Korean War was a typical example. With the perceived imminent invasion of American forces via Korea, the Communist government launched a *kangmei yuanchao* (resisting America and assisting Korea) campaign that mobilized popular energy, coupled with nationalistic rhetoric, to defend the nation. Since the end of the Cold War, China has been under heavy pressure from Western countries. In particular, Sino-U.S. relations have become entangled over many issues, such as human rights, intellectual property rights, trade deficits, weapons proliferation, and especially the Taiwan issue. The voice of containment of China reached its peak in the Western media when China launched military exercises in the Taiwan Strait following President Lee Teng-hui's visit to the United States in May 1995. Suspicion of the United States' intentions prevailed among many Chinese people after the United States sent two aircraft carrier battle groups to protect Taiwan in March 1996.

The Constraint on Chinese Nationalism

Nationalism is a double-edged sword. While nationalism may be used by the regime to replace the discredited Communist ideology as a new base of legitimacy, it may also cause a serious backlash and place the government in a hot spot facing challenges from both domestic and international sources. It is not hard for pragmatic leaders to realize that the Boxer Rebellion–style xenophobia that prevailed during the Cultural Revolution may cause more harm than good to the Communist regime. The destructive effects may set a limit on the utility of nationalism to Chinese leaders.

Domestically, rising nationalism has broken the taboo surrounding Chinese foreign policy and run into a criticism of current policy, especially the seemingly too "soft" stance toward the United States and Japan. It is certainly an embarrassment for the Chinese leadership when the authors of *The China That Can Say No* openly claim that "we need China's Zhirnovsky" and propose to take back Taiwan by force at any cost and to assume a confrontational approach to the United States and Japan, while China's modernization continues to depend on cooperation with these two countries. When the nationalist issues have fired up the general public about the conduct of the United States and Japan, it often does more harm than good to the Communist government. The CCP rose to power partly because it gained nationalistic credentials by fighting the Japanese invasion during the war. Now being widely criticized as too chummy with Japan, while it has failed to provide more compensation for wartime injuries, laid claim to Tiaoyu Islands, and allegedly waged economic imperialism by flooding China with Japanese products, the party suffers from losing some of its dwindling legitimacy. The Chinese leadership has worried about the patriotic movement against Japan that may evolve into a protest movement against the

Chinese government itself by those people who are jobless or angry about corruption in the government. In addition, coming at a time when China urgently needs Japanese trade and investment, the government does not want to see the surge of nationalism jeopardizing Sino-Japanese economic relations. Ironically, when Chinese patriots blamed Japan for economic imperialism, the *China Daily* reported that "Chinese products have edged their way into the Japanese market so steadily." According to the report, the trade between China and Japan hit a record $57.47 billion in 1995, up 19.9 percent from 1994. Of the total, China's exports soared 31.7 percent to stand at $28.46 billion. China's trade deficit was narrowed. While Chinese statistics claimed a $540 million trade deficit with Japan, Japan reported a $13.99 billion trade deficit with China in 1995.[31]

Another domestic cost that pragmatic nationalism has to consider is the possibility that separatists may take over nationalist appeals to challenge the very basis of the multinational state of the PRC. The rising nationalism has appealed largely to the Han history and nation. This could be dangerous due to the existence of at least three different Chinese nations. The first is the PRC, officially defined by state nationalism. The second, defined by ethnic nationalism, is the PRC's Han nation, composed of the core Han population, distinct from non-Han nations within the PRC. The third, a product of ethnic nationalism and the vagaries of Chinese political and migratory history, consists of the PRC plus the compatriots (*tongbao*) of Taiwan, Hong Kong, and Macao presently under different political authorities.[32] Ethnic division has always been a source of tension since the founding of the PRC. In spite of the dominance of state nationalism, ethnic nationalism, particularly Tibetan nationalism, with stirrings among the Uighurs, other Muslims, and Mongols, has never been suppressed. No doubt many Han Chinese, and some minorities, accept the idea that a Chinese identity is shared among all the nationalities, but most of the movement toward integration of this nation results from assimilation of the non-Han into Chinese culture. Although the Chinese government has granted minorities various kinds of special representation and autonomy and have been relatively effective in resisting open espousal of Han Chinese ethnic nationalism, all manner of manifestations of ethnic identity continue to endure. Ethnic consciousness in China is often conflictual and violent, most notably in Tibet where ethnic nationalism has triumphed over the state's version of Han-Tibetan relations. Ethnic nationalism among minorities makes it clear that some may choose greater differentiation and autonomy over integration into the PRC nation, a trend that on all sides can only strengthen awareness of the distinctiveness and dominance of the core Han Chinese nation. The third nation of compatriots is also troublesome. For many years after 1949, the gulf between the PRC and the three territories was so great that the nation existed only as a legal fiction. Growing contact with the mainland in recent years has rekindled awareness of one Chinese nation. For Hong Kong and Macao, reunification is now a virtual reality. However, Taiwan shows no signs of yielding to the PRC version of reunification while the Taiwan independence movement has gained momentum in recent years. Nationalism in the case of second and third Chinese nations is a rising challenge to the PRC government.

Internationally, the rise of Chinese nationalism has coincided with a presumption in the international community that nationalism has assumed an irrational and dangerous quality that might distort a state's true interests and threaten other states in the late twentieth century. Many analysts, sobered by two centuries of imperialism, revolution, and war, have taken a negative view of nationalism. Hans J. Morgenthau's classic study found nationalism of the late twentieth century "essentially different from what traditionally goes by that name and what culminated in the national movements and the nation-state of the nineteenth century." According to Morgenthau, "Traditional nationalism sought to free the nation from alien domination and give it a state of its own," while the nationalism of the late twentieth century "claims for nation and one state the right to impose its own valuations and standards of action upon all the other nations."[33] Because it is thought to unite and strengthen a nation internally, embracing traditional nationalism is often thought the right course for a country fighting for unity and independence. Because it claims universal dominance and enhances misunderstanding of external forces, the new crusading nationalism is often seen as a bad thing for one state strong enough to impose its will on others. In this context, although there is a tendency to see the absence of nationalism in imperial China as a fatal weakness, the new tide of Chinese nationalism has caused anxiety in Asia and the rest of the world in the 1990s. The rise of Chinese nationalism is seen as "a potent force in a country that is striving to shake off its image as the sick man of Asia and regain ancient glory."[34] Chinese military exercises in March 1996 may have fired a shot across the bow of Taiwanese independence, but it, in the meantime, set off alarms all over East Asia, causing a series of moves that were against China's national interests. The Japan-U.S. relationship was strengthened. The American military withdrawal from Okinawa was shelved temporarily. Indonesia was drawn closer to Australia and protested China's claim to gas fields in the South China Sea. The Philippines strengthened its military and improved relations with Taiwan.

Nationalist Rhetoric and Pragmatic Foreign Policy: A Case Study

Balancing the positive side and the negative backlash, the Chinese leadership has been very cautious and ambivalent toward nationalism. The Chinese government has never officially endorsed nationalism. The sentiments of the Chinese people are not described as nationalistic but patriotic. In the vocabulary of the Chinese Communists, "nationalism" carries a derogatory connotation and is used to refer to parochial and reactionary attachments to nationalities.[35] The official line defines the Chinese nation always as the total population of China comprising fifty-six nationalities, and loyalty to the nation must mean loyalty to the state. It is particularly interesting to take note that while pragmatic Communist leaders have consciously cultivated nationalism as a new glue to unite the nation against the so-called "splitting" (fenhua) and "Westernization" (xihua) of China by Western countries, strong nationalist rhetoric has often been followed by prudential policy actions in foreign affairs.

This behavior pattern is particularly clear in China's dealing with their relationship with the United States. Pragmatic Chinese leaders have tried very hard to assure that the Sino-U.S. relationship would not be dictated by nationalist rhetoric. Managing to control the expression of anti-American nationalism during the two crises of the embassy bombing and the EP-3 midair collision between a U.S. Navy EP-13 spy plane and a Chinese jet fighter is a good example of the pragmatic policy in practice.

The relationship with the United States is the most crucial and important one in all of China's foreign relations. It is also the most frustrating foreign policy challenge for China's leaders. While China's pragmatic leaders have hoped to establish and maintain a "friendly and cooperative relationship" with the United States, the unwieldy superpower holding the key to China's future of economic modernization, they are also concerned that rising nationalism would evolve into a criticism of Chinese foreign policy, especially its seemingly "soft" stance toward the United States. In this case, pragmatic leaders have tried to avoid the danger of falling victim to the nationalism that they themselves have cultivated and to ensure that Chinese foreign policy is not dictated by the emotional voice of university students and liberal nationalist intellectuals. It is not at all surprising that pragmatic Chinese leaders have described nationalism as a force that must be "channeled" in its expressions, including restraining or even banning students from holding anti-American demonstrations.

In particular, the crises caused by the accidental bombing of the Chinese embassy by the United States on May 8, 1999, and the midair collision between a U.S. Navy EP-13 spy plane and a Chinese jet fighter in the South China Sea on April 1, 2001, showed to pragmatic leaders the damage that uncontrolled nationalism could cause to China's national interest.

In the wake of the U.S. bombing of the Chinese embassy that killed three Chinese journalists, both Chinese government officials and average Chinese people found it impossible to comprehend how the North Atlantic Treaty Organization (NATO) forces led by the United States could have bombed the Chinese embassy in Belgrade by mistake. The Chinese official media carried blanket coverage of the bombing and highly emotive stories on the Chinese victims in virulent anti-U.S. language on the first couple of days after the bombing. In response, university students, spontaneously as well as organized by the university authorities, poured into the front of the U.S. embassy in Beijing and consulates in other cities, throwing eggs and stones to express their anger at the U.S.-led NATO actions. Sympathetic to the students, the police units guarding the embassy did not at first make any move to stop the demonstrations. The Chinese leadership apparently did not anticipate the vehemence of the student protests. The physical damage to the U.S. embassy and consulates spoke of the dangers of playing with nationalist fire. China's crucially important relationship with the United States could be permanently damaged by virulent nationalism unleashed in China. The price would be China's reform and economic growth, and "China would be seen as a rogue state that must be contained" by the United States and other Western countries.[37]

This situation was obviously not in the interest of the Chinese leadership, which sought to maintain stability as the domestic policy priority and to retain a cooperative relationship with the United States as the foreign policy priority. As a matter of fact, pragmatic leaders had tried very hard to improve the relationship with the United States after the Tiananmen incident in 1989. Publicly, Beijing took a strong position defending its stand against sanctions imposed by Western countries. Deng Xiaoping accused America of intervening in its domestic affairs and told his foreign visitors in September 1989, "China is not afraid of sanctions, which in the long run will backfire at those imposing them."[38] Privately, however, pragmatic leaders tried to avoid a confrontational policy against the United States and other Western countries.

This nonconfrontational principle produced positive results. Although President Clinton linked China's MFN status with its human rights record in the first year of his presidency, he was forced to reverse his position the next year and eventually proposed a "strategic partnership" with China in his second term. Pragmatic leaders in Beijing certainly did not want to see the embassy-bombing incident damaging this important relationship, whether or not the bombing was seen as a considerable provocation from the United States. As a result, they had to assume the difficult task of cooling down nationalist fury by calling for a reasoned response. Two days after the incidental bombing, China's vice-president Hu Jintao made a televised speech in which, while extending government support to students' patriotism, he appealed for calm and warned against extreme and destabilizing behavior. Frenzied demonstrations quickly ran out of steam. A tight police cordon was put up around the embassy where U.S. Ambassador James Sasser and his staff had been under virtual siege for almost four days.

When an increased police presence outside the U.S. embassy showed that the Chinese government was determined to prevent protests that might agitate the United States, the Chinese media stopped showing pictures of demonstrations on the streets and instead featured organized protests in schools and workplaces. In the meantime, *The People's Daily* reported that various Western countries had issued advisories against traveling to China, hurting tourism and foreign investment. In this case, although the government continued to demand that the United States perform a thorough investigation and promptly publish the results and punish those responsible, the official Xinhua agency promptly listed apologies by U.S. President Bill Clinton and other NATO leaders, and state television carried Clinton's public apology. Meeting with the visiting Russian envoy, Viktor Chernomyrdin, on May 11 to discuss the Kosovo crisis, President Jiang Zemin stated that life in China should now return to normal and that it was time to turn over a new page in the name of economic necessity: "The Chinese people have expressed their strong indignation in various forms. . . . This has demonstrated the enthusiasm, will and power of the great patriotism of the Chinese people. . . . The whole country is now determined to study and work harder, so as to develop the national economy continuously, enhance national strength, and fight back with concrete deeds against the barbaric act of U.S.-led NATO."[39]

Less than one month after the bombing, Beijing restrained from vetoing the G8 proposal, which had to gain approval from the UN Security Council, to end the Kosovo crisis when Yugoslav leaders announced their decision to accept its conditions.[40] This was striking to some observers because, after the bombing, Beijing threatened all kinds of restrictions on UN Security Council activity but now did nothing. To express goodwill to U.S. leaders, President Jiang sent his congratulations to the U.S. women's soccer team after its July 10 victory over the Chinese team in the Women's World Cup final. This congratulations was reported prominently in the Chinese media and was said to be a signal that Beijing was ready to move past the incident. Beijing's pragmatic policy paid off. Six months after the incident, China and the United States reached a historical agreement on the terms of China's accession to the WTO.

The midair collision between a U.S. Navy EP-3 surveillance plane and a Chinese fighter jet over the South China Sea on April 1, 2001, created another crisis that once again highlighted the possibility of a popular explosion of nationalist sentiment. The Chinese pilot, Wang Wei, killed in the collision, was quickly declared a "martyr of the revolution" and praised as a heroic defender of the motherland. China held the twenty-four U.S. crew members at PLA military facilities on Hainan Island for eleven days and accused the pilot of breaking the law by making an emergency landing at a Chinese airbase without permission. While the collision took place in international airspace about 50 miles outside China's territorial waters, China claimed an exclusive economic zone that extends 230 miles out to sea, and asserted that the plane had no right to conduct surveillance there. Beijing sought an apology from Washington, but the White House declined to apologize, saying the collision was an accident.

The spy plane incident reinforced many Chinese people's suspicion of the United States as a careless bully that threw around its weight without considering the views or feelings of people from other nations. With this new incident, many Chinese people were angered over American spying, saddened by the death of a pilot, and frustrated by President Bush's unwillingness to apologize. At the root of their complaints was a sense of wounded national pride. China had suffered at the hands of foreigners before and was not prepared to suffer again. Many university students threatened that there would be larger demonstrations if the government released the crew members before the United States made an apology.[41]

In response to rising nationalist sentiments, while talking stiffly, pragmatic leaders followed a two-pronged policy, as they did not want to see a repeat of the anti-American demonstrations one year earlier. It was reported that President Jiang Zemin very quickly laid down several tough principles on how to handle the standoff at a Politburo emergency meeting moments after the collision. While there is no way for outsiders to confirm what decisions were made at this meeting, some of the principles sound plausible. One was that the U.S. side "should offer a written apology for using its military aircraft to ram China's aircraft, resulting in a loss of a pilot, and for entering China's airspace and landing at China's airport without permission." Another was that the U.S. side "should stop its military spying and provocative activities along China's coast." A further principle was that "China should adopt measures and be prepared against the U.S. side using the air collision incident to blackmail China politically, economically, and militarily, resulting in tension in Sino-U.S. relations and even local confrontation."[42] However, in addition to these tough principles, the meeting also emphasized that "the leadership must protect itself from criticism by ordinary Chinese by not appearing weak before the 'hawkish' new Bush administration. Yet, at the same time, there should be no repeat of the anti-U.S. demonstrations after the 1999 NATO bombing of the Chinese embassy in Belgrade."[43]

Subsequent events proved that this emphasis was crucial in guiding the actions of Beijing's leaders during the crisis. On the one hand, Beijing's public stance was particularly uncompromising on the demand that the spy plane crew would only be released after a formal apology by the U.S. government to the Chinese people. Commentaries in major media attacked U.S. "neohegemonism" and extolled the patriotism of the Chinese people. On the other hand, pragmatic leaders ensured that the government rather than the public set the tone in determining how to deal with the United States. Beijing wanted to refrain from high-profile actions that might provoke a military confrontation with the United States. The leadership moved to censor vocal anti-American sentiment that had been pouring in on the Internet and, to some extent, in the state-run media. In Beijing and other major cities, while people's anti-American emotions ran high, the government allowed no demonstrations outside U.S. missions and no intimidation of foreign communities. To show his confidence in resolving this incident, President Jiang Zemin did not cancel an earlier scheduled foreign trip and left for a state visit to Latin America on April 4, 2001, four days after the collision. Vice President Hu Jintao was put in charge of an emergency team to handle the crisis.

Following this two-pronged policy, pragmatic leaders maintained control over the situation in Beijing's streets throughout the ordeal. Worrying about the possibility of losing control, these leaders were eager to find a face-saving solution for both sides to get out of the crisis. For this purpose, they took a flexible position under the veneer of toughness and eventually hammered out a so-called "diplomacy of apology." In an interview with CNN, the Chinese ambassador to the United States, Yang Jiechi, hinted that the United States should at least say "sorry" after doing harm to China. When U.S. Secretary of State Colin Powell expressed "regrets" over the loss of the Chinese pilot for the first time on April 4, the Chinese foreign ministry promptly responded by saying this was a "step in the right direction," while still insisting on a full apology. The next day (April 5), at a press conference in Chile, President Jiang provided a further hint by saying that it was normal for two people who had bumped into each other to say "excuse me." The U.S. side at this time also switched to a pragmatic position. President Bush expressed his regrets on the same day and, on Aril 9, Powell used the word "sorry" for the missing Chinese pilot and aircraft. Beijing squeezed again for something stronger than "sorry" in the next two days. When Washington said "very sorry" and indicated that it could not go any further, certainly not to the level of a full apology, Beijing accepted the "very sorry" as a close equivalent to an apology on April 10 and released the crew on the next day.

It was a testimony to the pragmatic leaders' tactical flexibility that the Chinese official media was instructed to translate Powell's expression of "very sorry" as *"baoqian,"* which is one word different from but has almost identical meaning as *"daoqian,"* the Chinese expression of "apology" that Beijing demanded initially. Although the United States did not make a full apology to China, pragmatic Chinese leaders interpreted the expression of being "very sorry" as a full apology and the American expressions of "regret" and "sorry" that meant in most instances only for the loss of the pilot and aircraft as meant for the whole incident. As a Western journalist suggested, "China stressed that it had forced the United States to admit its faults, as it was implicit in the usage of the character '*qian*,' which is both in *daoqian*, the apology demanded by Chinese leaders, and in *baoqian*, or deep excuses, the word used in the American statement to the Chinese." This was a face-saving solution not only for China but also for the United States. As the journalist indicated, President Bush "underscored that it did not give China the precise apology Beijing had demanded, had brought its people home, and thus was no longer subject to either possible Chinese blackmail or internal pressures over its difficulties in handling a difficult situation."[44]

Like President Bush, pragmatic Chinese leaders trumpeted the success largely for a domestic audience as they did not want to let nationalism get out of hand and hurt both the Communist state and the Sino-U.S. relationship. They declared that "China had won a victory at the stage" (*jieduan xing chengguo*) in a *People's Daily* editorial on April 11, the day when Beijing released the U.S. crew. The editorial told the Chinese people, "Our persistent struggle forced the U.S. government to change its tough and unreasonable attitudes at the beginning of the incident and finally apologized to the Chinese people. . . . This struggle extended justice, defeated the hegemonism, defended our country's sovereignty and national dignity, and demonstrated the big-nation spirit (*daguo fengmao*) of

China in defending world peace and fearless in the fact of great power threat." In the meantime it stressed the importance of maintaining a cooperative relationship with the United States: "What direction will the Sino-US relationship go is crucially important [*zhiguan zhongyao*] for the whole world. . . . The improvement of relationship between China and the U.S. is not only in the interests of the two countries but also to the advantage of world peace and stability."[45]

Apparently while pragmatic leaders did not alter their tough rhetoric for domestic reasons, they did almost everything they could from their perspective to avoid confrontation and maintain the framework of long-term cooperation with the United States during the two crises. This seemingly contradictory strategy of talking tough but acting in a calculated manner showed that pragmatic leaders were willing to move forward to rebuild and expand interactions with the West in general and the United States in particular, given their understanding of China's vital interests. They recognized that China's modernization inevitably depended on adapting to the modern world and required maintaining cooperation with the United States and other Western countries.

Conclusion

Pragmatic nationalism is assertive in international orientation and is particularly powerful when China's national interests or territorial integrity are in jeopardy. But it has not made China's international behavior particularly aggressive. That is why David Shambaugh characterized Chinese nationalism as "defensive nationalism," which is "assertive in form, but reactive in essence."[46] Defensive nationalism has made Chinese leaders very assertive in defending China's national interests, particularly on the issues concerning national security and territorial integrity, as seen in the continuing drama of PRC efforts to regain Hong Kong, Macao, and Taiwan, which ensure a steady diet of nationalistic themes in the official media. The same defensive nationalism has also prevented China from assuming international responsibilities beyond its immediate concerns and capabilities.

To be sure, pragmatic leaders have displayed a consistently *realpolitik* worldview. Nevertheless, their preferred ends have predominantly remained the defense of their own political power and the preservation of China's territorial integrity. They have played up a history of painful Chinese weakness in the face of Western imperialism, territorial division, unequal treaties, invasion, anti-Chinese racism, and social chaos, because eliminating "the century of shame and humiliation" is at the heart of a principal claim to CCP legitimacy. The regime's legitimization has always been based on its ability to defend China's national independence against foreign enemies. Even Mao, who pursued a nativist policy, claimed legitimacy not on his ability to carry out international aggression but on his success in defending China. Mao relied on deterrence and defense in depth in an attempt to wall out enemies. A three-fronts strategy formulated by Mao in the 1960s typically manifested this defensive strategy.[47]

Chinese defensive nationalism is calculated and often has a strong moral appeal. In case of border conflicts with neighboring countries, the PLA has not always taken territorial occupation as an ultimate objective. Several times, the PLA enacted the drama of unilateral withdrawal after gaining ground in the first series of skirmishes. The bloody sacrifice aimed at telling the opponent of the moral superiority of the PRC and the secularity of its state sovereignty. The best known example of unilateral withdrawal appeared in the Sino-India war of 1962 when the PLA demonstrated its ability to defend the territory and withdrew 40 kilometers after forcing the Indian troops back 40 kilometers. Another example is the Sino-Vietnam war of 1979 when the PLA suggested a truce and general cessation of hostilities after reaching the city of Liangshan, before suffering further casualties and totally losing the aura of invincibility it had gained in the otherwise analogous Indian border conflict seventeen years earlier.[48] Similarly the military exercises in 1995 and 1996 in the Taiwan Strait suggest no intention of the PLA to occupy Taiwan. The objective was to demonstrate determination to halt the tendency of Taiwan separatism and foreign intrusion.

After the end of the Cold War, while Chinese leaders found themselves facing a less favorable international environment, there were no immediate military threats to China's security. Thus, pragmatic leaders in Beijing could afford to avoid a confrontational policy against the United States and other Western countries. In November 1990, Beijing abstained during the UN Security Council's vote on the use of military force against Iraq, thus freeing the way for Operation Desert Storm led by the United States. According to a Chinese scholar, when America's power and influence were amplified by its victory in the Gulf War and the formal demise of the Soviet Union in 1991, there were great pressures on the Beijing leadership to launch an ideological campaign against Western political ideas and the Soviet leaders' betrayal of socialist principles. However, pragmatic leaders, particularly Deng Xiaoping, argued that "China's power and interests did not allow a confrontational relationship with Western countries."[49] As Deng was quoted on his twenty-four-character principle for handling world affairs after the Tiananmen incident in 1989: "Observe developments soberly, maintain our position, meet challenges calmly, hide our capacities and bide our time, remain free of ambition, never claim leadership." This principle later evolved into an official sixteen-character principle for dealing with Sino-U.S. relations: "enhancing confidence, reducing troubles, expanding cooperation, and avoiding confrontation," which was delivered by President Jiang Zemin at his meeting with U.S. President Clinton in Seattle in 1993. This defense strategy may sound deceptive, but it is in line with the essence of pragmatic nationalism and serves China's national interests. Using nationalism as a defensive weapon, pragmatic leaders encouraged the surge of nationalism in response to the rising voice of containment against China in the Western media after the Taiwan Strait crisis of 1995–1996. They also let nationalist sentiment decline when they saw an improvement of the Sino-U.S. relationship marked by the state visit of Jiang Zemin to Washington in October 1997.

Notes

1. Erica Strecker Downs and Philip C. Saunders, "Legitimacy and the Limits of Nationalism: China and the Diaoyu Island," *International Security* 23, no. 3 (Winter 1989–1999): 114–146; Allen Whiting, "Assertive Nationalism in Chinese Foreign Policy," *Asian Survey* 23, no. 8 (August 1983): 913–933 and "Chinese Nationalism and Foreign Policy after Deng," *The China Quarterly*, no. 142 (June 1995): 295–316; Michael Oksenberg, "China's Confident Nationalism," *Foreign Affairs* 65, no. 3 (1986–1987): 504.

2. Ying-shih Yu, "Minzu zhuyi de jiedu" (Interpretation of Nationalism), *Minzhu Zhongguo* (Democratic China), no. 35 (June–July 1996); Samuel P. Huntington, *The Clash of Civiliza-*

tions and the Remaking of World Order (New York: Simon & Schuster, 1996), p. 229; James R. Lilley, "Nationalism Bites Back," *New York Times,* October 24, 1996; and Richard Bernstein and Ross H. Munro, "The Coming Conflict with America," *Foreign Affairs* 76, no. 2 (March/April 1997): 19.

3. For one excellent description on how the CCP used the Great Wall for patriotic education purpose, see Arthur Waldron, "Scholarship and Patriotic Education: The Great Wall Conference, 1994," *The China Quarterly,* no. 143 (September 1995): 844–850.

4. Quoted from Deng Xiaoping Theory Research Center, Chinese Academy of Social Sciences, "Wusi Yundong de lishi yiyi" (Great Historic Significance of May Fourth Movement), *Guangmin Ribao* (Guangmin Daily), April 26, 1999 p. 2.

5. Zhongguo Gongchandang Zhongyang Weiyuanhui (the CCP Central Committee), "Aiguo Zhuyi Jiaoyu Dagang" (Guidelines for Education in Patriotism), *Renmin Ribao* (People's Daily), September 6, 1994, p. 1.

6. Francis Fukuyama, *The End of History and the Last Man* (New York: Free Press, 1992); Samuel Huntington, "The Clash of Civilization," *Foreign Affairs* 72, no. 3 (Summer 1993) and *The Clash of Civilizations*; and Richard Bernstein and Ross H. Munro, *The Coming Conflict with China* (New York: Alfred A. Knopf, 1997).

7. Wang Jisi, ed., *Wenming yu Guoji Zhengzhi: Zhongguo Xuezhe Ping Huntington de Wenming Chongtulun* (Civilization and International Politics: Chinese Scholars on Huntington's Clashes of Civilization) (Shanghai: Shanghai Renmin Chuban She, 1995); Shi Zhong (Wang Xiaodong), "Weilai de Chongtu" (Future Conflicts), *Zhanlue yu Guanli,* no. 1 (1993): 46–50; Shi Zhong, "Zhongguo xiandaihua mianlin de tiaozhan" (The Challenges to China's Modernization), *Zhanlue yu Guanli,* no. 1, 1994; Guan Shijie, "Cultural Collisions Foster Understanding," *China Daily,* September 2, 1996, p. 4.

8. Shi Zhong, "Zhongguo de minzu Zhuyi yu zhongguo de weilai" (Chinese Nationalism and China's Future), *Huaxia Wenzhai* (China Digest), 1996.

9. Song Qiang, Zhang Zangzang, Qiao Bian, *Zhongguo Keyi Shuo Bu* (Beijing: Zhonghua Gongshang Lianhe Chuban She, 1996), p. 199.

10. Song Qiang, Zhang Zangzang, Qiao Bian, Tang Zhengyu, and Gu Qingsheng, *Zhongguo Hai Shi Neng Shuo Bu* (The China That Still Can Say No) (Beijing: Zhongguo Wenlian Chuban Gongsi, 1996); Zhang Xueli, *Zhongguo Heyi Shuo Bu* (How China Can Say No) (Beijing: Hualing Chuban She, 1996); Tong Chuan, *Renminbi Keyi Shuo Bu* (Chinese Currency Can Say No) (Beijing: Zhongguo Chengshi Chuban She, 1998).

11. Among them are Hong Yonghong et al., *Zhong Mei Jun Shi Chong Tu Qian Qian Hou Hou* (U.S.-China Military Confrontations: Before and After) (Beijing: Zhongguo Shehui Chuban She, 1996); Xi Laiwang et al., *Da Yang Ji Feng: Liang Ge Shi Jie Da Guo De Bo Yi Gui Ze* (The Oceanic Wing: The Games of Two World Class Nations) (Beijing: Zhongguo Shehui Chuban She, 1996), 2 volumes; Zhang Shan and Xiao Weizhong, *E Zhi Tai Du: Bu Cheng Nuo Fang Qi Wu Li* (Stop Taiwan from Independence: No Promise on Not Using Force) (Beijing: Zhongguo Shehui Chuban She, 1996); Chen Feng, Zhao Xingyan, Huang Zhaoyu, Yang Mingjie, and Yuan Xiqing, *Zhongmei Jiaoliang Daxiezhen* (A Depiction of Trials of Strength between China and the United States) (Beijing: Zhongguo Renshi Chuban She, 1996).

12. Xu Zhuwang, Liu Yi, and Li Quanxing, *Mao Zedong Sixiang yu Zhongguo de jueqi* (Mao Zedong Thought and the Rise of China) (Beijing: Zhongguo Renmin Chuban She, 1993).

13. Among these books, see Ho Shudong, ed., *Yidai Juren Mao Zedong* (A Giant, Mao Zedong) (Beijing: Zhongguo Qingnian Chuban She, 1993); Zhang Yun, *Zhongguo Lishi Mingyun de da jueze,* (The Great Decison over the Historical Fortune of China) (Shanghai: Shanghai Renmin Chuban She, 1994); Xue Xingmin, ed., *Zai Mao Zedong Shenbian* (On the Side of Mao Zedong) (Beijing: Zhonggong Zhongyang Dangxiao Chuban She, 1993); Qi Pengfei, Wang Jing, eds., *Mao Zedong yu Gonghegu Jiangshuai* (Mao Zedong and the Marshals and Generals of the Republic) (Beijing: Hongqi Chuban She, 1993).

14. Stuart R. Schram, *Mao Zedong* (Beijing: Hongqi Chuban She, 1987).

15. Xu Zhuwang, Liu Yi, and Li Quanxing, *Mao Zedong Sixiang yu Zhongguo de jueqi* (Mao Zedong Thought and the Rise of China) (Beijing: Zhongguo Renmin Chuban She, 1993), p. 77.

16. For an English study of this "Mao Fever," see Edward Friedman, "Democracy and 'Mao Fever,'" *Journal of Contemporary China*, no. 6 (1994): 84–95.

17. Although some scholars believe that rudimentary nations and protonationalism are to be found in the premodern world, most scholars argue that only when the nation-state is constituted, do nations and nationalism come into existence, either as an expression of this nation-state or as a challenge to it on behalf of a future state. Ernest Gellner, *Nations and Nationalism* (Ithaca, NY: Cornell University Press, 1983), and Eric J. Hobsbam, *Nations and Nationalism since 1780* (Cambridge, UK: Cambridge University Press, 1990).

18. Fong-ching Chen, "Chinese Nationalism: A New Global Perspective," *The Stockholm Journal of East Asian Studies* 6 (1995): 4.

19. Benjamin I. Schwartz, "Culture, Modernity, and Nationalism—Further Reflections," in Tu Wei-ming, ed., *China in Transformation* (Cambridge, MA: Harvard University Press, 1993), p. 247.

20. S. Robert Ramsey, *The Languages of China* (Princeton, NJ: Princeton University Press, 1989), p. 3.

21. Martin E. Marty, "Fundamentalism as a Social Phenomenon," *Bulletin, The American Academy of Arts and Sciences* 42, no. 2 (November 1988): 15.

22. Harry Harding, "China's Changing Roles in the Contemporary World," in Harry Harding, ed., *China's Foreign Relations in the 1980s* (New Haven, CT: Yale University Press, 1984), p. 188.

23. Mao Zedong, "On the People's Democratic Dictatorship," *Selected Works of Mao Tse tung*, vol. IV (Beijing: Foreign Language Press, 1969), p. 415.

24. Wang Jisi, "Pragmatic Nationalism: China Seeks a New Role in World Affairs," *The Oxford International Review* (Winter 1994): 30.

25. Andrew J. Nathan and Robert S. Ross, *The Great Wall and the Empty Fortress: China's Search for Security* (New York: W. W. Norton, 1997), pp. 32–33.

26. *Renmin Ribao* (People's Daily), January 31, 1995, p. 1.

27. Ever since the publication of Hobsbawm and Ranger's influential volume, the concept of "invention of tradition," whether historical or cultural, has come into vogue. (Eric Hobsbaw and T. Ranger, eds., *The Invention of Tradition* [Cambridge, UK: Cambridge University Press, 1983]). Gellner's "Potato Principle"—roughly that groups will look back historically to periods when they were mainly farmers to justify the control of land in an urban and industrial age—even shows how territory itself is imagined (Ernest Gellner, "Nationalism in the Vacuum," in Alexander Motyl, ed., *Thinking Theoretically about Soviet Nationalities* [New York: Columbia University Press, 1992]). According to this theory, there is no predetermined cultural and territorial boundaries of nations.

28. Michael Hunt, "Chinese National Identity and the Strong State: The Late Qing Republican Crisis," in Lowell Dittmer and Samuel S. Kim, *China's Quest for National Identity* (Ithaca, NY: Cornell University Press, 1994), p. 63.

29. Liu Ji, "Making the Right Choices in Twenty-first Century Sino-American Relations," *Journal of Contemporary China* 7, no. 17 (1998): 92.

30. Xiao Gongxin, "Zhongguo Minzu Zhuyi de Lishi yu Qianjing" (The History and Prospect of Chinese Nationalism), *Zhanlie yu Guanli*, no. 2 (1996): 62

31. Gao Bianhua, "Chinese Products Find a Home in Japan," *China Daily*), April 1, 1996, p. 2.

32. Townsend's study put forward a fourth Chinese nation, including other overseas Chinese who retain some idea, however attenuated, of dual nationality. The fourth nation cannot take unified political form, as most of its external members have primary obligations to non-Chinese states. Nonetheless, it has contributed both politically and economically to the PRC

and Taiwan, continuing to nurture the idea that important community bonds remain. See James Townsend, "Chinese Nationalism," *The Australian Journal of Chinese Affairs*, no. 27 (January 1992): 128.

33. Hans J. Morgenthau, *Politics among Nations: The Struggle for Power and Peace* (New York: McGraw-Hill, 1993), pp. 272–273.

34. Steven Mufson, "China's New Nationalism: Mix of Mao and Confucius," *International Herald Tribune*, March 20, 1996.

35. Article 50 of the Common Program of the Chinese People's Political Consultative Conference on September 29, 1949 stated that "nationalism and chauvinism should be opposed." See Theodore H.E. Chen, ed., *The Chinese Communist Regime: Documents and Commentary* (New York: Praeger, 1967), p. 34.

36. Lu Ning, "The Central Leadership, Supraministry Coordinating Bodies, State Council Ministries, and Party Department," in David M. Lampton, *The Making of Chinese Foreign and Security Policy* (Stanford, CA: Stanford University Press, 2001), p. 50.

37. Dali Yang, "In Bombing Aftermath, Cool Heads Must Prevail to Keep China Focused on Reforms," www.Chinaonline.com, May 10, 1999.

38. "Deng Xiaoping Met with Japanese Visitors," *Beijing Review* 32, no. 40, October 2–8, 1989, p. 5.

39. *Xinhua*, May 11, 1999.

40. Liu Weijun, Yin De An, "China Did Not Threaten to Veto the G-8 Proposal," *China News Digest*, June 3, 1999.

41. Craig S. Smith, "Students' Unease over Weakness Could Threaten Beijing's Leaders," *New York Times*, April 6, 2001.

42. There were different reports about this Politburo meeting. Most Chinese reports said the meeting set five principles but a CNN report said there were three principles. For the Chinese reports, see, for example, Wen Ren: "Hu Jintao Shicha Jiefangjun Sizongbu he guofangbu tingqu yijian" (Hu Jintao Visits Four General Headquarters of PLA and Defense Ministry to Solicit Opinions), *Taiyangbao* (The Sun Daily), April 6, 2001. "Beijing Shiyao Meiguo Daoqian" (Beijing Is Determined to Ask the US for an Apology), Duowei Xinwen She, April 6. For the CNN report, see Willy Wo-Lap Lam, "Analysis: Behind the Scenes in Beijing's Corridors of Power," www.cnn.com, May 9, 2001.

43. Ibid.

44. Francesco Sisci, "Reading the Tea Leaves," *Asia Times Online*, April, 18, 2001.

45. Editorial, "Ba Aiguo Reqing huawei qiangguo liliang" (Transform the Warm Emotion of Patriotism into the Power of Strengthening the Nation), *Renmin Ribao*, April 11, 2001, p. 1.

46. David Shambaugh, "Containment or Engagement of China," *International Security* 21, no. 12 (Fall 1996): 205.

47. First-line defenses were built on China's west, north, and east, on China's long, exposed border with the Soviet Union and Soviet-supported Mongolia and facing the U.S. military bases in a Pacific island chain. Second lines of defense were prepared in defensible mountain regions behind the first. A third front, the secure militarized hinterland, was prepared in China's southwest, not far from Indochina. China moved most of its military industry into the third front hinterland, away from both the Soviet border and the coast.

48. For one study of this symbolic use of force by the PRC in border conflicts, see Jonathan R. Adelman and Chih-yu Shi, *Symbolic War: The Chinese Use of Force, 1840–1980* (Taipei, Taiwan: Institute of International Relations, 1993).

49. Wang Jisi, "Pragmatic Nationalism: China Seeks a New Role in World Affairs," *The Oxford International Review* (Winter 1994): 29.

Part Two

Ideology, Strategic Culture,
and Pragmatism

5

Pragmatic Calculations of National Interest

China's Hong Kong Policy from 1949–1997

Lau Siu-kai

Until 1978, when the Chinese Communist Party (CCP) decided to shift its major mission from class struggle to economic reconstruction, the People's Republic of China's (PRC) foreign policy was described by some scholars as primarily driven by political ideology, to the extent that rational calculation of power and interests were relegated to a secondary position. According to one scholar during the Maoist era, the themes that summed up Beijing's foreign policy were national unity, socialist revolution, export of Communist ideology, anti-Americanism and pro-Sovietism, and restoration of Chinese primacy in Asia.[1] For the sake of facilitating large-scale mobilization of the Chinese people for socialist reconstruction, Mao even deliberately fostered a high degree of tension between China and the West so as to create a siege mentality among the people and whip up nationalist fervor. As a result, the role of ideology in China's foreign policy was further fortified. Such understanding of Chinese foreign policy under Mao can also be found in studies on the relationship between the CCP and the United States during the Chinese civil war in the second half of the 1940s and the decision of the PRC to enter the Korean War.[2]

Since 1978 and under the direction of Deng Xiaoping, the emphasis on developing the economy and opening the door to foreign investment, technology transfer, trade, and training has propelled China's foreign policy toward pragmatism and realism. The central considerations have become peace and security.[3] On the other hand, renewed attention has been given to patriotism and nationalism. Antihegemonism and fraternity with the Third World continue to be the cornerstones of China's new foreign policy.

While the above interpretation of Chinese foreign policy since the establishment of the PRC is to a certain extent correct, the way China handled Hong Kong is problematic. By not recovering Hong Kong in 1949 or shortly thereafter, and by allowing it to remain in the hands of Britain—China's arch-imperialist foe in modern Chinese history—China made a significant exception to its fundamental policy lines. In China's Hong Kong policy, not only was ideology insignificant, but China was even prepared to go to the extent of ignoring accusations of inconsistency and hypocrisy by critics and foes alike.[4] The high degree of rationality and pragmatism displayed in China's approach to Hong Kong contrasts starkly with the overall thrust of its foreign policy. China's policy toward Hong Kong does not necessarily call for a major revision of the ideological interpretation of

Chinese foreign policy since 1949, but it definitely requires us to develop a more nuanced and comprehensive understanding of the way China defined its national goals, pursued its national interests, and handled its relations with nations of strategic importance.

Leaving Hong Kong in the Hands of Britain

The reason why the PRC did not recover Hong Kong by force or diplomacy in 1949 and instead allowed the territory to continue as a British colony is still a mystery. No official explanation has been given. Statements by Chinese leaders and officials on this matter are without exception sketchy, cryptic, and tongue-tied. Some Chinese officials hint obliquely at national security as the chief reason why resumption of sovereignty over Hong Kong did not take place in 1949, but they normally fail to elaborate. Post hoc explanations by outside observers however abound, which normally ascribe the benefits reaped by China after the event to account for it, attributing perhaps far too much foresight to the Chinese leaders, particularly Mao Zedong. And Chinese officials have not been innocent of giving force to such "explanations" by their own accounts. In exploring the rationale behind China's act to allow Hong Kong to remain a British possession, I have to rely on scattered evidence and my contact with Chinese officials. Hopefully a more plausible case can be made.

In the first place, it appears that sometime before the establishment of the PRC on October 1, 1947, the intention not to change the status quo of Hong Kong was already quite evident. It also came at a time when the PRC was critically dependent on Soviet support. The intention was against the wish of Joseph Stalin, the paramount leader of the Soviet Union.[5] According to Michael Yahuda, "Evidence from Soviet Archives suggests that as early as January 1949 (three months before the crossing of the Yangtze River) Mao Zedong had already decided to defer the seizure of the two remaining Western colonies of Hong Kong and Macao."[6] Another scholar placed the time of the decision as early as November 1948.[7] When the People's Liberation Army (PLA) reached the border between China and Hong Kong, not only did it not take any initiative militarily against the colony, it even worked to maintain peace in the area.[8] Even though China at that time did not issue any official statements with regard to Hong Kong, the editorials in two leftist papers in the colony as well as a report by the Xinhua news agency on Hong Kong hinted at China's intention to maintain the status quo there.[9]

Why did China fail to recover Hong Kong at a time when doing so would definitely enhance the Communists' prestige both domestically and abroad for sticking to their well-known principles of anti-imperialism and nonrecognition of the unequal treaties imposed on the Chinese by dint of force? It is particularly intriguing as Hong Kong occupies a very important symbolic place in modern Chinese history, for the Treaty of Nanking, which ceded the territory to Britain, represents the beginning of a period of excruciating national humiliation. Piecemeal comments by Chinese leaders and officials, though not helpful as definitive explanations, are nevertheless useful as clues to seek better understanding.

Thus far the most revealing explanation of China's decision was given by Li Hou, former vice-director of the Hong Kong and Macao Office of the State Council, as follows:

> The decision of the Chinese leaders not to recover Hong Kong for a long time to come was based on the following considerations: (1) In an international situation character-

ized by sharp confrontation between two political camps, it was not possible to resolve the Hong Kong problem through peaceful method. Hong Kong could only be recovered by a resort to force. The British were well aware of the fact that they could not singly deal with China by their own power, so they would definitely bring the U.S. along to jointly defend Hong Kong. This scenario was certainly not what China wanted to see. In the eyes of the Chinese government and its leaders, it was better for Hong Kong to be left in the hands of Britain than to allow Britain to enlist the help of the U.S. to defend Hong Kong. (2) New China was just established; it had no diplomatic relations with many countries. Western countries, led by the U.S., were imposing economic embargo on China. Under these circumstances, maintaining the status quo of Hong Kong would allow it to serve as China's channel to the outside world, making it possible for China to obtain things which could not be obtained from other channels.[10]

It is quite obvious that Li emphasized the threat of the United States and national security as the two major considerations in China's policy. The possible economic role Hong Kong was subsequently to play in China's development was apparently of limited significance in the reckoning of China.

In view of the perception of the international situation by the Communist leadership on the eve of the establishment of the PRC, it would appear that the decision to allow Britain to continue to administer Hong Kong represented a rational attempt of the CCP to avert war with the West over a piece of land that—unlike the Korean Peninsula—did not pose a military threat to China. By allowing Hong Kong to remain a British colony, China deliberately manipulated the differences between Britain and the United States and ameliorated the pain inflicted on China by the latter's policy of containment.

In making sense of China's decision, the first issue that has to be dealt with concerns the possibility of war between China and Britain, and possibly also the United States, if China recovered Hong Kong by military force in 1949. Dick Wilson, for one, thought that "Chinese Communist troops had stopped at the border because China was not ready to absorb Hong Kong, and perhaps feared that such an action would provoke British and American retaliation before the civil war in China was even complete."[11] An analysis of the situation at that time would support the view that there was a probability of a Sino-British or even a Sino-American military showdown over Hong Kong if the PLA forced its way into the colony.

British determination to keep Hong Kong was obvious during World War II. Winston Churchill, the British prime minister, angrily repudiated U.S. President Franklin D. Roosevelt's request to return Hong Kong to China after the war in order to be made an internationalized free port.[12] In order to keep Hong Kong, Churchill had even considered making a secret deal with Stalin on postwar territorial arrangements in 1944.[13]

The determination of Churchill to remain in Hong Kong was shared by his Labor party successors after the war ended. In fact, in the Pacific war, Britain's concerns were basically defensive, being fully cognizant of the limited powers at its disposal. The war aims were defined as: hold on to India, "liberate" Singapore and Malaya, keep Hong Kong, and insulate Britain's Asian empire from any sort of international accountability.[14] Even though Hong Kong had proven to be a military liability during the war, the strategic and symbolic value of the colony was still substantial. Apparently, while Britain could not afford to reassert its prewar position in China, the retention of Hong Kong was important to the empire, to Britain's role as a major world power, and to the future British economic interest in East Asia.[15]

As a result of the perceived importance of Hong Kong, Britain did take steps to militarily fortify the colony and to make preparations to deal with emergencies.[16] How far Britain was willing to go to defend Hong Kong militarily was difficult to gauge, as British resolve was never subjected to a test. In any case, the position of Britain not to let go of Hong Kong without a fight, might have made some impression upon the Chinese Communists. As far as the CCP was concerned, the possibility of American intervention on the side of Britain was the more important consideration.

In fact, as the Pacific war drew to a close, Roosevelt's confidence in China ever becoming a great and "responsible" power was shaken by America's experience with China and Generalissimo Chiang Kai-shek.[17] As the end of World War II ushered in the Cold War, Roosevelt's successor, Harry S Truman, took a hard-line approach to Maoist China as part of a global strategy of opposing the spread of communism.[18] Moreover, in contrast with Roosevelt's sentimental attachment to China, Truman's attitude toward the Chinese was one of aloofness. In any event, in the extreme anti-Communist atmosphere in the United States and in light of the strident views of the China lobby, it was difficult for Truman to take a conciliatory approach to the PRC.[19]

American support for the Nationalists in the Chinese civil war, though coming short of direct military involvement, was seen as a serious threat by the CCP. When complete victory was near in late 1948 and early 1949, CCP leaders were very concerned about Washington sending its troops to China to rescue the Nationalist regime.

In this context, Communist leaders' suspicions of American motives in Hong Kong were not without grounds. As a matter of fact, as the Nationalists were on their way to total defeat, Hong Kong assumed a new importance in the eyes of the United States. Concern about the possibility that the Communists would attack and overwhelm Hong Kong in the closing days of the Chinese civil war brought about military cooperation between Britain and the United States.[20] Though the United States had never publicly committed itself to the military defense of Hong Kong, on several occasions in 1949 and thereafter, it hinted at such a possibility with the explicit purpose of deterring the PRC from attacking the colony.[21] It is difficult to gauge the extent to which the possibility of war with Britain and the United States had influenced the PRC's decision not to recover Hong Kong in 1949, but apparently such a possibility had entered into the mind of the Communist leadership.

Britain, while making military preparations for a possible Communist attack on Hong Kong, had also taken steps to mollify the Chinese Communists in order to safeguard Britain's extensive interests in China and to keep Hong Kong. After the PRC was established, against the opposition of the United States, the British government on January 5, 1950, withdrew recognition from the Nationalist government, and the next day it accorded *de jure* recognition to the Communist regime.[22] Britain's move prompted a flurry of recognitions of the PRC between January 6 and January 18 by Norway, Ceylon, Denmark, Israel, Afghanistan, Finland, Sweden, and Switzerland.

These goodwill moves by Britain did not fail to impress the Communist leaders. In addition, China very early on noticed the differences in national interests between Britain and the United States and intended to exploit them to promote its national security. As early as 1949, during the first conference of the Chinese People's Political Consultative

Conference, Mao Zedong proposed employing peaceful methods to resolve historical problems, including the Hong Kong problem. For the sake of improving China's diplomatic situation, Zhou Enlai proposed the policy of "advancing Sino-British relationship and promoting peaceful cooperation," exploiting the contradictions between the United States and some Western nations to build a united front in favor of keeping peace and promoting trade.[23] According to these calculations, allowing continued British rule in Hong Kong would drive a wedge between Britain and the United States and hence ease the harshness of the anti-Communist policy of the latter.

China's decision not to recover Hong Kong in 1949 might also have something to do with the CCP's mistrust of Stalin and the Soviet Union. Mao's deep-seated personal antagonism toward Stalin could be traced back to the days of the Jiangxi Soviet, in the early 1930s. Mao blamed Stalin and the Comintern for the CCP's early disasters. During World War II, Stalin nurtured his vision of a *realpolitik* partnership with the Western allies and sided with the Chinese Nationalists, and, until late in the Chinese civil war, considered the CCP's cause as hopeless.[24] Only after the beginning of the Cold War, when all chances for reconciliation with the West were lost and he faced the need to seek new allies, did Stalin begin to develop a strategic relationship with the Chinese Communists.[25] Yet despite this, Mao was deeply repulsed by Stalin's territorial ambitions in China.[26] Privately, Mao even had fonder impressions of the Americans than the Soviets.[27]

Therefore, notwithstanding Mao's "lean to one side" policy and anti-American rhetoric, even before 1949, he seemed to have wished to develop a normal relationship with the United States when the PRC was established.[28] In 1944–1945, Mao wanted to create the best possible balance of power for China in the international environment, leaning neither toward Moscow nor toward Washington, but using the assistance of both powers to reconstruct a country devastated by the Japanese occupation and the civil war.[29]

In the end, despite the failure of Mao's efforts to win over the United States and his policy of total dependence on the Soviet Union, his hatred and suspicion of Stalin had never abated. Mistrust of the Soviet Union might have led to the wish to not foreclose all possibilities of an outlet to the West in the future. The Soviet factor might thus have had a modicum of influence on China's policy toward Hong Kong in 1949.

Long-term Consideration, Full Utilization

When the Chinese Communists decided to maintain Hong Kong's status quo in 1949, it might have only been an expedient and short-term move in view of the national security needs of the new regime. Later on, it was apparent to the Communist leaders that this move did bear fruit and was beneficial to China. What was more important was that the international situation since 1949 continued to be unfavorable to China, hence any attempt by China to recover Hong Kong would have jeopardized its national interests.

The Chinese leadership defined the international environment as threatening and turbulent, hence it refrained from taking risks over Hong Kong. The Sino-Soviet relationship, which was the cornerstone of Chinese foreign policy, though superficially cordial, was in reality strained because of Mao's deep resentment stemming from the Soviets' refusal to treat China as an equal.[30] It completely broke down in the late 1950s when Stalin's successor, Nikita Krushchev, reversed his approach to the West by seeking to reach a "peaceful coexistence" with the United States. The turnabout of the Soviet Union reflected Krushchev's fear of nuclear war and desire to divert Soviet resources to peaceful Socialist construction.[31] The looming possibility of a Soviet-U.S. collusion against China and the fear of total containment by Western and socialist imperialists presented the Communist leadership with the greatest national security threat since 1949.

The Sino-American relationship reached the nadir after the Korean War in the early 1950s and began to thaw only in the early 1970s, when Richard Nixon visited Beijing in 1972. Since then, it improved because of the two nations' mutual interests in containing the Soviet Union. But formal recognition of the PRC did not come until 1979. Even then, the Taiwan issue remained unresolved and continued to haunt the Sino-American relationship. Therefore, apprehension about American military intervention in Hong Kong was still in the mind of the Chinese leaders.

While the economic value of Hong Kong to China was not obvious in 1949, the industrial takeoff of Hong Kong in the late 1950s made Hong Kong of critical economic importance to China, which was reeling from the trade embargo imposed by the West and the termination of Soviet aid resulting from the Sino-Soviet confrontation.

Allowing Britain to remain in Hong Kong did bring about a certain degree of conflict between the foreign policies of Britain and the United States, but in view of the fact that in the postwar period Britain had to depend on the United States for economic revival and military assistance, the influence of the former on the latter was limited.[32] British efforts did, however, produce some talks between China and the United States on practical and minor matters and alleviated, to a limited degree, American harshness toward China.[33]

The sense of insecurity from the CCP rendered the continuation of the status quo of Hong Kong after 1949 inevitable. Gradually, China came up with an explicit and stable policy toward Hong Kong, the thrust of which was instrumentalism and pragmatism. The strategic goals were to avoid international conflicts and to create a peaceful environment for China's socialist construction.[34] The evolution of this policy was officially described in the following passage:

> In 1959, referring to the impatience of a minority of comrades over Hong Kong, Mao Zedong pointed out that from the perspective of overall global strategy, "it is better not to recover Hong Kong in the meantime. We are not in a hurry. The current [status of Hong Kong] is advantageous to us." In 1960, the Chinese government, after drawing conclusions from experiences in the last decade, put forth the policy of *"changqi dasuan, chongfen liyong"* [long-term consideration, full utilization] to guide work on Hong Kong and Macao. This policy means that China would adopt long-term consideration as far as Hong Kong's future was concerned. In the foreseeable future, China would not adopt policies which would drastically change Hong Kong's *status quo*. At the same time, Hong Kong's special status would be fully utilized to serve China's socialist construction and diplomatic strategy.[35]

Under the principle of "long-term consideration, full utilization," China's long-term policy toward Hong Kong was eventually established. The high degree of stability of this policy can be seen in the fact that it was adhered to even during the Cultural Revolution, when ideological fanaticism threatened to push anti-imperialism to the extreme.[36]

This Hong Kong policy of China had several interrelated components. First and foremost, Hong Kong had been part of the territory of China since ancient times. It was occupied by Britain because of the unequal treaties forced upon China, but China did not recognize these treaties. When the conditions were ripe, the Hong Kong problem would be resolved in a peaceful manner through negotiations; however, until that happened, the status quo of Hong Kong would be maintained.[37] Second, the resolution of the Hong Kong problem was a matter entirely within the sovereignty of China, it had nothing to do with the commonly understood problem of the colonies. In a letter to the United Nations dated March 8, 1972, China demanded that the United Nations inform its relevant committee that Hong Kong and Macao did not at all fall under the ordinary category of colonial territories and should not be included in the UN's list of colonial territories covered by the declaration on the granting of independence to colonial countries and peoples.

These two principles constituted the foundation of China's Hong Kong policy. Following from these were a series of conditions imposed upon Britain as the prerequisites for continued British rule in Hong Kong. One condition established that prior to 1997 Hong Kong was a problem to be resolved only between China and Britain, and no third parties would be allowed to interfere in the matter. Aside from referring to other nations, "third parties" also included the people of Hong Kong, whose anti-Communist sentiments were well known to the CCP. Another crucial demand was that Hong Kong should not be steered toward independence. Any moves in Hong Kong that smacked of independence would not be tolerated. These moves apparently included political reforms that would result in democracy or self-governance. Britain should not allow Hong Kong to become a Chinese Nationalist base of subversion or become a military base of foreign powers against the PRC. A Soviet presence in Hong Kong was not welcome in view of the hostilities between China and the Soviet Union. Britain should also prohibit activities in Hong Kong that aimed at undermining the prestige of the PRC. Furthermore, the safety of Chinese officials should be protected and China's economic interests in Hong Kong should not be obstructed.[38]

Implicit in China's Hong Kong policy was that the territory should remain economically valuable to China. While it is obvious that China's decision not to recover Hong Kong in 1949 was largely due to national security considerations in the political and military sense, the growing economic importance of Hong Kong to China has since loomed increasingly large as a factor in China's policy to maintain Hong Kong's status quo as a British colony.[39]

Generally speaking, Britain was able to satisfy China's demands in an adroit and low-key manner, though problems occasionally arose to produce conflicts in Sino-British relations[40] and not all Chinese wishes were met.[41] Yet all signs showed that China was pleased with the situation in Hong Kong. In addition, China took every step possible to ensure that Hong Kong enjoyed economic prosperity and political stability under British rule.[42]

The 1997 Problem, China's Hong Kong Policy, and Unresolved Dilemmas

As discussed above, before the late 1970s, China's Hong Kong policy was based on national security considerations chiefly defined in physical terms. This policy was rational and instrumental in nature. Since the late 1970s, the domestic and international environment of China fundamentally changed. By 1978, the Cultural Revolution had ended, and China had regained a semblance of political stability. China's international situation also improved tremendously as the result of the establishment of diplomatic relations with the

United States and the détente between China and the Soviet Union. Accordingly, the definition of the national interests of China was revised. Under the new definition, which stressed the primacy of economic development, the importance of Hong Kong to China escalated. At the same time, the expiry of the lease on the New Territories (representing more than 90 percent of the land area of the colony) in 1997 called for serious thinking on the future of Hong Kong.

After the arrest of the "Gang of Four" and the marginalization of the Leftists, Deng Xiaoping consolidated his power and became China's paramount leader. Deng Xiaoping's political line was established during the Third Plenary Meeting of the Eleventh Central Committee of the CCP held in December 1978. The three major tasks laid down for the 1980s and 1990s for the CCP were domestic economic construction, national unification, and the maintenance of world peace. In essence, political struggle and world revolution were no longer in vogue. China's relations with the West should be improved so that Western capital, markets, and technology could be utilized to promote China's development. The focus on national unification meant that the return of Hong Kong to China became a salient item on the CCP agenda. And, after joint discussions by the Hong Kong and Macao Office and the Foreign Ministry chaired by Liao Chengzhi, the Chinese leader most experienced in Hong Kong affairs, China decided in the first half of 1981 to resume sovereignty over Hong Kong in 1997.[43] This official position was arrived at before China entered into formal negotiations with Britain over the territory in late 1982. China's decision to recover Hong Kong in 1997 was based on the reasoning that due to the expiry of the New Territories lease, Britain had perforce to return Hong Kong to China at that time. Any delay in recovering Hong Kong would create complications afterward and might make peaceful resolution of the matter difficult.[44] By early 1982 at the latest, the basic policy for post-1997 Hong Kong, though still inchoate, was already laid down. After further considerations, the policy of "one country, two systems" (OCTS) (*yiguo liangzhi*) was formally established as the means to resolve the Hong Kong problem.

The primary considerations underlying the "one country, two systems" policy were: (1) Hong Kong's economic value to China should be retained; (2) China's relations with Britain and other Western countries should not be affected; (3) Hong Kong as an economic bridge between China and the West should be maintained; (4) Hong Kong should not become a security threat to China; and (5) Hong Kong's return should facilitate the reunification of Taiwan with China.

The essence of the OCTS policy represented the attempt by China to preserve Hong Kong's prosperity and stability without the British. The key elements in the policy included:

1. The Hong Kong Special Administrative Region (HKSAR) would be set up on July 1, 1997;
2. The HKSAR would be placed directly under the central government and, aside from foreign and defense affairs, would enjoy a high degree of autonomy;
3. The HKSAR had the powers of administration and legislation. Its judicial power, including that of final adjudication, would be independent;
4. The HKSAR government would be made up of the inhabitants of Hong Kong;
5. The preexisting social and economic institutions as well as the way of life of Hong Kong would remain unchanged;

6. Hong Kong would retain its free port status and remain as a separate customs territory;
7. Hong Kong would continue to be an international financial center;
8. Hong Kong would enjoy independent finances;
9. Hong Kong could establish mutually beneficial economic relations with Britain and other countries;
10. Hong Kong could develop economic and cultural relations with other countries;
11. Hong Kong would be responsible for the maintenance of its public order. The defense of the HKSAR would be the responsibility of the PLA; and
12. The OCTS policy would remain unchanged for fifty years. Eventually, these elements were incorporated into the Basic Law, the miniconstitution of Hong Kong, after its reversion to China.

On the whole, China's policy for post-1997 Hong Kong was a pragmatic and rational policy based on self-interests, yet from the beginning China failed to grasp the momentous ramifications of the policy. China wanted to remove the British but at the same time leave the rest of the status quo in Hong Kong unchanged. That China was so optimistic about achieving the two goals simultaneously in the early 1980s has to do with China's mechanical understanding of Hong Kong as a society, its overestimation of the political passivity of the people of the place, and its overconfidence in Sino-British cooperation and in British capacity to control the situation in Hong Kong in the transitional period. However, it transpired that once it was a known fact that British rule in Hong Kong was going to end in 1997, a Pandora's box was opened, and China's key premises about Hong Kong were in shambles. In retrospect, China was ill prepared for the fundamental changes in Hong Kong that were inaugurated by the end of colonial rule. Consequently, China's policy in the transitional period was in disarray.

One big premise of China's policy was that since Britain was willing to surrender Hong Kong to China in a peaceful manner, and since Britain had tremendous interests in Hong Kong and in China, Britain should be willing to cooperate with China fully in the transitional period and grant China de facto veto power over major British initiatives before the handover. China insisted that Britain should and could continue to run Hong Kong in the same manner as in the past. Undoubtedly, China was not fully cognizant of the changed political situation of the British rulers who were on the way out. By the same token, China could not sympathize with the need of the British to initiate political reforms in order to stabilize colonial rule in its last days. Nor could China understand the imperative of Britain to depart Hong Kong with honor. In fact, alongside China's optimism about British cooperation was a lingering suspicion of British intentions. The persistence of Britain in introducing political reforms in Hong Kong had greatly exacerbated China's fears of British conspiracies.[45] It was indubitably true that Britain did want to groom popular leaders through political reforms to guarantee the realization of Hong Kong people governing Hong Kong with a high degree of autonomy as understood by it, yet the self-interests of Britain to maintain effective rule before 1997 counseled it against radical moves. The demand of China that Britain take absolute control of Hong Kong before its return to China eventually clashed with the British political agenda to leave with glory. As a result, the two countries moved from uneasy collaboration to intense confrontation when the Tiananmen incident in 1989 proved to be the last straw. The progressive breakdown of Sino-British cooperation in the transitional period created immense instabilities, uncertainties, and disruptions in Hong Kong.[46]

Another crucial premise was that British cooperation before 1997 was indispensable to a smooth transition. This is somewhat ironic in view of the fact that China repeatedly depreciated the role of Britain in Hong Kong's success and trumpeted instead the contributions of China and the Hong Kong people. This premise was based on the consideration that Britain would still be in full control of Hong Kong before the handover, and that it was risky to work with the preponderantly anti-Communist people there, who were averse to the return of Hong Kong to China. Because of China's ambivalence toward Britain, the reaction of China to the abrupt shift to a confrontation policy by the British in 1992 was a mixture of anger and dismay. The ambiguity displayed by China in the early stage of Sino-British confrontation over Hong Kong and the continued hope on the part of Chinese officials for Britain to reverse its policy underlined the existence of the premise of "indispensability." This premise prompted China to depend on behind-the-scenes negotiations with Britain over Hong Kong issues. Mobilization of mass support in Hong Kong was considered as not only unnecessary but also risky. During the transitional period, China failed to build rapport between itself and the people of Hong Kong. When confrontation with Britain occurred shortly before 1997, China tried to coerce Britain into cooperation, and in the process further alienated public opinion, which largely sided with the departing regime. The gap between the Chinese government and the people of Hong Kong remained wide throughout the transitional period.

Yet another questionable premise was that despite the anxieties and fears generated among the Hong Kong people by the 1997 problem, they would remain politically subservient notwithstanding their mistrust of the CCP and fears of the future. To the people of Hong Kong, however, Britain had served as a shield against the Communist regime since 1949. The departure of the British would expose them to the caprice of the CCP and the thought of it was tormenting. The people were disposed to seek whatever forms of political protection that were available. The support, though mild and shallow, given by the Hong Kong people to political reform as a means to secure their future alerted China to their anti-Communist intentions. The strong support and intense sympathy by the Hong Kong people to the prodemocratic protesters in Beijing in 1989 alarmed the Chinese leadership to the extent that not only the political motives of the Hong Kong people were reinterpreted, but also that Hong Kong was perceived as a base of subversion. Chinese leaders and officials repeatedly admonished the Hong Kong people not to turn the place into a "political city" (zhengzhi chengshi), and not to allow "the well water to offend the river water" (heshui pu fan jingshui).[47]

The incorporation of antisubversion clauses into the Basic Law in 1990 and the tightening up of China's Hong Kong policy widened the chasm between China and Hong Kong and greatly weakened public confidence in the territory.

Reliance on British cooperation coupled with mistrust of British motives made Sino-British relations extremely fragile and unstable during the transition. A major victim of China's zigzag policy toward Britain emerged during the cultivation of local political leadership. Local leadership was underdeveloped under British colonial rule. The prospect of Hong Kong people ruling Hong Kong theoretically should have provided fertile soil for the rise of popular leaders in the territory. However, Sino-British rivalries were not congenial to the rise of local leaders. On the contrary, by vying with each other for the support of local people in their confrontation, China and Britain inexorably divided the meager pool of local political leaders, widened the gap between leaders and people, discredited leaders in the public eye, and aggravated political apathy among the masses.[48] The paucity of local political leaders who were simultaneously trusted by China and popular with the people inevitably threatened the implementation of Hong Kong people ruling Hong Kong.

In addition to the dearth of local political leadership in Hong Kong was the problem of legitimacy for the political leaders entrusted by China to run Hong Kong after 1997. China's mistrust of the political intentions of the people of Hong Kong naturally gave rise to a leadership selection process subject to Chinese influence. This process of leadership formation was bound to meet with public resistance. Moreover, China's conception of the political needs of Hong Kong had made the problem of legitimacy even more intractable. By defining Hong Kong as a capitalist society, China instinctively determined that after the British were gone, the bourgeoisie should naturally become the ruling class.[49] This logic was also derived from the imperative to maintain the confidence of the bourgeoisie in Hong Kong. China's political reasoning gave the bourgeoisie the political power and status they, as political dependents, did not enjoy under colonial rule. By turning the bourgeoisie into a governing class, the status quo in Hong Kong was fundamentally altered. Signs of class conflict appeared in the transitional period, portending more ominous social conflicts in the post-1997 period.[50]

As a corollary to the emphasis on the bourgeoisie, China paid insufficient attention to the aspirations and apprehensions of the growing middle class in Hong Kong. In fact, China was alienated by the Westernized and cosmopolitan values held by the growing middle class. At the same time, China saw the middle class not as an autonomous class, but as a class dependent on the bourgeoisie.[51] This disparaging view of the middle class and the lack of sympathy with their aspirations strained relations between the CCP and the middle class in Hong Kong. The alliance between Britain and the middle class to undertake political reform in Hong Kong further weakened the latter's relations with China. The alienation of the middle class greatly hampered China's effort to shape public opinion in its favor during the transition.

Finally, the scheduled end of British rule in Hong Kong changed the international environment of the place. As Hong Kong would soon be part of China and not part of the Western world, the relations between Hong Kong and the West had undergone changes. Western governments had repeatedly expressed interests in the way Hong Kong was treated by China, which aroused Chinese suspicions of Western interference in China's internal affairs. China was also worried about Western governments using Hong Kong as a base of subversion against Chinese socialism, or turning the place into a bargaining chip to their advantage. For the sake of shielding China from the "Hong Kong threat," China adopted institutional and legal safeguards to forestall external forces from turning Hong Kong against China. And, in doing so, China further undermined the Hong Kong people's confidence in their future.

Conclusion

Throughout the period of 1947–1997, China's Hong Kong policy evinced a pragmatic approach to a problem that was intrinsically highly emotional and ideological. A study of China's approach to Hong Kong, particularly in the period of 1949–1978, provides material for a more nuanced understanding of its foreign policy, which is far less irrational or ideologically driven than commonly believed. National security considerations led China to leave Hong Kong in British hands for another half century, and this decision, in retrospect, was highly beneficial to China. It was only when China decided to recover Hong Kong in 1997 that this pragmatic approach met with difficulties.

Admittedly, the policy of "one country, two systems" was an ingenious creation to resolve the Hong Kong problem, which would definitely benefit both China and Hong Kong, and there was tremendous sincerity on the part of China to make that policy a success. Nevertheless, China's understanding of the reverberations unleashed by the end of British rule was minimal. This lack of understanding or even misunderstanding had served China poorly during the transitional period, making the transition unnecessarily rough, particularly in the last five years. The problems that arose in the transitional period are bound to continue to affect the Hong Kong Special Administrative Region (HKSAR). The essence of these problems originated from the simplistic calculation of China that the role played by Britain in Hong Kong could be easily superseded by a set of promises and arrangements while the rest of the status quo remained undisturbed. And, true to its obstinately adhered principle that Hong Kong was a matter between China and Britain, China only permitted Britain to be its interlocutor in Hong Kong affairs. China's simplistic reasoning, in turn, produced overconfidence and, when reality did not match expectations, to frustration and scapegoating. Consequently, China's behavior during the transition was erratic and in many aspects detrimental to Hong Kong.

Having said that, the "one country, two systems" policy was undoubtedly a product of rational imagination. China had to take great risks in pursuing this policy, which included, *inter alia*, allowing a politically dangerous capitalist enclave to exist in a socialist country, furnishing the West a lever against China, and creating jealousies among the people on the mainland. Obviously, China, by taking these risks, must have come to the conclusion that the gains to national security accrued from the policy warranted the costs incurred.

Notes

1. Thomas W. Robinson, "Chinese Foreign Policy from the 1940s to the 1990s," in Thomas W. Robinson and David Shambaugh, eds., *Chinese Foreign Policy: Theory and Practice* (Oxford, UK: Clarendon Press, 1994), p. 556.

2. See, for example, Tang Tsou, *America's Failure in China, 1941–1950* (Chicago: University of Chicago Press, 1963), and Chen Jian, *China's Road to the Korean War: The Making of the Sino-American Confrontation* (New York: Columbia University Press, 1994).

3. Robinson, "Chinese Foreign Policy," p. 569.

4. Several instances illustrate the political embarrassment sustained by China by leaving Hong Kong in British hands. (1) India's march into the Portuguese colony of Goa in December 1961 prompted the Soviet press to condemn the "urinal of colonialism" along the China coast. (2) As Sino-Soviet relations deteriorated in the early 1960s, the Soviets constantly

taunted China for failing to expel the British from Hong Kong. In September 1964, a meeting in Moscow of the World Youth Forum affronted the Chinese intentionally by including Hong Kong and Macao in a resolution on the elimination of colonies in Asia. (3) In early 1963, the Communist party of the United States criticized the PRC for acquiescing to British imperialism. See, for example, Nancy B. Tucker, *Taiwan, Hong Kong, and the United States, 1945–1992* (New York: Twayne, 1994), p. 213, and Kevin P. Lane, *Sovereignty and the Status Quo: The Historical Roots of China's Hong Kong Policy* (Boulder, CO: Westview Press, 1990), pp. 65–66.

5. Michael Yahuda, *Hong Kong: China's Challenge* (London: Routledge, 1996), p. 46.

6. Ibid., p. 45.

7. According to James T.H. Tang, as early as November 1948, the Chinese Communists had indicated to British authorities that they would not attack Hong Kong. The head of the New China News Agency, Qiao Mu, told the Reuter correspondent in Hong Kong that the Communists had no intention of changing the status quo of the colony. And, according to a former Xinhua News Agency employee in Hong Kong, a few days after October 1, 1949, Qiao Mu relayed the following message from Zhou to the agency's staff members in the colony: "We are not taking Hong Kong back, but it does not mean that we are abandoning or retreating from Hong Kong." See his "World War to Cold War: Hong Kong's Future and Anglo-Chinese Interactions, 1941–55," in Ming K. Chan, ed., *Precarious Balance: Hong Kong between China and Britain, 1842–1992* (Hong Kong: Hong Kong University Press, 1994), pp. 116–117.

8. See Yu Shengwu and Liu Shuyong, eds., *Ershi Shiji de Xianggang* (Hong Kong in the Twentieth Century) (Hong Kong: Qilin Shuye, 1995), pp. 197–198.

9. Ibid., pp. 196–197.

10. Li Hou, *Huigui de Licheng* (The Path of Hong Kong's Return) (Hong Kong: Joint Publishing, 1997), p. 47.

11. Dick Wilson, *Hong Kong! Hong Kong!* (London: Unwin Hyman, 1990), p. 65.

12. In the 1941 Atlantic charter, while both Roosevelt and Churchill upheld the inalienable right of all peoples to self-determination, Churchill however took this to refer only to the peoples of Europe. "I have not become the King's First Minister in order to preside over the liquidation of the British Empire," Churchill defiantly proclaimed. Speaking to the American ambassador to China in 1945, the prime minister said in regard to Hong Kong, "never would we yield an inch of the territory that was under the British flag." In Yalta, Churchill insisted that "nothing would be taken away from England without a war" and specifically referred to Hong Kong. See William Roger Louis, *Imperialism at Bay: The United States and the Decolonization of the British Empire, 1941–1945* (Oxford, UK: Clarendon Press, 1977), pp. 7, 285; John Keay, *Last Post: The End of Empire in the Far East* (London: John Murray, 1997), p. 200; Chan Lau Kit-ching, *China, Britain and Hong Kong, 1895–1945* (Hong Kong: Chinese University Press, 1990), pp. 293–323; John J. Sbrega, "The Anticolonial Policies of Franklin D. Roosevelt: A Reappraisal," *Political Science Quarterly* 101, no. 1 (1986): 65–84; Clive Ponting, *Churchill* (London: Sinclair-Stevenson, 1994), p. 691; and Robin Renwick, *Fighting with Allies: America and Britain in Peace and at War* (New York: Times Books, 1996), pp. 64, 108. For a sample of the Chinese view of the matter, see Yu Qun and Cheng Shuwei, "Meiguo de Xianggang Zhengce (1942–1960)" (The Hong Kong Policy of the United States [1942–1960]), *Lishi Yanjiu* (Historical Research) (in Chinese), no. 3 (1997): 53–66; and Liu Cunkuan and Liu Shuyong, "1949 nian yiqian zhongguo zhengfu shoufu xianggang de changshi" (Attempts by the Chinese Government as Recovering Hong Kong Before 1949), *Lishi Yanjiu* (Historical Research), no. 3 (1997): 5–18.

13. "The Soviet Union should have 'effective rights at Port Arthur,'" said Churchill. More important, "any claim by Russia for indemnity at the expense of China, would be favorable to our resolve about Hong Kong." The Hong Kong issue led Churchill quickly to instruct that no agreements be reached with the United States to oppose a "restoration of Russia's position in

the Far East." See Warren F. Kimball, *Forged in War: Roosevelt, Churchill, and the Second World War* (New York: William Morrow, 1997), p. 287.

14. Ibid., p. 328.

15. Tang, *America's Failure*, p. 114.

16. Yu Shengwu and Liu Shuyong, eds., *Ershi Shiji*, pp. 175–180.

17. Roosevelt had no faith in the willingness of either Chiang Kai-shek or Mao Zedong playing a "responsible" role in world peace. He thought that "three generations of education and training was required before China could be a serious [political] factor." With no American force in East or Southeast Asia, and with China needing "tutelage" before it could play the role of responsible policeman, Roosevelt's only option was to have the Europeans reclaim their empires. See Kimball, *Forged in War*, p. 304.

18. Thomas J. Christensen, *Useful Adversaries: Grand Strategy, Domestic Mobilization, and Sino-American Conflict, 1947–1958* (Princeton, NJ: Princeton University Press, 1996), p. 8.

19. See David McCullough, *Truman* (New York: Touchstone, 1992), pp. 742–744. Also see Richard G. Powers, *Not without Honor: The History of American Anticommunism* (New York: The Free Press, 1995), pp. 228–229.

20. As London reinforced the garrison in Hong Kong, the U.S. Army placed a liaison officer in British defense headquarters, the first such peacetime appointment by Britain. See Tucker, *Taiwan, Hong Kong*, p. 200.

21. Ibid., pp. 200–209.

22. On the issue of recognition of the PRC, the hand of Britain was forced by the anxiety of India to establish a friendly relationship with China. In fact, on December 30, 1949, India extended formal recognition to the PRC, and Pakistan followed suit on January 5, 1950.

23. Guowuyuan Gangaobangongshi Xianggangshehuiwenhuasi (Bureau of Hong Kong's Society and Culture, Hong Kong and Macao Affairs Office, State Council), *Xianggang Wenti Duben* (Reader on the Hong Kong Problem) (Beijing: Zhongyang Dangxiao Chubanshe, 1997), p. 24.

24. See *inter alia* Li Hou, *Huigui de Licheng*, pp. 116–117; Harrison E. Salisbury, *The New Emperors: China in the Era of Mao and Deng* (Boston: Little, Brown, 1992), pp. 14–15, 84–85; Vladislav Zubok and Constantine Pleshakov, *Inside the Kremlin's Cold War: From Stalin to Khrushchev* (Cambridge, MA: Harvard University Press, 1996), p. 6; and Warren I. Cohen, *America's Response to China: A History of Sino-American Relations* (New York: Columbia University Press, 1990), pp. 150–176.

25. Zubok and Pleshakov, *Inside the Kremlin's Cold War*, p. 57.

26. During World War II, Stalin was persistent in regaining the Russian sphere of influence in Xinjiang, Outer Mongolia, Manchuria, and Korea. He expected to "obtain" Manchuria by virtue of its liberation by the Red Army from the Japanese. In November 1944, the Soviets supported a separatist rebellion in Xinjiang and since that time controlled the area. See Zubok and Pleshakov, *Inside the Kremlin's Cold War*, pp. 33, 56; and Salisbury, *The New Emperors*, p. 85.

27. According to Mao's personal doctor, though Mao adopted the "lean to one side" approach to the Soviet Union, he still had an admiration for the technology, dynamism, and science of the United States and the West. His propensity to "lean to one side" was always tempered by a recognition that the Soviet Union was not the only potential source of lessons in revitalization. Mao also reportedly said that the United States and the Soviet Union were different in the sense that the former never occupied Chinese territory. Mao believed that America's intentions in China had always been relatively benign. While Great Britain, Japan, and Russia had imperialistic designs and became deeply involved in China's internal affairs, the United States had remained aloof. See Zhisui Li, *The Private Life of Chairman Mao: The Inside Story of the Man Who Made Modern China* (London: Chatto & Windus, 1994), pp. 124, 514, 566.

28. According to a report sent to Moscow by Petr P. Vlasov (Vladimirov) from his headquarters in Yenan, Mao in 1944 tried very hard to come to terms with America. Mao was even

prepared to "soften" Marxism, and the works of the corresponding period reflected this. In fact, since December 7, the Communists became extremely friendly toward the United States, not only on the surface, but even in their educational programs. In the quarrels between the United States and the Chinese government, the Communists persistently sided with the Americans. When the Japanese warned China of the dangers of the embrace of American imperialism, the Communists defended the United States against the charges. In 1944, July 4th was celebrated in Yenan with tremendous enthusiasm and fulsome praise for Roosevelt, whose policies were lauded as expressions of the great tradition of freedom and democracy in the United States. See Zubok and Pleshakov, *Inside the Kremlin's Cold War*, pp. 142–143.

29. Ibid., p. 213.

30. After the establishment of the PRC, the Soviet Union assumed that China was its ideological inferior and a junior partner in the international Communist revolution. Stalin treated Mao arrogantly. See, for example, Dmitri Volkogonov, *Stalin: Triumph and Tragedy* (New York: Grove Weidenfeld, 1991), pp. 538–541. The Soviet Union also made clear to China that it would retain its "interests" in China (a base in Port Arthur, railroad access to it, and exclusive rights in Manchuria and Xinjiang). In the Sino-Soviet Treaty of 1950, a supposedly equal treaty reluctantly signed by Stalin, several secret agreements were imposed on Mao that were extremely embarrassing to China. One agreement prohibited foreigners from living in Manchuria and Xinjiang and encouraged joint Sino-Soviet economic concessions. Another agreement, with no expiration date, allowed Soviet troops to move to Port Arthur across Manchuria at any time, and without forewarning Chinese authorities. In an agreement on intelligence cooperation Stalin asked Mao to set up a joint global network of espionage among Chinese living abroad. See Zubok and Pleshakov, *Inside the Kremlin's Cold War*, p. 61.

31. Ibid., p. 201.

32. Renwick, *Fighting with Allies*, pp. 149, 157, 235, 237; and Christensen, *Useful Adversaries*, p. 135.

33. See Sa Benren and Pan Xingming, *Ershi Shiji de Zhongying Guanxi* (Sino-British Relationship in the Twentieth-Century) (Shanghai: Shanghai Renmin Chubanshe, 1996), pp. 340–341, 354; and Huang Hongzhao, *Zhong Ying Guanxi Shi* (The History of Sino-British Relationship) (Hong Kong: Kaiming Shudian, 1990), pp. 260–263.

34. Huang Yi, *Xianggang Wenti he Yiguo Liangzhi* (The Hong Kong Problem and One Country Two Systems) (Beijing: Dadi Chubanshe, 1990), p. 15.

35. Guowuyuan, *Xianggang Wenti Duben*, p. 25.

36. See Yu Changgeng, "*Zhou Enlai Yaokong 'Fanying Kangbao' Neimu*" (Inside Stories of the Remote Control by Zhou Enlai of the "Anti-British, Resist Suppression"), *The Nineties*, no. 316 (May 1996): 70–76, and no. 317 (June 1996): 92–98.

37. See the editorial of *The People's Daily*, March 8, 1963.

38. See, *inter alia*, Robert Cottrell, *The End of Hong Kong: The Secret Diplomacy of Imperial Retreat* (London: John Murray, 1993), p. 27; Wilson, *Hong Kong! Hong Kong!* pp. 65–66; and Xu Bin, *Xianggang Huigui Fengyun* (The Winds and Clouds of Hong Kong's Return in 1997) (Changchun: Jilin Sheying Chubanshe, 1996), p. 47. Some of these demands were given by Zhou Enlai to Sir Alexander Grantham, who paid a private visit to Beijing in October 1955 and met with Zhou "unofficially" for a three-hour conversation.

39. See, for example, Norman Miners, *The Government and Politics of Hong Kong* (Hong Kong: Oxford University Press, 1995), pp. 32–42; and Guowuyuan, *Xianggang Wenti Duben*, pp. 52–57.

40. See Lane, *Sovereignty and the Status Quo*, pp. 70–78; and Sa Benren and Pan Xingming, *Ershi Shiji*, pp. 367–370.

41. For example, China's request to station an official representative in Hong Kong was rejected by Britain, who was afraid of the coexistence of two power centers. The result was that a news agency—the Hong Kong branch of the New China News Agency became the de facto representative of China in the colony. Even so, Britain was willing to give it de facto recogni-

tion only in the late 1970s. See Liang Shangyuan, *Zhonggong zai Xianggang* (The Chinese Communist Party in Hong Kong) (Hong Kong: Wide Angle Press, 1989), pp. 131–134, 167–168.

42. See Qi Pengfei, *Yichu Yiluo: Xianggang Wenti Yibaiwushiliu Nian (1841–1997)* (Sunrise, Sunset: Hong Kong Problem for One Hundred and Fifty-Six Years [1841–1997]) (Beijing: Xinhua Chubanshe, 1997), pp. 309–349.

43. See Li Hou, *Huigui de Licheng*, pp. 72–73.

44. See Huang Yi, *Xianggang Wenti*, pp. 13–14. According to Huang Wen-fang, a retired senior official with the Hong Kong branch of the New China News Agency—the de facto Chinese consulate in Hong Kong, China initially had reservations about recovering Hong Kong in 1997. If Britain was willing to officially recognize that Hong Kong was part of China and that China had sovereignty over the place, there might be a possibility that Hong Kong would remain administered by Britain. There is no way to corroborate Huang's argument, but in any case Britain had not done anything along that line. See Huang's *Zhongguo dui Xianggang Huifu Xingshi Zhuquan de Juece Licheng yu Zhixing* (China's Renewed Exercise of Sovereignty over Hong Kong: The Process of Policy-making and Policy Implementation) (Hong Kong: Hong Kong Baptist University, 1997), pp. 8–9.

45. Deng Xiaoping had expressed his suspicions back in the early 1980s to Prime Minister Margaret Thatcher and Foreign Secretary Geoffrey Howe. He was worried that Britain would instigate serious disorders in Hong Kong during the transition. He expressed his hope that the following would not happen: (1) the Hong Kong dollar becoming unstable; (2) the colonial government spending the revenue from land sales irresponsibly; (3) the colonial government creating fiscal burdens for the HKSAR government; (4) the colonial government grooming a separate political leadership and imposing it upon the HKSAR; and (5) British capital taking the lead in fleeing Hong Kong. See Deng Xiaoping, *Lun Xianggang Wenti* (On the Hong Kong Problem) (Hong Kong: Joint Publishing, 1993), pp. 1–3, 9–10.

46. See Lau Siu-kai, "Decolonization à la Hong Kong: Britain's Search for Governability and Exit with Glory in Hong Kong," *The Journal of Commonwealth and Comparative Politics* 34, no. 2 (July 1997): 28–54.

47. See, for example, Deng Xiaoping, *Lun Xianggang*, pp. 29, 33; and Zhao Rui and Zhang Mingyu, *Zhongguo Lingdaoren Tan Xianggang* (Chinese Leaders Talked About Hong Kong) (Hong Kong: Ming Pao Press, 1997), pp. 16, 29–31, 34, 36, 64, 185–186, 188, 196.

48. See Lau Siu-kai, "Democratization and Decline of Trust in Public Institutions in Hong Kong," *Democratization* 3, no. 2 (Summer 1996): 158–180.

49. The equation of capitalism with bourgeois rule was typical of Marxian analysis: A former director of the Hong Kong branch of the New China News Agency, Xu Jiatun, had unmistakably pointed out that the "political regime of the future Special Administrative Region should be composed chiefly of the bourgeois class, but with the participation of the proletariat. While it should be a regime reflecting the unity of different social classes, it however is basically bourgeois in nature." And, "the thinking of the [CPP] center is: (1) the bourgeoisie should have substantial influence on the regime of Hong Kong, (2) there should be political parties [to reflect the interests of] not merely the individual elements of the bourgeois class but the [politically] organized bourgeoisie." See Xu Jiatun, *Xu Jiatun Xianggang Huiyi Lu* (Xu Jiatun's Recollections on Hong Kong) (Taibei: Lianhe Bao, 993), pp. 142, 190.

50. Lau Siu-kai, "The Fraying of the Socioeconomic Fabric of Hong Kong," *The Pacific Review* 10, no. 3 (1997): 426–441.

51. See Xu Jiatun, *Xu Jiatun*, pp. 132–133.

6

Soldiers, Statesmen, Strategic Culture, and China's 1950 Intervention in Korea

Andrew Scobell

Is China a rational, peaceful, and defensive-minded power or an irrational, bellicose, and expansionist state? Are Chinese soldiers more hawkish than Chinese statesmen? Do China's generals hold different views on the use of force from civilian leaders? Do they exhibit tendencies similar or different from their counterparts in other countries?

Episodes such as the 1995–1996 saber rattling in the Taiwan Strait and the rising clamor in the United States about the "China threat" highlight the need for dispassionate scholarship to place these events in context.[1] One trend on which many analysts who study the People's Republic of China (PRC) agree is the growing influence—particularly in foreign affairs—of the People's Liberation Army (PLA) in the post–Deng Xiaoping era.[2] There is no consensus, however, on what this increasing influence means or will mean in practice. During the military exercises on the coast of China and the missile tests in the seas off Taiwan, there were widespread reports of a split in Beijing between PLA figures advocating a harsh line and civilian leaders pushing a more moderate approach. However, some experts have questioned the reliability of these accounts.[3]

The question of the PRC's propensity to use force is not simply a policy issue but is significant theoretically. Strategic culture and civil-military relations have become central, if controversial, topics in China studies. Civil-military relations have emerged as a critical dimension for analyzing post-Deng China,[4] and strategic culture has recently been adopted in several important monographs that study China's foreign policy.[5] Yet both areas are fraught with controversy as scholars disagree on the conceptualization of military politics in the People's Republic and on the fundamental tradition and orientation of Chinese statecraft.

Civil-military relations, the study of the military's role in domestic politics and foreign policy, are now an essential ingredient in understanding the dynamics of China's internal politics and in assessing Beijing's foreign relations after Deng Xiaoping. The PLA is now widely assumed to have a major impact on political outcomes, but analysts are often at a loss to demonstrate this in specific cases, particularly in foreign policy.[6] Moreover, some scholars of Chinese military politics contend that the PLA does not even constitute a separate coherent interest group,[7] and Chinese soldiers do not have a perspective distinct from that of Chinese statesmen. Civil and military leaders in the People's Republic are often conceived of as being a monolithic elite—forged during a protracted armed struggle for power—that embraces nearly identical views on most matters. When differences of opinion are evident, they are perceived as occurring along factional lines that transcend the civil-military dichotomy.[8]

Strategic culture, defined here as a persistent system of values held in common by the leaders or group of leaders of a state concerning the use of military force, was employed during the Cold War to explain national differences in the strategic outlook and behavior of the two superpowers. More recently strategic culture has been used to examine Chinese foreign policy in the Ming and Qing dynasties, the Republican era, and Mao's China.[9] Prevailing scholarly opinion has been that China possesses a weak martial tradition especially in comparison to Japan. Moreover, China is widely believed to have a cultural predisposition to seek nonviolent solutions to problems of statecraft, exemplified by the thinking of sages such as Sun Zi and Kong Zi, and to be defensive-minded, favoring sturdy fortifications over expansionism and invasion.[10] This assumption has recently been challenged by Alastair Iain Johnston who argues that, on the contrary, China possesses a stark realist strategic culture that views war as a central feature of interstate relations.[11]

While civil-military relations are thought to have an important impact on, or to be a key element of, strategic culture,[12] the precise nature of the relationship has not been explained. Strategic culture entails the study of the "mindset" of political actors.[13] Hence one might expect that distinctive types of belief systems are likely to be based on differences in an actor's past experience, training, and environment. Thus, military leaders may likely hold perspectives distinct from those of political leaders. Indeed, studies indicate that soldiers in the United States and other countries hold belief systems and attitudes that are distinct from those of nonmilitary officials.[14] Moreover, other research suggests that soldiers tend to respond differently from their civilian counterparts to decisions to commit troops to combat.[15]

This study analyzes the attitudes of civilian and military leaders toward China's decision in 1950 to intervene in Korea in order to determine how these two groups compare in their basic orientations toward the initiation of hostilities. The results of the exercise hold significant implications about the nature and impact of China's strategic culture.

Why Korea?

The Korean case is particularly valuable not only because analyses of China's strategic disposition in this instance vary radically, but also because the perspectives of soldiers and statesmen have not been fully examined, and the primary focus has been on the dominant figure of Mao Zedong. Today, with no single leader as dominant as Mao or Deng, such an approach is of limited utility. While the reliability and availability of information on the views of military and civilian leaders regarding the PRC's use of coercive diplomacy in more recent instances (such as against Taiwan in 1995–1996) is extremely limited, such data is readily available for earlier episodes. The richest and most extensive sources now available are those on China's intervention in Korea.[16]

Each decision to commit troops has its own specific context, and the case of China's entry into the Korean War is no exception. In 1950, the decision-making apparatus was dominated by a single paramount leader, Mao Zedong. Beijing's leaders perceived a major threat to China from the considerable military might of the United States, and in addition may have held a predisposition to resort to military means when confronted with a challenge.[17] The purpose of this chapter, however, is to assess the views of individual Chinese soldiers and statesmen regarding the use of force—most of the recent literature on the

Korean War does not consider this—rather than focus exclusively on Mao or study the decision-making process.[18]

The fundamental goal here is to determine whether a pattern of attitudes among political and military elites is discernable. Such a finding would be suggestive of whether one can speak of a "Chinese strategic culture," a basic but crucial initial step in assessing China's strategic orientation in the early twenty-first century.[19] This chapter does not seek to identify the possible sources of a Chinese strategic culture nor to demonstrate the impact of strategic culture on the views or behavior of actors. Rather, the focus is strictly on attitudes in order to discern whether a pattern exists.

Interpretations of the logic behind China's decision to enter the Korean conflict have shifted from emphasizing Beijing's reluctance and caution to stressing Beijing's enthusiasm and recklessness. Until the late 1980s, studies of Chinese intervention in Korea highlighted Beijing's attempts at deterrence and gradual escalation leading ultimately to the Chinese People's Volunteers (CPV) crossing the Yalu River.[20] In stark contrast, however, studies published in the 1990s, utilizing the wealth of newly available primary sources, underscore Beijing's impetus to pursue armed conflict.[21] Moreover, the views of political and military officials tend not to be examined in any systematic fashion.[22] In fact all these studies tend to be Mao-centric, giving limited attention to the thinking and influence of other Chinese leaders.[23]

This study briefly sketches the context of China's intervention in Korea in October 1950, then examines the attitudes and actions of senior Chinese civilian and military leaders, and finally, factoring in the findings of other scholars, makes some preliminary assessments about the strategic dispositions of elites in post-1949 China.

The Context

When the first units of CPV troops crossed the Yalu River into North Korea under the cover of darkness on the night of October 19, 1950, it marked the culmination of a protracted and deliberative decision-making process. Beijing had been caught by surprise when forces of the Korean People's Army (KPA) launched a dramatic strike against South Korea across the Thirty-eighth Parallel on June 25. While the Chinese were aware of Kim Il Sung's goals, having provided him with some fifty thousand to seventy thousand Korean troops from the PLA between late 1949 and the mid-1950s, Pyongyang did not keep Beijing appraised of its invasion plans or notify Beijing of the attack across the Thirty-eighth Parallel until June 27.[24] Still the Chinese had months to assess the significance and threat posed by the presence of U.S. troops on the Korean Peninsula and the positioning of the U.S. Seventh Fleet in the Taiwan Strait. The Chinese responded prudently in mid-July by creating a Northeast Frontier Defense Army (NEFDA) with headquarters in Manchuria. Initial Chinese indignation over the insertion of the U.S. fleet and the awareness of the involvement of U.S. ground forces in the Korean conflict turned to considerable alarm by early September, even before the Inchon landings in midmonth, when intelligence reports indicated KPA units were overextended and in danger of being destroyed by an enemy counteroffensive.[25]

After the overwhelming success of the Inchon operation in mid-September, UN Command (UNC) troops under the direction of General Douglas MacArthur continued northward beyond the Thirty-eighth Parallel toward the Yalu River. In late September and early October 1950, Beijing issued stern public warnings to Washington and transmitted other private warnings via New Delhi that China would intervene if these troops did not pull back. Beijing's gradual escalation of rhetoric and military preparation appears evident, particularly the explicit warnings to UNC forces not to cross the Thirty-eighth Parallel, up until the time the Chinese actually intervened in mid-October.[26] However, while Chinese leaders considered the possibility of intervention as early as August,[27] this is not to say that the decision was solidified at an early date, that intervention was inevitable, or that the leadership in Beijing spoke with one voice.[28]

Political Elite: Sustained Ambivalence

In fact, China's civilian leadership pondered the matter of whether to enter the Korean conflict for months before finally acting. The process was tortuous, and there was widespread reluctance as well as significant opposition within the Chinese elite to becoming embroiled in Korea. Nevertheless, as key leaders became convinced of the necessity of intervention, dissenters were gradually won over or silenced.

The Paramount Leader

The opinion of China's top civilian leader, Mao Zedong, of course, was critical in Beijing's decision to dispatch troops to Korea. Evidence of the extent of the chairman's preoccupation with the situation in Korea and his considerable involvement in the most minor details of Beijing's response is overwhelming.[29] The decision was a difficult one for Mao who, although having reached a decision in August that, in principle, China would probably have to intervene in Korea,[30] wavered on the actual decision until virtually the last minute. According to acting Chief of General Staff, Nie Rongzhen, Mao "pondered deeply for a long time and from many different angles before he finally made up his mind."[31] According to many accounts, the chairman had all but made up his mind on October 1 after receiving an urgent telegram from North Korean leaders Kim Il Sung and Pak Hon Yong requesting that Chinese troops intervene. In an enlarged Politburo Standing Committee meeting the following day, Mao stated that China should intervene but left undecided the timing of the operation and the appointment of a military commander.[32]

It was another week before Mao finally issued the formal order (on October 8) after convincing himself and other top officials, following days of deliberation and discussion with Chinese civilian and military leaders, that China had no choice but to intervene. It was also on this date that Mao informed Kim Il Sung of China's decision to assist Korea.[33] Nevertheless, after receiving an October 10 telegram from Zhou Enlai in Moscow reporting that Stalin had decided against providing air support for the intervention, on October 12, Mao instructed Commander Peng Dehuai of the CPV to postpone the intervention. Mao directed Peng and Gao Gang, the CCP boss of Manchuria, to hurry to Beijing for urgent consultations.[34] During the period October 11–13, Mao did not sleep, pondering "a most difficult decision."[35] Finally, on October 13, Mao reaffirmed his decision.[36]

The Diplomat

China's foremost diplomat, PRC Premier and Foreign Minister Zhou Enlai, firmly supported intervention.[37] Judging by his statements to India's ambassador in China, K. M. Panikkar, and speeches about Korea, he consistently took a hawkish view.[38] While it is debatable whether these words actually reflect his own thinking, it is very likely that these views are his own.[39] According to CCP Central Committee member Bo Yibo, as early as August 1950, Zhou strongly favored intervention in Korea.[40] A military attaché named Chai Chengwen reports briefing Zhou in Beijing on September 1, 1950, upon Chai's return from Korea. According to him, he gave the premier a sober and ominous assessment of the military situation. After listening attentively, Zhou asked him pointedly: "In case the situation suddenly worsened—if we had to dispatch troops to intervene in Korea—what difficulties do you envision?" Chai took this to mean that Zhou was favoring intervention and, if the military situation deteriorated, China would enter the conflict.[41]

Another account, citing an unnamed high-level Communist source, notes Zhou firmly supported intervention in the face of serious opposition from others.[42] Still, one can discern even in Zhou hints of reservations about the prudence of dispatching Chinese forces to Korea. The foreign minister seemed to harbor some degree of doubt about the prospect of a Korean adventure.[43] It was Zhou who reportedly insisted on issuing a final warning to the UNC via Ambassador Panikkar in the early morning hours of October 3.[44] Certainly, this was in part a propaganda ploy, but it also held out the last minute possibility of averting Chinese military action if the UNC heeded the warning.

Doubts in the Party

There is evidence of considerable opposition within the CCP leadership about the wisdom of sending Chinese troops to Korea. The PRC had barely celebrated its first anniversary on October 1, 1950. Five months earlier the CCP Central Committee ordered the demobilization of 2.4 million soldiers to be accomplished in two phases.[45] Many Chinese looked forward to a period of peace in which the country could focus on economic development after decades of protracted armed struggle.

There appears to have been nothing short of a "high-level policy debate" on the merits of intervention.[46] According to one account, "some comrades" feared that China was completely unprepared for war with the most powerful country in the world and would probably be defeated. At a Politburo meeting on October 4, these individuals argued that it would be best to postpone intervention for a few years so that China could properly prepare itself.[47] When the meeting resumed the next day, these opponents of intervention remained but seem to have been won over by the shrewd arguments of Peng Dehuai and Mao.[48] The prospect of a relatively unsophisticated peasant army with little armor, artillery, and no air force or navy to speak of against the most technically advanced armed forces in the world possessing nuclear weapons was extremely daunting. Accordingly, for China's top leaders: "The policy decision to dispatch troops to resist America and aid Korea was one of the most difficult . . . of their lives."[49]

One bloc of opposition identified by many scholars is the economic bureaucracy.[50] Chen Yun, an economic planner and Politburo member seems to have opposed a war. The tone and content of a November 15, 1950, report about the country's finances, made to a national economic conference on the impact of China's involvement in Korea, supports the notion that Chen opposed the war. Chen's mood is noticeably unenthusiastic and matter-of-fact.[51] Still, economists appear to have been convinced eventually by Mao, Zhou, and others that China must intervene.[52]

"Most of the old Communists," including senior CCP leader Dong Biwu, reportedly opposed China's entry into the war until the PRC was more firmly established.[53] Liu Shaoqi, a member of the Politburo Standing Committee, and Rao Shushi, the leader of China's eastern region, were also reportedly against intervention.[54] There was allegedly opposition from some other civilian leaders including Gao Gang. It appears that Gao initially opposed sending troops to Korea but eventually decided to support the move.[55]

Military Elite: Deep Reluctance

Military men proved extremely reticent to commit troops to the peninsula and were ultimately convinced as to the wisdom of the move only after the sustained and forceful arguments of Mao and later CPV Commander Peng Dehuai.

Peng Dehuai

General Peng Dehuai had strong reservations about the wisdom of intervening in Korea. After the outbreak of war in June 1950, Peng saw Korea mainly as a Soviet-American conflict, one that the Chinese should keep an eye on but in which they should avoid involving themselves. In August 1950, he told top party and army leaders in the northwest region the following:

> The Korean peninsula is divided in two, both mutually antagonistic, the problem is relatively complicated, related to problems in bilateral relationships between the Soviet Union and the United States. [As] the North Korean People's Army is fighting in the south, the U.S. will not sit back and ignore it. This possibly creates a problem, and our country also ought to be prepared.[56]

In early October 1950, Peng was urgently summoned to Beijing from Xian where he was engaged in reconstruction work in northwest China. He came straight from the airport on October 4 to the top leadership compound at Zhongnanhai, arriving at about 4 P.M. to attend a Central Committee meeting already in progress where the merits of sending Chinese troops to Korea were being debated. Peng, who was expecting to be quizzed about his work on the northwest frontier, had no idea that the subject would be Korea. He sat and listened in silence to the debate. When the meeting adjourned for the day, he found it impossible to sleep in his comfortable hotel room with the possibility of war with the United States on his mind. While he later told his military subordinates in Shenyang on October 10, shortly after arriving to assume command of the CPV, that he believed the decision to intervene in Korea was correct, Peng clearly had serious reservations about the move and agonized over his decision.[57]

Peng wrote in his memoirs that, although he said naught during the October 4 meeting, his inclination at the time was that "troops should be [*yinggai*] sent to rescue Korea." But this was a tentative, preliminary reaction. Peng goes on to explain it was only during the course of a sleepless night that he concluded intervention was "essential" (*biyao*). He recalled: "Having straightened out my thinking, I gave my support to the Chairman's wise decision."[58] Having made up his mind that China had no choice but to intervene, Peng acceded without hesitation to Mao's request that he assume command of the troops being readied to go to Korea.[59]

When Mao asked Peng's opinion on October 5, Peng replied:

> Chairman, last night I got almost no sleep. I couldn't stop thinking about what you said. I realized this was a question of fusing internationalism with patriotism. If we only stress the difficult aspect, don't consider the critical implications of American troops bearing down on the Yalu River, not to mention the uncertainties this holds for the Democratic Republic of Korea—which are linked—then can we not see that the defense of our country's northeast frontier is also directly threatened. Is dispatching troops advantageous or not? After thinking it over, I endorse Chairman Mao's wise policy decision to send troops to Korea.
>
>   . . . [I]f [we] allow the enemy [the United States] to occupy the entire Korean peninsula the threat to our country is very great. In the past when the Japanese invaded China they used Korea as a springboard. First they attacked our three eastern provinces, then using these as a springboard, they launched a large scale offensive against the interior. We cannot overlook this lesson of history. We must fight the enemy now, we cannot hesitate.[60]

Once he had convinced himself of the necessity of intervention, Peng could without hesitation accept Mao's October 5 offer to make him commander. He later told staff officers in Shenyang, "I, Peng Dehuai, do not know how to say the word 'no.'"[61] Once convinced, Peng took on the role of a key advocate of intervention in the face of continued opposition in the October 5 meeting.[62] Given Peng's reputation for forthrightness and frankness, if the general had concluded that intervention was wrong he would undoubtedly have stated his opinion as he did on other occasions much to his detriment.[63]

Peng's deliberations highlight what seems to have been a considerable dilemma for many senior officers: Their political loyalty was torn between a sense of their Chinese identity on one hand and a sense of global class-consciousness on the other. Many senior soldiers remarked that important to their support of intervention was the linkage of patriotism with the cause of global communism. Peng remembered lying awake on the night of August 4–5, 1950, pondering whether it was in China's best interests to intervene in Korea or not. He recalled: "Again and again I turned over in my mind the Chairman's remarks. . . . I came to realize that his instruction combined internationalism with patriotism."[64] Phrases linking both internationalism and patriotism appear frequently in military memoirs. While the repetition could be dismissed as the parroting of official propaganda, judging from the context in which the phrases are mentioned, it is no mere platitude. The phrase seems to have great significance to the Chinese veterans of that war. It is usually mentioned reverently, sometimes emotionally.[65] Initially many soldiers were either opposed to or very reluctant to get involved in Korea: They viewed intervention as counter to China's best interests. Their feelings of "proletarian solidarity" with the oppressed people's of the world, particularly the strong ties forged with the Korean Communist troops who fought side by side with them during the anti-Japanese war and the Chinese civil war,[66] did not seem sufficient at the outset of hostilities on the peninsula to dispose them toward intervention. Rather the basic rationale behind China's eventual involvement came down to defense of the motherland. Mao shrewdly linked the two concepts so as to make support of the war not just a patriotic crusade to stir every Chinese heart but a politically correct cause that would satisfy dedicated Communists.[67]

Nie Rongzhen

Others, including acting Chief of General Staff Nie Rongzhen, also had reservations about intervening in Korea. While Nie does not specifically say in his memoirs that he opposed the idea of intervention, neither does he say he vocally endorsed it from the start. Implicit in his discussion of the deliberations is his own ambivalence. Nie wrote vaguely of the doubts of unnamed "comrades" and alludes to his own doubts: "Whether to fight that war was a question on which no one could easily make up his mind."[68] By the end of September he was convinced that if U.S. forces kept advancing toward the Yalu, China would have no alternative but to intervene. On September 25, during a dinner in Beijing with K. M. Panikkar, Nie told the Indian diplomat, "in a quiet and unexcited manner that the Chinese did not intend to sit back with folded hands and let the Americans come up to their border." He continued soberly: "We know what we are in for, but at all costs American aggression has to be stopped."[69]

Zhu De and Xu Xiangqian

By October, PLA Commander in Chief Zhu De had also apparently decided China must intervene.[70] During the critical period when the debate over whether to enter Korea raged, PLA Chief of General Staff Xu Xiangqian was at a sanitarium in the coastal resort of Qingdao recuperating from an illness. Because of his incapacitation, his duties were assumed by his deputy Nie Rongzhen. When Xu returned to Beijing in late October, he labeled the policy on Korea "wise."[71]

Determined Military Dissent

Some soldiers openly opposed China's involvement in a war on the Korean Peninsula. One such general was Lin Biao.

Lin Biao

Until very recently Lin was in official disgrace. Many scholars agree that Lin was a shrewd and clever, if not brilliant, tactician who appeared to suffer from self-doubt and a "strong streak of pessimism."[72] Although claims that he opposed Mao on Korea suggest character assassination, this stand is consistent both with what we know about Lin's character and circumstantial evidence, and is corroborated by many different sources.[73] While one account speculates that Lin was "an enthusiastic proponent" of intervention in Korea, this view was based on fragmentary and faulty evidence and the erroneous assumption that Lin Biao had served as the first commander of the CPV before being replaced by Peng in 1951.[74]

The evidence suggests that Mao first asked Lin to command the CPV.[75] An important question is why Mao waited so long before deciding to select Peng Dehuai (October 2), and waited several more days before actually asking him (October 5). This delay is understandable if Mao had assumed that Lin would take the job but at the last minute begged off on grounds of ill health. Lin declined Mao's request to lead the CPV but left open the possibility of stepping in later if he recuperated quickly.[76] Only in late September did Mao learn that Lin would not be available and spent two "days and nights" before selecting Peng. Mao apparently also considered General Su Yu, who had been charged with masterminding the invasion of Taiwan, to lead the CPV.[77]

Chai Chengwen, a military attaché posted to the hastily activated Chinese embassy in Pyongyang, gives an account of a conversation with Lin in early September. Chai had just returned with an ominous sounding report on the military situation in Korea, indicating that while on the surface things appeared to favor the KPA, North Korean forces were overextended and would be particularly vulnerable to a sudden American counteroffensive. According to Chai, Lin asked him "bluntly," "If we don't dispatch troops and told them [North Korea's leaders] to head for the mountains and wage guerilla warfare, would that be okay?"[78] Mao reportedly told the October 2 Politburo meeting that, in a conversation several days earlier, Lin had raised serious doubts about the wisdom of intervention. Indeed, Mao said, Lin insisted that involving the country in Korea would only hurt China.[79] Foremost among Lin's concerns was the possibility that the United States would use nuclear weapons.[80] Another account states that Lin opposed dispatching troops both at the Central Committee meeting chaired by Mao on October 4 and also at a meeting of the Central Military Commission chaired by Zhou Enlai two days later. At this second meeting Lin reportedly declared, "The United States is highly modernized. Furthermore, it possesses the atomic bomb. There is no guarantee of achieving victory [against the United States]. This issue should be considered with great care by the central leaders."[81]

Zhou criticized Lin's attitude saying the matter had already been decided. Zhou and Lin were eventually selected to go to the Soviet Union to facilitate arms transfers, the latter since Lin was already scheduled to travel there to receive medical treatment.[82]

Lin Biao's reaction to Mao's request that he command the Chinese force in Korea contrasts dramatically with Peng's response. Lin, pleading illness, declined his superior's request. Many accounts report Lin "said he was ill" (*shuo bing*), or used sickness as "an excuse" (*jiekou, jiegu* or *tuoci*).[83] It is likely that he was feigning illness to shirk duty in Korea; although, it is also possible that he really was sick.[84] The refusal to obey, which could have been considered insubordination, was necessarily couched in terms of poor health since this offered an honorable out, short of flat refusal for, despite Mao's polite phrasing, it was an order not a request.

Where the likelihood of character assassination comes into play is over the scornful tone of many accounts suggesting that Lin was using sickness as an excuse to avoid a difficult mission that he personally opposed. There is the clear inference in several accounts that Lin acted out of cowardice.[85] Far from qualifying as shameful behavior, however, this was accepted face-saving etiquette for a senior official (military or civilian) strongly opposed to a major policy. From Mao's words and actions, he seemed to have believed that

Lin was really ill. Mao was deeply concerned and went so far as to personally order a top physician to treat him.[86] Mao told the October 2 meeting of a conversation in which Lin insisted he was ill and suffered from insomnia. Lin was "afraid of wind, light, and noise, aiya!" Mao exclaimed. He continued, "he [Lin] has the three afraids—how could he possibly command troops!" Mao later recounted this story to Peng.[87]

Indeed Lin, like several other military figures including PLA Chief of General Staff Xu Xiangqian, had a record of chronic ill health, and others, including Nie Rongzhen, suffered serious bouts of illness at different times, probably brought on by overwork.[88] In the final analysis Lin's affliction is best viewed as a "political illness" (*zhengzhi bing*), and this is the way many senior Chinese leaders perceived it.[89] Thus, Lin's response was a convenient and time-honored excuse to avoid an escalation of intra-elite conflict over policy.

Other Military Opposition

General Ye Jianying, like Lin Biao, also reportedly opposed dispatching troops to Korea. Even after UNC successes following Inchon, Ye appeared to believe that it was unnecessary for China to become involved. He told a gathering in Guangzhou that the KPA should take to the mountains and gave no indication that he believed China should consider dispatching troops.[90] Generals He Long and Su Yu allegedly also opposed intervention in Korea.[91] Su, the man charged with masterminding the invasion of Taiwan, appeared to be an extremely cautious commander. Speaking in February 1950, Su was circumspect about the prospects for the invasion of Taiwan; he was adamant that China needed more time to prepare.[92] Liu Bocheng reportedly opposed intervention in Korea because it would mean postponing indefinitely the invasion of Taiwan.[93]

Despite initial reservations, most military leaders became convinced that intervention was the best course of action. Many were persuaded by Peng Dehuai after he became adamantly convinced of the correctness of this course. Even those skeptical or opposed concurred with the decision out of strong personal loyalty to Mao and Peng, or they kept silent.

Officers in the Field

Field commanders in the northeast initially seemed genuinely enthusiastic about the prospect of fighting in Korea, but they soon exhibited extreme caution and sought to postpone intervention. Some units had been moved to the northeast as early as mid-July 1950, after the NEFDA had been formed in response to the American military involvement in Korea. Deng Hua, commander of the Thirteenth Army Corps, and his deputy, Hong Xuezhi, had been preparing for action since they received orders on August 5 to be ready to move by early September. These orders were superseded by orders dated August 15 to be ready before the end of September.[94]

After the rapid deterioration in the military situation following the Inchon landings, Pak Hon Yong, KPA commander and North Korean interior minister, arrived in Andong, China, near the Yalu to brief local PLA commanders on the military situation in Korea. He reportedly asked the officers for Chinese troops, and the NEFDA commander Deng Hua and others sought to reassure him, saying they would pass the request onto Beijing and stood ready to move as soon as they received the order to go to Korea. Deng Hua, Hong Xuezhi, and the other commanders all agreed that the advancing American troops posed a direct threat to China and expressed a readiness to confront this threat head on.[95] Hong recalled thinking on October 2, "The circumstances of the Korean military situation had already swiftly worsened, requiring China to assist promptly by dispatching troops. Our Thirteenth Corps leaders and troops were already well prepared. All we needed was the order from CCP Chairman Mao and we would immediately move into Korea."[96]

The units preparing to cross the Yalu were from the Fourth Field Army and had been under the command of Lin Biao. Yet officers such as Deng Hua and Hong Xuezhi gave no indication that they were disappointed at not being led by their former commander. In fact, Deng and Hong were both enthusiastic about the news that Peng had been selected to lead them. Peng's reputation for being a no-nonsense, battle-hardened general directly affected their outlook on the impending mission.[97] On October 10 when Peng, after arriving in Shenyang to assume command of the CPV, asked Deng what he thought about intervention, Deng replied that he agreed completely with Peng's viewpoint. He firmly held that China had no choice but to dispatch troops. Hong Xuezhi also told Peng he believed: "We should resolutely dispatch troops."[98]

Nevertheless, many troop commanders in the prospective field of operations were deeply concerned about the nuts and bolts of the actual intervention. Their concerns reveal a large measure of doubt as to the wisdom of prompt intervention. At a meeting of commanders at the corps level and above in Shenyang on October 9, after Peng and Gao Gang explained Beijing's order to dispatch them to Korea, there was a chorus of concerns, "Various officers raised questions, they were most worried that troops were being sent abroad to fight without air support."[99] General Liu Zhen, commander of the infant air arm of the CPV stated candidly that after being appointed commander, he surveyed the years of his military career and acknowledged that while he had commanded infantry, artillery and armored units, he had never commanded air units. Liu admitted air warfare and organization were a "mystery." He said, "Neither I nor any of the other cadres had any experience organizing or commanding air combat operations and there was no shortage of problems." Liu also noted the obvious: "in our levels of tactics and technology we were way, way below those of our enemy."[100] Considering the enemy China was facing and the paltry resources available, building an air force was a daunting task. Liu could only draw strength from the fact that the PLA had long struggled successfully against overwhelming odds, learning military strategy, tactics, and combat through trial and error in battle. Liu recalled, "I nevertheless had a resolute thought running through my mind over and over again. The cause of the revolution had all along developed out of nothing, gone from small to big, developed as a brutal, difficult, death-defying struggle."[101] Chai Chengwen, the first military attaché at the fledgling PRC embassy operating in Pyongyang in mid-1950, enthusiastically supported intervention. He was excited about the prospect.

On September 1, 1950, after Chai briefed Zhou Enlai on the Korean situation Chai recalls thinking: "Sending Chinese troops was the correct move—all that remained was the question of the order."[102]

Despite these prointerventionist opinions, Deng, Hong, and their fellow officers in the theater of operations soon grew ambivalent about engaging in a mission beyond China's borders—something with which they had no experience. During a meeting with Peng Dehuai on October 10, Deng, Hong, and other senior officers from the Thirteenth Army Corps all insisted that Peng query Beijing's decision to send two army corps across the Yalu. They argued that such a force was simply "too few" (*tai shao*), and urged Peng to ask Beijing to approve a doubling of the initial expeditionary group to four corps. Peng agreed, requested the increase in troop strength, and Mao approved it. But the next day (October 11), Deng and Hong warned Peng, "Even if we initially send four corps into Korea, this still would be insufficient manpower. If we advance these four corps, who will protect the rear? We don't have any troops to protect the rear area, [so] how can our supplies be guaranteed? We must designate another corps to protect our rear."

And again Peng concurred.[103]

Then on October 17, Deng and Hong sent a cable to Peng and Gao suggesting intervention be postponed until the spring of 1951. The CPV field officers were growing increasingly perturbed; the lack of air support and the onset of winter raised grave concerns about the wisdom of entering Korea in late 1950. The telegram from Deng Hua, Hong Xuezhi, and other commanders declared, "It is our opinion that we are not yet fully prepared, and our troops have not undergone thorough political indoctrination. We suggest it would be more suitable with the onset of winter to postpone the operation until the spring."[104]

Peng and Gao transmitted these concerns to Mao who ordered Peng and Gao back to Beijing and put the operation on hold for twenty-four hours.[105] A final hurdle, the extreme caution of CPV leaders, was overcome during an October 18 meeting in Beijing when Mao and Zhou Enlai, the latter just returned from Moscow with promises of Soviet military assistance and air support from Stalin, insisted on forging ahead with intervention.[106] Mao promptly issued orders for CPV units to cross the Yalu the following evening and Peng returned to his headquarters to inform his commanders that the decision to enter Korea was final.[107]

Conclusion

While Mao certainly dominated the decision-making on Korea in 1950, other figures, including soldiers, played major roles. Focusing on individual leaders' cautions against viewing intervention as preordained or China's path toward Korea as inexorable once the subject of possible entry was raised. It is clear that despite key factors that favored Chinese intervention, including Mao's underlying propensity to dispatch troops, many senior leaders, Mao included, still harbored great ambivalence and others, such as Lin Biao, were adamantly against such a move. While Mao's thinking was undoubtedly critical and his resolve carried the day, the considerable discussion regarding intervention constituted nothing less than a major policy debate in which the views and concerns of many leaders were aired and weighed. Particularly important in winning over doubters and preserving party unity were Zhou Enlai and Peng Dehuai. Moreover, soldiers (e.g., Peng, Lin, and Nie Rongzhen) seemed to have been more reluctant than their civilian counterparts (e.g., Mao and Zhou) to intervene in Korea.

The analysis here is consistent with research by other scholars about the dispositions of elite Chinese figures regarding the deployment of force that reveals that soldiers seem to be no more hawkish than their civilian counterparts. Indeed, Jong Sun Lee concluded that, on Korea, military men tended to be consistently dovish throughout the deliberation and preparation for intervention.[108] Furthermore, in at least three instances other than Korea, the Taiwan Strait (1958), India (1962), and Vietnam (1964–1965), it seems at least some top soldiers opposed the use of force.[109] In each case the opinion of China's top civilian leaders (Mao Zedong, 1949–1976; and Deng Xiaoping, 1978–1997) appears to have carried the day, while the military view has never seemed to be the determining factor in decisions to initiate hostilities. However, this could change without the presence of a dominating civilian leader. While China's generals may remain reluctant to advocate war, they may engage in brinkmanship, particularly on matters, such as Taiwan, which they believe threaten core issues of national sovereignty and vital strategic interests.[110] Such brinkmanship could escalate into a full-blown conflict. While in earlier crises Chinese military leaders, far from being the bellicose figures they are sometimes painted, actually seem to have been leery of initiating hostilities,[111] this may be changing if reports over PLA hawkishness on Taiwan are confirmed.

Although the evidence is inconclusive, the overwhelming predominance of the land component in the PLA and the small size of its air and naval arms suggest that in the past the Chinese military may have been significantly less hawkish than the armed forces of other states that possess more substantial air and naval capabilities.[112] This too may be changing as China's navy and air force are undergoing modernization and expansion. The apparent belligerence of military figures over Taiwan might be explained as a result of the PLA's improving operational capabilities combined with the emotional issue of national unification.

The findings of this study suggest that the attitudes of Chinese soldiers and statesmen conform roughly to the pattern found by Richard Betts in his study of the attitudes of American leaders during the Cold War.[113] Betts found that military figures tended to be no more hawkish than civilian leaders and, in fact, often were more dovish. However, once a country has embarked on the path of war, soldiers, especially commanders in the theater of operations, become eager for combat. Still, on the eve of combat, officers in the field exhibit considerable caution and conservatism on strategy and tactics and a desire to deploy maximum force.

There are variations based on a soldier's branch of the service, position in the military hierarchy, and whether or not he is in the high command or in the theater of operations. This chapter supports the contention by Paul Godwin that on issues of national security and foreign policy, Chinese soldiers and statesmen hold basic dispositions generally consistent with the findings of Betts.[114]

These results suggest the existence of a "Chinese military mind," comparable to that identified in the United States, distinct from the thinking of Chinese civilian leaders. A distinction along civil-military lines has significant implications for Chinese strategic culture. First, this division suggests there may not be one clearly dominant Chinese strategic culture. Perhaps neither the parabellum strategic culture identified by Johnston nor the Confucian-Mencian strategic culture of conventional scholarly wisdom operates alone. It is possible that both may operate on a more equal and competitive footing than scholars have thus far been willing to recognize.[115] Second, civil-military relations may exert considerable influence on strategic culture, particularly in mediating its impact on the actual deployment of force. Indeed, this raises considerable questions about whether researchers can actually demonstrate a clear, direct, and causal relationship between strategic culture and specific instances of the use of military force.[116]

If future studies bear out the findings of this chapter, then China's strategic disposition may seem less uniquely Chinese and hold more in common with those of other countries. Further research will enable us to gain a more nuanced understanding of China's propensity for war making and saber rattling.

Notes

1. For some examples of China threat scholarship, see Richard Bernstein and Ross H. Munro, *The Coming Conflict with China* (New York: Alfred A. Knopf, 1997); Caspar W. Weinberger and Peter Schweitzer, *The Next War* (Washington, DC: Regnery, 1996).

2. See, for example, Michael D. Swaine, *The Military and the Political Succession in China: Leadership, Institutions, Beliefs* (Santa Monica, CA: Rand, 1992); Michael D. Swaine, *The Role of the Chinese Military in National Security Policymaking* (Santa Monica, CA: Rand, 1996).

3. See for example, Ting Chen-wu, "Hawks Dominate China's Policy on Taiwan," *Hsin Pao* (Hong Kong Economic Journal), March 14, 1996, p. 9, cited in Foreign Broadcast Information Service, *Daily Report: China*, March 21, 1996, pp. 11–13; Matt Forney, "Man in the Middle," *Far Eastern Economic Review*, March 28, 1996, pp. 14–16. For a direct challenge of the reliability of these reports, see June Teufel Dreyer, "The Military's Uncertain Politics," *Current History* 95, no. 602 (September 1996): 258–259. For a sophisticated, comprehensive analysis of the most recent strait crisis that incorporates civil-military relations, see You Ji, "Making Sense of War Games in the Taiwan Strait," *Journal of Contemporary China* 6 (1997): 287–305.

4. See, for example, sources cited in note 2.

5. Leading scholars of Chinese foreign and security policy have identified strategic culture as a fruitful area of research. See, for example, Robert S. Ross and Paul H.B. Godwin, "New Directions in Chinese Security Studies," in David Shambaugh, ed., *American Studies of Contemporary China* (Armonk, NY: M. E. Sharpe, 1993), p. 145. For some monographs that utilize a strategic culture or cultural approach, see Shu Guang Zhang, *Deterrence and Strategic Culture: Chinese-American Confrontations, 1948–1958* (Ithaca, NY: Cornell University Press, 1992); Jonathan Adelman and Chih-Yu Shih, *Symbolic War: The Chinese Use of Force, 1840–1980* (Taipei: Institute for International Relations, 1993); Alastair Iain Johnston, *Cultural Realism: Strategic Culture and Grand Strategy in Chinese History* (Princeton, NJ: Princeton University Press, 1995).

6. See, for example, Swaine, *The Role of the Chinese Military*. In the past, scholars tended either to argue that the nature and degree of PLA influence in foreign policy-making was difficult to discern or assert that it was "virtually nonexistent." On the former perspective, see A. Doak Barnett, *The Making of Foreign Policy in China* (Baltimore, MD: Johns Hopkins University Press, 1985), pp. 96–104; on the latter, see Gerald R. Segal, "The Military as a Group in Chinese Politics," in David S.G. Goodman, ed., *Groups and Politics in the People's Republic of China* (Armonk, NY: M. E. Sharpe, 1983), pp. 83–101 [quote on p. 96]. See also Gerald Segal, "The PLA and Chinese Foreign Policy Decision-Making," *International Affairs* 57, no. 3 (Summer 1981): 449–466.

7. See, for example, Segal, "The Military as a Group"; June Teufel Dreyer, "Civil-Military Relations in the People's Republic of China," *Comparative Strategy* 5, no. 1 (1985): 27–49.

8. William H. Whitson with Chen Hsia Huang, *The Chinese High Command: Military Politics, 1927–1971* (New York: Praeger, 1973); Lucian W. Pye, *The Dynamics of Chinese Politics* (Boston: Oelgeschlager, Gunn, and Hain, 1981); Segal, "The Military as a Group"; Dreyer, "Civil-Military Relations." It should be noted that some scholars do contend there is a

distinct military perspective. These include Harlan Jencks, Ellis Joffe, and, to an extent, Paul Godwin.

9. Alastair Iain Johnston, "Cultural Realism and Strategy in Maoist China," in Peter J. Katzenstein, ed., *The Culture of National Security: Norms and Identity in World Politics* (New York: Columbia University Press, 1996), pp. 216–270. See also the sources cited in note 5.

10. See, for example, F. F. Liu, *A Military History of Modern China, 1924–1949* (Princeton, NJ: Princeton University Press, 1956), esp. pp. 281–283; Edward L. Dreyer, "Military Continuities: The PLA and Imperial China," in William W. Whitson, ed., *The Military and Political Power in China in the 1970s* (New York: Praeger, 1972), pp. 3–24; Frank A. Kierman, Jr. and John K. Fairbank, eds., *Chinese Ways in Warfare* (Cambridge, MA: Harvard University Press, 1974) esp. John K. Fairbank, "Introduction: Varieties of the Chinese Military Experience"; Adelman and Shih, *Symbolic War*, esp. chap. 2.

11. See Johnston's *Cultural Realism* and "Cultural Realism."

12. See David R. Jones, "Soviet Strategic Culture," in Carl G. Jacobsen, ed., *Strategic Power: USA/USSR* (London: St. Martin's Press, 1990), pp. 35–49, and Desmond Ball, "Strategic Culture in the Asia-Pacific Region," *Security Studies* 3, no. 1 (Autumn 1993): 47, 62–66.

13. Alastair Iain Johnston, "Thinking about Strategic Culture," *International Security* 19, no. 4 (Spring 1995): 45. See also David J. Elkins and Richard E.B. Simeon, "A Cause in Search of Its Effect, or What Does Political Culture Explain?" *Comparative Politics* 11, no. 2 (January 1979): 128.

14. Samuel P. Huntington, *The Soldier and the State: The Theory and Politics of Civil-Military Relations* (Cambridge, MA: Belknap Press, 1957); Morris Janowitz, *The Professional Soldier: A Social and Political Portrait*, rev. ed. (New York: The Free Press, 1971).

15. Richard K. Betts, *Soldiers, Statesmen and Cold War Crises*, 2nd ed. (New York: Columbia University Press, 1991).

16. Scholars can now utilize participant autobiographies and participant interviews. Moreover there are numerous scholarly monographs, articles on party and military history, and collections of telegrams and speeches all published in China.

17. See, *inter alia*, Chen Jian, *China's Road to the Korean War* (New York: Columbia University Press, 1994); Sergei N. Goncharov, John W. Lewis, and Xue Litai, *Uncertain Partners: Stalin, Mao, and the Korean War* (Stanford: Stanford University Press, 1993); Shu Guang Zhang, *Mao's Military Romanticism: China and the Korean War, 1950–1953* (Lawrence: University Press of Kansas, 1995).

18. This is not to say that there has not been any coverage of the perspectives of specific Chinese elite figures. Rather, it means that the study of individual views (other than Mao's) toward Chinese intervention in Korea has yet to be the focus of an article or monograph published in English. For a strategic culture approach that focuses exclusively on Mao, see Johnston, "Cultural Realism and Strategy in Maoist China." For some recent studies of policy-making and decision-making in China, see Carol Lee Hamrin and Suisheng Zhao, eds., *Decision-Making in Deng's China: Perspectives from Insiders* (Armonk, NY: M. E. Sharpe, 1995); Kenneth Lieberthal and David M. Lampton, eds., *Bureaucracy, Politics and Decision Making in Post-Mao China* (Berkeley and Los Angeles: University of California Press, 1992).

19. Johnston, "Thinking About Strategic Culture," p. 54.

20. Allen S. Whiting, *China Crosses the Yalu: The Decision to Enter the Korean War* (Stanford, CA: Stanford University Press, 1968); John Gittings, *The World and China, 1922–1972* (New York: Harper and Row, 1974), chap. 9; Lawrence S. Weiss, "Storm around the Cradle: The Korean War and the Early Years of the People's Republic, 1949–1953," Ph.D. diss., Columbia University, 1981, chap. 3; Melvin Gurtov and Byung Joo Hwang, *China under Threat* (Baltimore, MD: Johns Hopkins University Press, 1982), chap. 2; Gerald Segal, *Defending China* (London: Oxford University Press, 1985), chap. 6. See also Adelman and Shih, *Symbolic War*, chap. 10, which, although published in 1993, is based on older secondary sources.

21. Thomas J. Christensen, "Threats, Assurances, and the Last Chance for Peace: Lessons

of Mao's Korean War Telegrams," *International Security* 17, no. 1 (Summer 1992); Goncharov, Lewis, and Xue, *Uncertain Partners*; Chen Jian, *China's Road to the Korean War*; Shu Guang Zhang, *Mao's Military Romanticism*; Michael M. Sheng, "Beijing's Decision to Enter the Korean War: A Reappraisal and New Documentation," *Korea and World Affairs* 19, no. 2 (Summer 1995): 294–313.

22. A noteworthy exception is the analysis of Jong Sun Lee. This perceptive study, however, is based on a limited array of mostly secondary sources. See Lee, "Attitudes of Civilian and Military Leaders Toward War Initiation: Application of Richard Betts' Analysis of American Cases to Other Countries," Ph.D. diss., Ohio State University, 1991, chap. 3.

23. One of the most nuanced, comprehensive, and circumspect analyses of China's decision to intervene appears in William Stueck, *The Korean War: An International History* (Princeton, NJ: Princeton University Press, 1995), pp. 98–103.

24. Chen, *China's Road to the Korean War*, pp. 110–111, 134.

25. Xu Yan, *Diyici jiaoliang: kangMei yuanChao de lishi huiyi yu fansi* (The First Trial: A Historical Retrospective and Review of the Resist America and Aid Korea War) (n.p. Zhongguo Guangbo Dianshi Chubanshe, 1990), pp. 19–20; Chai Chengwen and Zhao Yongtian. *Banmendian tanpan* (Panmunjom Negotiations) (Beijing: Jiefangjun Chubanshe, 1989), pp. 77–78.

26. Whiting, *China Crosses the Yalu*.

27. Chen, *China's Road to the Korean War*, p. 13.

28. Zhang Xi, "Peng Dehuai shouming shuaiyuan kangMei yuanChao de qianqian houhou" (The Complete Story of Peng Dehuai's Appointment to Lead the Resist America, Aid-Korea War), *Zhonggong dangshi ziliao*, no. 31 (1989): 111–159; Michael H. Hunt, "Beijing and the Korea Crisis, June 1950–June 1951," *Political Science Quarterly* 107, no. 3 (1992): 475–478. But one cannot assert that had the United States heeded China's warnings, or at least taken the possibility of Chinese intervention more seriously, China would not have intervened. China's leaders decided that, if they intervened, their aim was nothing less than the complete expulsion of UNC troops from the peninsula. Hope of a cease-fire or negotiated settlement in late 1950 after China intervened was based on the erroneous assumption that the lull in fighting in November was a deterrent signal to the UNC. See Christensen, "Threats, Assurances."

29. This is clear not only from most accounts, but also from the contents of telegrams sent by Mao including those sent to Soviet leader Joseph Stalin. See Mao Zedong, *Jianguo yilai Mao Zedong wengao* (Mao Zedong's Manuscripts since the Founding of the Republic), vol. 1, *September 1949–December 1950* (Beijing: Zhongyang Lishi Wenjian Chubanshe, 1987); for English translations of key telegrams, see Mao Zedong, "Mao's Dispatch of Chinese Troops to Korea: Forty-Six Telegrams, July–October 1950," trans. Li Xiaobing, Wang Xi, and Chen Jian, *Chinese Historians*, no. 1 (Spring 1992): 63–86.

30. Zhang Xi, "Peng Dehuai shouming," p. 119–131.

31. Nie Rongzhen, *Inside the Red Star: The Memoirs of Marshal Nie Rongzhen*, trans. Zhong Renyi (Beijing: New World Press, 1988), p. 635; Jonathan Pollack, "The Korean War and Sino-American Relations," in Harry Harding and Yuan Ming, eds., *Sino-American Relations, 1945–1955: A Joint Reassessment of a Critical Decade* (Wilmington, DE: Scholarly Resources, 1989), p. 219; Chen Jian, "China's Changing Aims During the Korean War, 1950–1951," *Journal of American-East Asian Relations* 1, no. 1 (Spring 1992): 23.

32. The two-day sequence version is from Zhang Xi, "Peng Dehuai shouming," pp. 123–125; Chen Jian, "China's Changing Aims," p. 17; Hunt, "Beijing and the Korean Crisis," p. 460. Another account gives the dates as October 4 and 5. See Hong Xuezhi, *KangMei yuanChao zhanzheng huiyi* (Memoirs of the War to Resist America and Aid Korea)(Beijing: Jiefangjun Wenyi Chubanshe, 1990), pp. 18–19. Other reports say Mao decided on October 2. Chai Chengwen and Zhao Yongtian, *KangMei yuanChao jishi* (Chronicle of the Resist America, Aid Korea War)(Beijing: Zhonggong Dangshi Ziliao Chubanshe, 1987), p. 56. The significance of October 2 is further underlined by the fact that it was also on this date that Mao sent a telegram

to Stalin informing him of China's intention to intervene militarily in Korea.

33. Zhang Xi, "Peng Dehuai shouming," pp. 140–142.

34. Chen Jian, "China's Changing Aims," pp. 21–22; Zhang Xi, "Peng Dehuai shouming," pp. 148–149.

35. Zhai Zhihai and Hao Yufan, "China's Decision to Enter the Korean War: History Revisited," *China Quarterly* 121 (March 1990): 111.

36. Zhang Xi, "Peng Dehuai shouming," pp. 149–151; Hong Xuezhi, *KangMei yuanChao*, p. 24; Chen Jian, "China's Changing Aims," pp. 21–22.

37. Shu Guang Zhang, "In the Shadow of Mao: Zhou Enlai and New China's Diplomacy," in Gordon A. Craig and Francis L. Loewenheim, eds., *The Diplomats, 1939–1979* (Princeton, NJ: Princeton University Press, 1994), pp. 346–347, 352–355.

38. Although Zhou varied in the harshness of his terminology, his tone is consistent. Thus, for example, his speech printed in *Renmin Ribao* on September 30, 1950, is far more virulent than his remarks to Panikkar in the early morning hours of October 3. *Zhou Enlai waijiao wenxuan* (Selected Diplomatic Works of Zhou Enlai) (Beijing: Zhongyang Wenxian Chubanshe, 1990), pp. 24, 25. His tough tone on Korea is traceable to at least late August from the report he gave the Central Military Commission at that time. Chen Jian, *China's Road to the Korean War*, pp. 149–150.

39. Qiao Guanhua, one of Zhou's colleagues in the Ministry of Foreign Affairs, insists that Zhou was a hawk on Korea. While the source for this is not wholly reliable, its depiction of Zhou is entirely consistent with the other sources cited above. See Russell Spurr, *Enter the Dragon: China's Undeclared War against the U.S. in Korea, 1950–51* (New York: Henry Holt, 1988), p. xx. Zhou's strong, sometimes harsh words could have been an effort to stave off any accusation that he was soft on American imperialism, but Zhou's words on Korea are consistent with his Hobbesian outlook on the world. See Dick Wilson, *Zhou Enlai: A Biography* (New York: Viking, 1984), pp. 189–191, 295, 300. It is also possible that Zhou could have supported intervention to side with Mao in order to avoid incurring the chairman's wrath. Indeed Zhou's political longevity was due in large part to expediently siding with Mao on many issues. See Roderick MacFarquhar, *The Origins of the Cultural Revolution*, vol. 1, *Contradictions among the People, 1956–1957* (New York: Columbia University Press, 1974), pp. 7–9; Shu Guang Zhang, "In the Shadow of Mao." However, Zhou did not seem to shy away from disagreeing with Mao when he felt strongly about an issue. Wilson, *Zhou Enlai*, pp. 297–298.

40. Bo Yibo, *Ruogan zhongda jueche yu shijian de huigu* (Reflections on Certain Important Decisions and Events), vol. 1 (Beijing: Zhonggong Zhongyang Dangxiao Chubanshe, 1991), p. 43.

41. Chai Chengwen and Zhao Yongtian, *Banmendian tanpan* (Panmunjom Negotiations) (Beijing: Jiefangjun Chubanshe, 1989), pp. 77–78.

42. Carsun Chang, *The Third Force in China* (New York: Bookman Associates, 1952), p. 286.

43. Chen Jian, *China's Road to the Korean War*, p. 281, n. 78.

44. Shu Guang Zhang, "In the Shadow of Mao," p. 354.

45. In the first phase, the army would be cut from 5.4 million to 4.0, and then down to 3 million during the second phase. Xu Yan, *Diyici jiaoliang*, p. 13; Zhai Zhihai and Hao Yufan, "China's Decision to Enter the Korean War," p. 99. Certainly this is best viewed as a "restructuring" of the PLA. Chen Jian, *China's Road to the Korean War*, pp. 95–96. This was initiated in order to economize and respond to China's changing security needs at the end of the civil war on the mainland and during the preparations for the invasion of Taiwan. Goncharov, Lewis, and Xue, *Uncertain Partners*, pp. 148, 152. While this should not be interpreted as a step toward Chinese disarmament, nor should it be read as preparation for war with the United States as one scholar suggests. Chen Jian, *China's Road to the Korean War*, p. 96.

46. Pollack, "The Korean War," pp. 218, 220; Hunt, "Beijing and the Korean Crisis," pp. 460–

461. See also Zhai Zhihai and Hao Yufan, "China's Decision to Enter the Korean War," p. 104.

47. Hu Guangzheng, "Yingming de juece, weida de chengguo—lun kangMei yuanChao zhanzheng de chubing canzheng juece" (Brilliant Policy Decision, Great Achievements—on the Policy Decision to Dispatch Troops to Fight the Resist America, Aid Korea War) *Dangshi yanjiu* no. 1 (February 1983), p. 34; Zhang Xi, "Peng Dehuai shouming," p. 132.

48. Zhang Xi, "Peng Dehuai shouming," p. 136.

49. Xu Yan, *Diyici jiaoliang*, p. 20.

50. Gurtov and Hwang, *China under Threat*, p. 55; Weiss, "Storm around the Cradle," pp. 80–82; Segal, *Defending China*, p. 105.

51. Chen Yun, *Chen Yun wenxuan, 1949–1956* (Selected Works of Chen Yun, 1949–1956) (Beijing Renmin Chubanshe, 1982), pp. 111–112. David Bachman argues that Chen was less than enthusiastic about the war but not strongly opposed. See David Bachman, *Chen Yun and the Chinese Political System*, Chinese Research Monograph no. 29 (Berkeley: Institute of East Asian Studies and Center for Chinese Studies, University of California, 1985), p. 34. Red Guard documents insist Chen was against intervention. One charged: "Chen Yun took the attitude that to fight the war of resistance against the Americans and continue with economic construction was absolutely incompatible." Cited in Gittings, *The World and China*, pp. 182–183. See also Weiss, "Storm around the Cradle," p. 81. Given the limited evidence, however, a definitive answer cannot be given.

52. Bo Yibo stresses the persuasiveness of Mao and Zhou. While not explicitly revealing his own thinking, his account of the decision-making process strongly implies that he and others were won over by these convincing advocates of intervention. See *Ruogan zhongda jueche*, pp. 43–44.

53. Carsun Chang, *The Third Force*, p. 286.

54. Segal, *Defending China*, pp. 105–106; Weiss, "Storm around the Cradle," pp. 81–82.

55. Frederick C. Teiwes, *Politics at Mao's Court: Gao Gang and Party Factionalism in the Early 1950s* (Armonk, NY: M. E. Sharpe, 1990), p. 307, n. 39. Teiwes cites a 1955 report given by Deng Xiaoping. Zhai Zhihai and Hao Yufan also contend that Gao opposed intervention in Korea. "China's Decision to Intervene in Korea," p. 105. A 1967 Red Guard document also alleges this. See Union Research Institute, *The Case of Peng Teh-huai, 1959–1968* (Hong Kong: Union Research Institute, 1968), p. 154. Two other accounts suggest otherwise. According to one anonymous high level Communist source, Gao backed the decision to intervene. See Carsun Chang, *The Third Force*, p. 286. Reportedly, an aide to Peng Dehuai heard Gao remark in early September 1950 "[On Korea] I fully support him [Mao]. Spurr, *Enter the Dragon*, p. 85.

56. Zhang Xi, "Peng Dehuai shouming," pp. 120–121. Another account by a British journalist citing interviews with Chinese military officials, although differing in details, supports this view of Peng's thinking. See Spurr, *Enter the Dragon*, pp. 63–69.

57. Pollack, "The Korean War," p. 218.

58. Peng Dehuai, *Peng Dehuai zishu* (Peng Dehuai's Own Account) (Beijing: Renmin Chubanshe, 1981), pp. 257–258. The term *biyao* is translated as "necessary" in the English version of Peng's reminiscences but this does not capture the emphatic tone that is implied in the Chinese. See Peng Dehuai, *Memoirs of a Chinese Marshal: The Autobiographical Notes of Peng Dehuai (1898–1974)*, trans. Zheng Longpu (Beijing: Foreign Languages Press, 1984), pp. 473–474; Hong Xuezhi, *KanMei, yuanChao*, pp. 18–19. Yao Xu gives different dates (October 5 and 6—not October 4 and 5), but the rest of his account jibes with the other accounts cited here. Yao Xu, "KangMei yuanChao," p. 8. See also Pollack, "The Korean War," p. 218.

59. Zhang Xi, "Peng Dehuai shouming," pp. 133–135. Nie simply recalls that "At the [October 4] meeting, he [Peng] firmly supported Comrade Mao Zedong's proposal to dispatch troops to Korea." Nie Rongzhen, *Inside the Red Star*, p. 636. Nie's account probably has the date wrong and also does not note the deliberation that went into the decision. Some other accounts do not mention this either. See, for example, Hu Guangzheng, "Yingming de juece,"

p. 37; and Xu Yan, 1990, p. 23. Red Guard pamphlets accused Peng of opposing the Korea decision but this is not corroborated by any post–Cultural Revolution source. Cited in Gittings, *The World and China*, pp. 182–183; Weiss, "Storm around the Cradle," p. 81. According to one post-Mao account, this allegation is "completely false." Hu Guangzheng, "Yingming de juece," p. 37. Perhaps, aside from the obvious effort at character assassination, this charge has some basis in reality since, as the analysis here indicates, Peng was not enthusiastic about intervention until after some serious soul-searching on the night of October 4–5, 1950.

60. Zhang Xi, "Peng Dehuai shouming," pp. 133, 134.

61. Hong Xuezhi, *KangMei yuanChao*, p. 19.

62. Zhang Xi, "Peng Dehuai shouming," pp. 136–137; Chen Jian, *China's Road to the Korean War*, pp. 183–184.

63. Roderick MacFarquhar, *The Origins of the Cultural Revolution*, vol. 2, *The Great Leap Forward, 1958–1960* (New York: Columbia University Press, 1983), pp. 193–195; Frederick C. Teiwes, "Peng Dehuai and Mao Zedong," *The Australian Journal of Chinese Affairs*, no. 16 (July 1986): 81–98.

64. Peng Dehuai, *Memoirs of a Chinese Marshal*, pp. 473–474.

65. See, for example, Yang Chengwu, *Yang Chengwu huiyilu* (Memoirs of Yang Chengwu) (Beijing: Jiefangjun Chubanshe, 1990), pp. 338, 356; Nie Rongzhen, *Nie Rongzhen huiyilu*, p. 736; Liu Zhen, *Liu Zhen huiyilu* (Memoirs of Liu Zhen) (Beijing: Jiefangjun Chubanshe, 1990), p. 381.

66. See, for example, Bruce Cummings, *Origins of the Korean War*, vol. 2, *The Roaring of the Cataract: 1947–1950* (Princeton, NJ: Princeton University Press, 1990), chap. 11; Chen Jian, *China's Road to the Korean War*, pp. 106–109.

67. In fact, the eight-character slogan in the campaign to rally support for the war took the element of patriotism even further, linking it to the protection of one's family: "*KangMei yuanChao; baojia, weiguo*" (Resist America, Aid Korea; Protect Your Family, Defend Your Country).

68. Nie Rongzhen, *Inside the Red Star*, pp. 633, 634.

69. K. M. Panikkar, *In Two Chinas: Memoirs of a Diplomat* (London: Allen and Unwin, 1955; reprint ed. Westport, CT: Hyperion Press, 1981), p. 108. Nie Rongzhen, *Inside the Red Star*, p. 637. One scholar on the basis of this conversation labels Nie a proponent of intervention. Weiss, "Storm around the Cradle," p. 82.

70. Gittings, *The World and China*, p. 184.

71. "The Korean war situation was a matter of prime importance, and Zhu De, He Long, Luo Ronghuan, Nie Rongzhen, Ye Jianying, Li Xiannian and other comrades paid me visits and much of the contents of our conversations were on this situation." Xu Xiangqian, *Lishi de huigu* (Reflections on History), vol. 3 (Beijing: Jiefangjun Chubanshe, 1987), p. 798.

72. Pollack, "The Korean War," p. 223.

73. Xu Yan, *Diyici jiaoliang*, pp. 23–24; Chai Chengwen and Zhao Yongtian, *Banmendian tanpan*, p. 83; Zhai Zhihai and Hao Yufan, "China's Decision to Enter the Korean War," p. 105; Nie Rongzhen, *Inside the Red Star*, p. 636; Chen Jian, *China's Road to the Korean War*, pp. 153, 185.

74. The belief that Lin served as commander of the CPV was once widely accepted as fact because many of the Chinese units in Korea were from the Fourth Field Army, which had been commanded by Lin and captured CPV troops listed Lin as their commander-in-chief. See Jurgen Domes, *Peng Te-huai: The Man and the Image* (Stanford, CA: Stanford University Press, 1985), p. 61. On Lin as an advocate of intervention, see Whitson with Huang, *The Chinese High Command*, p. 329. On Lin as the first commander of the CPV, see ibid., and June Teufel Dreyer, *China's Political System: Modernization and Tradition*, 2nd ed. (Boston: Allyn and Bacon, 1996), p. 194. The latter account states that after major battlefield reverses, a "new" CPV commander was appointed: Peng Dehuai. While the author does not specifically name the "first" commander, by implication it must be Lin Biao.

75. Lin Qinshan, *Lin Biao zhuan* (Biography of Lin Biao), vol. 1 (Beijing: Zhishi Chubanshe, 1988), p. 71; Chen Jian, *China's Road to the Korean War*, pp. 173–174. There is some controversy about this. Michael Hunt argues that the choice of Lin is strange because Mao and Lin had a history of conflict over military strategy during the Civil War. Hunt, "Beijing and the Korean Crisis," p. 462, n. 29.

76. Lin Qinshan, *Lin Biao zhuan*, vol. 1, pp. 71–72.

77. Su Yu was a logical choice since he had been reassigned to command the NEFDA, which was created several months earlier. But he was ruled out, also because of ill health. Zhang Xi, "Peng Dehuai shouming," pp. 125–126, 135.

78. Chai Chengwen and Zhao Yongtian, *Banmendian tanpan*, p. 78.

79. Zhang Xi, "Peng Dehuai shouming," p. 126.

80. Goncharov, Lewis, and Xue, *Uncertain Partners*, p. 167. There was considerable concern among Chinese leaders that the United States might use the bomb either on the Korean Peninsula or against cities and/or military installations in China itself. Proponents of intervention downplayed this possibility, arguing that the bomb was unsuitable for use in Korea and the United States would be unwilling to risk triggering a nuclear response from the Soviet Union. These arguments seem to have calmed their fears. Ibid., pp. 164–167, 182.

81. Quote cited in ibid., p. 167. See also Xu Yan, *Diyici jiaoliang*, pp. 23–24.

82. Xu Yan, *Diyici jiaoliang*, pp. 23–24.

83. Chai Chengwen and Zhao Yongtian, *Banmendian tanpan*, p. 83; Zhang Xi, "Peng Dehuai shouming," pp. 135, 143; Hu Guangzheng, "Yingming de juece," p. 37; Yao Xu, "KangMei yuanChao de yingming juece" (The Wise Policy Decision to Resist America and Aid Korea), *Dangshi yanjiu*, no. 5 (October 1980), p. 8; Hong Xuezhi, *KangMei yuanChao*, p. 19. Hong's source is Peng who had a strong personal and professional rivalry with Lin and therefore good reason to imply that Lin was faking illness.

84. Certainly this is very possible. Nie Rongzhen remarked that the whole episode was "most peculiar" *(qiguai dehen)*, insisting he had "never seen him [Lin] so frightened of anything." *Nie Rongzhen huiyilu*, p. 736. In any event by the 1960s—if not before—Lin appears to have become a hypochondriac. See the bizarre account by one of Lin's secretaries: Zhang Yunsheng, *Maojiawan jishi: Lin Biao mishu huiyilu* (Maojiawan Account: The Memoirs of Lin Biao's Secretary) (Beijing: Chunqiu Chubanshe, 1988).

85. See, for example, Lin Qinshan, *Lin Biao zhuan*, vol. 1, pp. 65ff.

86. Ibid., vol. 1, pp. 68–73, 77–78.

87. Zhang Xi, "Peng Dehuai shouming," pp. 126, 135.

88. Frederick C. Teiwes, *Leadership, Legitimacy, and Conflict in China: From a Charismatic Mao to the Politics of Succession* (Armonk, NY: M. E. Sharpe, 1984), pp. 30, 106; MacFarquhar, *Origins of Cultural Revolution*, vol. 2, pp. 245–246. Nie Rongzhen suffered from stress and overwork, collapsing in late 1952. He remained acting chief of general staff, however, until near the end of the Korean War when he formally resigned the post to recuperate properly. Nie only resumed work again in late 1956 after three years of treatment. See Nie Rongzhen, *Inside the Red Star*, pp. 618, 659. Furthermore, according to one source, Mao wanted Su Yu to command the CPV but had to discount him because Su was seriously ill and recuperating in Qingdao. Zhang Xi, "Peng Dehuai shouming," p. 125.

89. Lin Qingshan, *Lin Biao zhuan*, vol. 1, p. 74.

90. Cummings, *Origins of the Korean War*, vol. 2, pp. 726–729. Ye allegedly voiced grave concerns about intervening in Korea from midsummer 1950 and warned Mao that mobilizing a force of sufficient strength to be effective in Korea would take months rather than weeks as the chairman believed. Spurr, *Enter the Dragon*, pp. 63–64, 66–67, 59–60.

91. Ibid., pp. 53–65.

92. Whiting, *China Crosses the Yalu*, p. 21.

93. James Chieh Hsiung, *Ideology and Practice: The Evolution of Chinese Communism* (New York: Praeger, 1970), p. 172.

94. Zhang Xi, "Peng Dehuai shouming," pp. 118–119.
95. Hong Xuezhi, *KangMei yuanChao*, pp. 8–9.
96. Ibid., p. 15.
97. Ibid., pp. 16–18.
98. Ibid., p. 19.
99. Zhang Xi, "Peng Dehuai shouming," p. 144.
100. Liu Zhen, *Liu Zhen huiyilu*, pp. 337, 338, 342.
101. Ibid., pp. 337.
102. Chai Chengwen and Zhao Yongtian, *Banmendian tanpan*, p. 78.
103. Hong Xuezhi, *KangMei yuanChao*, pp. 21, 22.
104. Zhang Xi, "Peng Dehuai shouming," p. 157.
105. Mao Zedong, *Jianguo yilai*, p. 567; Mao Zedong, "Mao's Dispatch of Chinese Troops to Korea," p. 75.
106. Zhang Xi, "Peng Dehuai shouming," pp. 157–158; Chen Jian, *China's Road to the Korean War*, p. 208.
107. Hong Xuezhi, *KangMei yuanChao*, chap. 3; Zhang Xi, "Peng Dehuai shouming," p. 159.
108. Lee, "Attitudes of Civilian and Military Leaders Toward War Initiation," pp. 65–67. Based on limited data, however, Lee also erroneously concluded that civilian leaders, including Zhou Enlai, were equally—if not more—dovish than soldiers.
109. On Taiwan, see Segal, *Defending China*, p. 134; on India, see ibid., pp. 153–154 and Gurtov and Hwang, *China under Threat*, p. 114; on Vietnam, see ibid., pp. 180, 183–185 and Segal, *Defending China*, pp. 169–70.
110. You Ji, "Making Sense of the War Games," esp. pp. 301, 229.
111. Nevertheless, it is impossible to say with any certainty where the views expressed by individuals cited in this chapter stem. Thus, for example, it is likely that some are the result of individual predilections (e.g., Lin Biao), while others probably flow from bureaucratic interests (e.g., Chen Yun).
112. Soldiers in the air force, and, to a lesser extent, those in the navy tend to be less reticent to commit forces to combat that those in the army. See Betts, *Soldiers, Statesmen and Cold War Crises*, pp. 116–142.
113. Ibid. This pattern has also been found in some but not all other countries. See Lee, "Attitudes of Civilian and Military Leaders Toward War Initiation."
114. Paul. H.B. Godwin, "Soldiers and Statesmen: Chinese Defense and Foreign Policies in the 1990s," in Samuel S. Kim, ed., *China and the World: New Directions in Chinese Foreign Relations* (Boulder, CO: Westview Press, 1989), pp. 181–202.
115. See Andrew Scobell, *China's Use of Military Force: Beyond the Great Wall and the Long March* (New York: Cambridge University Press, 2003). While Iain Johnston does acknowledge the existence of two strategic cultures, he argues that only one—parabellum—is actually influential. The other—Confucian-Mencian—is purely symbolic with essentially no impact according to him.
116. This particular challenge is even acknowledged by strategic culture adherents themselves. See, for example, Johnston, *Cultural Realism*, pp. 13–14, 52.

7

Traditional Chinese Military Thinking

A Comparative Perspective

Zhang Junbo and Yao Yunzhu

If traditional Chinese and Western military theories are examined from a philosophical perspective, it is not difficult to discern the enormous impact exerted on them respectively by the two distinct philosophical traditions—the Chinese and the Western. As cultural legacies, traditional military theories still play an important role in modern strategic thinking and military decision-making, which renders a comparative study in this respect highly necessary. The purpose of this chapter is to reveal some of the most outstanding differences between Chinese and Western traditional military theories and trace the divergence back to their respective philosophical roots.

"Justice" versus "Interests" (*Yi* and *Li*)

In a very broad sense, Chinese traditional military theories place more emphasis on upholding "justice" while Western theories focus more on pursuing "interests." This may be best demonstrated by the different approaches each takes in thinking and explaining the causes as well as the aims of wars.

Reflections on the causes of war have always been a major concern in military thinking. Such reflections can be traced back to the early stages of human history. Homer's *Iliad*, which gives a vivid description of life in ancient Greece, points to the contention over a beautiful woman as the cause of the bloody Trojan War. Behind this kind of straightforward version lies the interpretation that conflicts over interests, whether physical or sentimental in nature, trigger wars among human groups. There are legends and epics of the same period in China describing wars between Huangdi and Chiyuo, Zhuanxu and Gonggong, Shun and Xiang, Shun and Danshu, as well as Qi and Boyi. These wars were invariably depicted with a strong tendency to praise the moral winner and to castigate the moral loser, which implies that wars were thought to be triggered by moral disputes. These early reflections on the causes of war, simple as they seem to be, do take different approaches and reach different conclusions.

The earliest effort at explaining the causes of war in a logical and rational way was made in the classic period. While Herodotus inherited the traditional interpretation when discussing the Trojan War in his *History*, Thucydides, a later historian, dug out the real cause behind seemingly irrelevant events and explained that the Peloponnesian War was fermented by conflicts over economic gains between the Athenians and the Spartans, rather than disputes over three prostitutes. The Roman historian Appian also paid special attention to material interests in analyzing the origin of the Roman civil war. He pointed to the struggle over land between those who owned land and those who did not as the fundamental cause.

During the Spring and Autumn Period and the Warring States Period (770–221 B.C., inclusive), the classic period in China, Chinese scholars made the greatest contribution to war theories. Their studies of the causes of war revealed their preference for digging out moral reasons to explain reckless and bloody military activities. For example, most scholars justified Zhou Wu's usurpation of power from his lawful king because the latter had been notoriously tyrannical and oppressive. Zhou's launching of war against the king was praised as "Zhou Wu's revolution" in the classic era, for he himself later made a much more democratic and benevolent king. The war was legitimated as an action to redress the wrong. However, Chinese scholars never confined all their study efforts entirely within an ethical scope. When they argued that moral reasons were the major dynamics behind war, they did not hesitate to mention other factors. For instance, a classic divination book, *Zhou Yi*, ascribed the staging of war to conflicts over "access to food and drinks."[1] Wu Zi listed five factors giving rise to war: struggle for fame, disputes over interests, dissatisfaction of the general populace, internal unrest, and famine.[2]

By and large, the ancient Chinese held disputes over moral issues as the most important factor in causes of war while ancient Greeks and Romans stuck to interest conflicts when explaining the war ferment. The Chinese tended to think of war as an important part of human life and studied the phenomenon in the cultural framework of a highly ethical society. The Greeks and Romans, on the other hand, were more straightforward, practical, and held conclusions closer to the truth.

Similar disparities existed in reflections on the aims of war. Homeric epics described how war heroes in the West seized land, cities, jewelry, women, and slaves as trophies. Both admiration and appreciation were evident in those poetic lines. Earlier recognition and acceptance of material gains as the aim of war helped to legitimate some extreme atrocities in wars.

On the other hand, one can hardly find any assertions in the huge number of Chinese history books stating that material gains were legitimate aims of war. On the contrary, what one constantly comes across in those books is the condemnation of such assertions. In the Warring States period, the school led by Mo Zi argued that wars were fought to punish the immoral. Guan Zi asserted that wars were waged to uphold the ethical codes, that is, courtesy, justice, loyalty, and commitment (*li, yi, zhong,* and *xin*). Mencius, one of Confucius' most famous followers, claimed the aim of a justifiable war was to overthrow a tyranny, just like the purpose of Zhou Wu's revolution. His view was echoed by a later philosopher, Xun Zi, who approved of the use of force for "the purpose of stopping tyranny and getting rid of a dictator."[3] Wars with all their atrocities and bloodshed, could only be accepted for the purpose of pursuing or maintaining justice, although the promotion of interests was more often the true aim.

The difference in the reflections on the causes and aims of war can be traced to distinct philosophical traditions of China and the West. Traditional Chinese philosophy focused on the perfection of the inner world. Human behavior, desire, feelings, and aspirations were always explored and explained in an ethical framework. Social, political, economic, as well as military practices were all regulated with moral codes and judged accordingly. On the other hand, traditional Western philosophy portrayed man as the "only animal with reason" whose purpose in life was to know the outside world and to reveal the mystery of the universe with his rationality and wisdom, so as to control and dominate the universe with his brain power and muscle force. Hence, a meaningful and active life for human beings must be full of challenges, explorations, and adventures. Human achievements were measured, to a certain degree, by material gains seized from the outside world.

Different philosophical definitions of human life lead to different approaches in military thinking, which in turn gives rise to distinctive strategic traditions. The Chinese strategic tradition stressed the moral dimension in interpreting and responding to conflicts. Throughout Chinese history, political leaders and generals have been doctrinated to go to war for moral reasons as well as for economic interests. Military thinking in modern times is still heavily indented by this tradition. For instance, Mao Zedong, the man who has contributed most in the formation of the Chinese PLA's (People's Liberation Army) military doctrine, tends to judge military conflicts by moral standards. He firmly believes in the principle that if a state acquires *dao* (justice or moral strength), it would eventually win no matter how inferior it is in physical strength to its opponent. If a state does not acquire *dao*, it is doomed to failure no matter how powerful it might be. In the early 1960s when Mao still counted the United States as China's number-one enemy, he assessed the situation and assured the nation with these words: "Who is afraid of whom in today's world? It is not our people who are afraid of the U.S. imperialists, but vice versa. Those who have acquired the *dao* will have support from the majority of the world's people, whereas those who have lost the *dao* will have no one's support. U.S. imperialists are doomed to defeat. This historical trend is irresistible."[4]

The strong desire to uphold justice may have been the working force behind Mao's decisions to send the Chinese People's Volunteers across the Yalu River in the early 1950s and to provide military assistance to Vietnam in the late 1960s and early 1970s. In both cases, ideological affinity, a sense of duty to help the weaker party when it is unjustly invaded or bullied by a bigger power, combined with the need to safeguard national interests that were perceived to be at stake, justified the risk of a direct confrontation with the world's strongest nation. Moral reasons are not only taken into consideration in the war decision-making process but also counted upon as force multipliers in the conduct of war.

The People's War Doctrine, most systematically explored and successfully applied by Mao Zedong and still held as official military doctrine by the PLA, provides yet another good example of how the Chinese traditional military thinking continues to play a part in modern military theories. The basic logic of the People's War Doctrine goes as follows: A war fought for the interests of the people is undoubtedly a just war—that is, a war justified and necessitated by moral as well as other reasons. As such, the war is expected to win support from the masses, both domestically and internationally. People's support and participation in the war will eventually turn into unlimited power that will overwhelm the enemy in the end, even if the enemy might enjoy superiority at the beginning. Mao Zedong, reviewing the Korean War at the Conference of the Central Government Committee in September 1953, remarked, "The reason why we have been able to fight such an enemy as the U.S. imperialists, whose arms are many times superior than ours, and force them to come to a truce with us, lies mainly in the fact that for us it is a people's war which the

Chinese people have fought shoulder to shoulder with the Korean people."[5] Mao's strong belief in the just nature and unlimited potential of the People's War can also be sensed in his Telegraph to the South Vietnamese People on December 19, 1967, in which he claimed: "Any nation, no matter large or small, so long as it fights a People's War, will beat any enemy however powerful it might be."

The People's War Doctrine, though still playing a dominating role, is undergoing major revisions and adjustments. A defense debate began in the latter half of the 1980s and is still going on at present. One of the major issues under discussion is "national interests": How to define China's national interests? What are the long-term as well as the short-term interests of China? What is the best way to reconcile the two? As patriots and internationalists, how should the policy-makers balance the obligations of the two roles? It is interesting to note that while moral factors are expected to play a continuous role in the decision-making process concerning crisis response, a strong undercurrent goes in the reverse direction. Leading military theorists in the PLA today are calling for a more interests-oriented approach in crisis management. They have found a powerful supporter in none other than Deng Xiaoping, who explicitly referred to "national interests" as the "overriding consideration" in policy-making.[6] This change, subtle as it seems, may carry profound implications in the PLA's strategic thinking and war planning.

"Human Factor" versus "Weapon Factor" (*Ren* and *Wu*)

One of the major efforts made by both Western and Chinese traditional military theories is to explain the relationship between the "human factor" and the "weapon factor" in war and other military practices. Generally speaking, Chinese military tradition places more value on the human factor while Westerners give higher priority to the weapon factor. The difference can also be traced to philosophical roots.

Many Chinese scholars emphasized in their writings the importance of the human factor in war. Mencius' well-known maxim "[H]aving the right timing is not as good as having the favorable terrain; and having the favorable terrain is not as good as having mutually devoted people" (*tianshi buru dili, dili buru renhe*)[7] picked out "mutually devoted people" as the single most important strategic factor. The influence of this priority listing upon generations of political and military leaders can hardly be overestimated. A contemporary reformist, Shang Yang, advised his lord to pay special attention to the human factor in war by saying, "Brave soldiers win; cowardly people lose" (*Bing yong ze zhansheng; min buyong ze zhanbai*).[8]

These scholars went further to elaborate why human beings played a more important role than things and how to utilize the strategic assets to their full potential.

First, they argued that human beings were important as the directors of war. They possessed the intellectual power to understand the guiding rules of war and to conduct war in a proper way. They could develop right strategies and tactics for military activities. In his *Art of War* (Sun Zi Bingfa), Sun Zi suggested that in preparation for war, generals must "appraise it in terms of the five fundamental factors, make comparisons of the seven elements and assess its essentials." Here, Sun Zi asked the generals "to appraise, make comparisons and assess" the objective conditions of both sides, then make their own judgment and find the best way to defeat the enemy. Guan Zi also said that in war the most important tasks for generals were "to decide upon the right strategy, to make use of the best terrain, to form alliances and to start an operation at the right time."[9]

Second, ancient Chinese scholars argued that human beings were important as the operators of war. They had, at their command, the will as well as the physical power to carry out the strategies and tactics. They were active actors, while weapons were taken as passive actors, on the battleground, and the outcome of war depended very much on the way they used their initiatives. A Chinese military classic, *Hu Ling Jing*, listed the basic requirements that military personnel must meet if they were going to win on the battlefield. The requirements were resolve in decision making, speed in movement, secrecy in planning, prudence in advance and retreat, and coordination in the use of force and schemes.[10]

In war, how powerful an army can be depends on how well the human factor can function. Chinese military tradition has an abundance of literature expounding on this subject. Some of the more general principles concerning army building and personnel management are summarized below:

1. The Principle of Keeping a Smaller but Better Army (Jingbing)

Traditional theories think that the combat power of an army has as much to do with the quality of the people as with the quantity of them. If an army is well organized so that every individual soldier can fight most effectively, it would defeat a much larger but poorly organized enemy. Therefore, soldiers must be trained to know their job well and to know how to coordinate with each other on the battlefield. The concept of *jingbing* has occupied a very special place in Chinese military tradition ever since the Warring States Period and remains one of the basic tenets in today's PLA army building.

2. The Principle of Keeping a Well-Disciplined Army (Lubing)

In answering the king's question "What makes troops gain the victory?" a very famous counselor in the Warring States Period named Wu Zi said, "The number of the soldiers is not all that important. It is the proper discipline that enables them to win."[11]

Sima Qian in his famous *History* (Shiji) recorded that when the King of Wu asked if Sun Zi could show him the power of discipline by training his chamber maids into soldiers, Sun Zi accepted the challenge under the condition that he be given the full authority to do whatever he deemed necessary. Sun appointed two of the king's favorite concubines as leaders of the maids and beheaded them when the girls failed to follow his instructions for the third time. The maids were terrified and followed his orders word by word. Within days, Sun Zi turned the giggling girls into fierce and obedient warriors and handed them back to the king. The king was more than convinced and gave Sun Zi the command of his army. *Zhou Yi*, an earlier divination classic, also points out, "An army going to war must have good discipline, the absence of which is an ominous sign."[12]

3. The Principle of Boosting Morale (Jiqi)

The morale of an army can be measured by its willingness to fight and the bravery it demonstrates in war, which can make all the difference in the outcome of war. It is the duty of a general to "award the fearless and encourage the brave," because "with what [does] a general fight a war? Soldiers. And with what [do] soldiers fight a war? Morale. If the morale is high you can fight; if the morale is low you must retreat."[13]

4. The Principle of Unity in the Army (Yixin)

The concept that unity and cohesion in the ranks leads to well-synchronized and closely coordinated military operations has been fully accepted by traditional Chinese theories. Guan Zi argued, "Wherever you come from, far or near, you must think in the same way, so that no matter numerous or few, you can act as one person. If you act as one, you can advance when you attack and hold when you defend."[14] Later on, the importance of unity was also stressed in Lu's Spring and Autumn Annals (Lushi Chunqiu): "[I]f all units in the army think in the same way, orders will be carried out without hindrance; if orders are carried out without hindrance, the army can triumph in all wars."[15]

Compared with the Chinese, Westerners had an earlier and more comprehensive appreciation of the importance of the "weapon factor" in war. They tried more enthusiastically and conscientiously at applying new technologies to military uses. Weapons made of iron were extensively used by the Assyrians as early as 800 B.C. The Assyrians also invented vehicles equipped with bumping horns to break the city walls, tall mobile towers to move soldiers close to the city walls, and tunnels to get inside from the underground.

Maritime operations acquired strategic significance during the Greco-Persian War (492–449 B.C.) and the Peloponnesian War (431–404 B.C.). All belligerents were equipped with cleverly designed warships. The Athenians, by improving the hull design, fixing metal horns to the bow, and expanding the rowing crew, succeeded in obtaining better speed, mobility, and striking power, which helped their navy to crush the Spartans and gain the final victory.

Emphasis on the weapon factor in war helped to nurture an atmosphere in which scientists, technicians, and engineers were encouraged to apply any novel technology to military use. Even the great Greek scientist Archimedes tried his hand at military invention. He designed a giant grab that allegedly could seize enemy warships.

A brief discussion on the characteristics of Western and Chinese philosophical traditions might assist us to find the links between military thinking and philosophical exploration. The ancient Chinese used to take philosophy, politics, and ethics as a closely interwoven system. One of the long-held propositions of political philosophy is that people are the foundations of society (yimin weiben). Mencius expressed this idea in a most explicit way. He said that the general populace weighed heavier than kings in political considerations. In Xun Zi's well-known metaphor, kings were likened to a ship and people to the water. It was the water that had the power to either support or sink the ship. The same concepts were expressed in military literature as well.

When discussing the different roles of people and heroes, The Three Strategies argued that the heroes were the pillars of a state while the people were the groundwork upon which stand the pillars supporting the state. If the king wins the support of the people and the commitment of the heroes, he has obtained the key to victory of war.[16]

A central task of ancient ethical philosophy in China is to inspire people to cultivate the good in their nature and to discard the evil. It holds that the "heavenly laws" (tiandao), which refers to the laws of nature or even more broadly to the objective universe, is fundamentally in conformity with the "inner life" (neizai xingming), which means human nature. All the great thinkers in ancient China invariably agree that the highest achievement in human life is to reconcile human nature with heavenly laws and to reach the ultimate harmony of the two. Man has been endowed with everything he needs in his mind. By looking introspectively into his own mind, he can gain a better knowledge of the heavenly laws. By constantly cultivating and improving himself, he can compromise with nature and achieve·ultimate harmony. Driven by these notions, the military thinkers take human beings as the key factor in war. They place a much higher priority on how to utilize human potential in fighting a war than upon how to invent new weapons to arm the people.

Western philosophers have followed a different path, studying the relationship between man and nature. Their understanding of nature is more scientific and materialistic. What is more, they try to go beyond the known universe and explore the endless unknown. Instead of focusing on the inner world, they seek answers from solid facts in the outer world, which is mostly made of matter. One Greek philosopher suggested that water was the origin of everything in the world, which illustrates how early Western philosophical efforts were directed to matter, instead of the human mind, for explanations concerning the universe. Unlike the ancient Chinese who have looked upon man and the universe as an organic whole, ancient Westerners treated them as two distinctive systems. They tended to approach nature in a very objective manner, explaining it as it is and using it to their own benefit. Such a philosophy has helped to cultivate a culture in which exploration and utilization of the objective world are not only accepted but also encouraged, and in which no effort is spared in the quest for better material means to win a war.

Contemporary Chinese strategic thinking still gives priority to the human factor in war. Mao Zedong, for example, had serious doubts about the utility of nuclear weapons. His reference to the weapon as a "paper tiger" should not be taken simply as bragging or a helpless resignation at China's not owning such a weapon. Mao followed the same thinking as his ancestors. He always stressed the human factor and believed that as long as China demonstrated domestic harmony and a resolve to fight, no enemy, including the two superpowers, would risk a general war against China. In Mao's view, "Human beings are the most precious resource in the world. As long as therse are people, miracles can happen." Mao's emphasis on the mobilization of the masses, on the political indoctrination of the troops, and on the morale-boosting functions of political work, show his firm belief in human potential.

Chinese and Western perceptions of "deterrence" differentiate in this respect as well. Western deterrence theory was developed mainly in the 1950s and 1960s, spurred by the development of nuclear weapons, while the Chinese concept of deterrence can be dated back to the times of Sun Zi. The killing power of an advanced weapon has been considered by the West as the most important means of deterrence. From a Chinese point of view, however, the killing power of weapons is not decisive. China's huge human resources, its unbeatable national spirit, and its history of always denying final victory to any foreign invader, are taken as counterthreats to the use of even nuclear weapons. The rationale behind China's own development of nuclear weapons also differs from the West. The atom bomb is not an absolute weapon on which the survival and destruction of mankind hinges. It is only a more useful means that could be used when necessary. Therefore, it is very hard to subdefine China's deterrence theory as nuclear or conventional, because the Chinese have always considered the process in a comprehensive strategic framework, of which nuclear weapons are only one component.

In recent years, however, an increasing number of Chinese strategists began to rethink the interrelating roles of the human factor and weapon factor. The performance of high-tech weapons in the 1991 Gulf War served as an eye-opener to PLA leaders. While holding that human beings played more important roles than weapons in wars of older times, military theorists began to accept that new military technology may have changed the rule in modern wars. It seems that the tide is now turning in military thinking. Increased efforts on the acquisition of new weapons and military application of new technologies are to be expected.

"Stratagem" versus "Strength" (*Mou* and *Li*)

A more striking difference exists in the understanding of what constitutes the most decisive component in emerging as the victor in war. The Chinese tend to depend very much on stratagem (brain power, wisdom) and the Westerners on strength (muscle power, force). This distinction is even more outstanding and consistent than the other two discussed in the above sections.

In summing up the achievements of ancient Chinese military art, which reached its greatest splendor in the Spring and Autumn Period and the Warring States Period, a Chinese history compiled in the Han Dynasty (206 B.C.–A.D. 220) (*Han Shu*) classified military art into four categories: (1) application of stratagem (*quanmou*), (2) assumption of military posture (*xingshi*), (3) manipulation of ying and yang (the positive and negative) (*yingyang*), and (4) the skillful use of forces (*jiqiao*). Among the four, the application of stratagem was deemed the most important, because it "embraces the assumption of military posture, the manipulation of ying and yang as well as the skillful use of forces." The history went on to specify "application of stratagem" as "to defend in an orthodox way and to attack in an unorthodox way: to achieve a goal by using scheme first and using force second."

From the ancient Chinese point of view, war represented more of a confrontation of wisdom than a force between two opposing sides. In the process of a war, the deployment and exploration of armed forces, the development of strategy and tactics, and the flexible application and timely adjustment of these strategies and tactics were very much the business of the human mind. The uncertainty in war mainly resulted from the uncertain responses the human mind made in the chaos of war. The uncertainty, which the ancient Chinese termed as "trick situations" (*guidao*) and considered as a defining characteristic of war, could only be reduced by wisdom as seen in the responsiveness, flexibility, perceptiveness, and vision of the human mind. Therefore, the Chinese military tradition concentrated on the study of the development and application of stratagem, the utmost achievements of which were embodied in Sun Zi's writings. Sun defined the "acme of skill" not as "winning one hundred victories in one hundred battles," but as "subduing the enemy without fighting." Seeing that "as water has no constant form, there are in war no constant conditions," he summed up the basic principle of stratagem application as "modifying tactics in accordance with the enemy situation" and praised those who were able to do so as "divine." In commenting on Sun Zi's writings, one later theorist said, "If one does not have a cunning character, he is unable to use stratagem; and if he is unable to use stratagem, he can by no means subdue his enemy."[17]

In short, the significance of stratagem cannot be overemphasized in Chinese military tradition. One more example is that a great portion of *Wujing Qishu* (Five Classics and Seven Books), a collection of martial classics compiled by later scholars, is devoted to the study of the development and the adoption of stratagem in military activities.

In Chinese history, there are certainly no lack of examples where the weak defeated the strong and the few defeated the many. Victories won by stratagem have always been highly valued. Furthermore, stratagem, loosely defined as brilliant use of wisdom, has been considered the most dynamic and determining instrument that can be utilized to reduce the inherent uncertainty in war ("*guidao*" in Chinese or "fog of war" in Clausewitzan language). The fact that wisdom was such a highly valued attribute is evident in the fact that most great figures in Chinese military legends are not generals in helmets and armored suits who led the troops on the battlefields; rather they are the army counselors who spoke key words into the ears of the commanders-in-chief at the critical moment. The most well-known figure is Zhuge Liang, during the period of the Three Kingdoms (280–220 B.C.), who, with a feather fan in hand, directed his troops in a wheelchair.

Unlike Chinese culture, the adoration of physical power has been built into Western culture from the very beginning of the civilization. While Chinese legends reveal a civilization that admired the saintly virtues of their early rulers, Greek mythology tells of a people who worshiped the amazing and well-defined divine power of the gods and goddesses of Olympia.

The ancient Greek civilization and the whole Mediterranean civilization developed actually as a civilization of city-states. These city-states were formed originally as military fortresses surrounded by marketplaces. Later, fortresses and markets were merged into cities. People in quest of fame or material gain gathered in the cities, and those who possessed the power to fight against social as well as natural odds emerged as rulers of the city-states. From Greek mythology to the rise of city-states, the appreciation of physical strength and the worship of power have been molded into the Western mentality.

Affected by this mentality, the Western answer to the question of what wins a war differs substantially from the Chinese answer. Westerners think that physical strength or the extension of physical strength (weapons) is the most important variable that wins a war. The sword has been a symbol of strength in Western culture. The duel between medieval knights was not a contest of brain power in which schemes and tricks were used as much as possible but a contest of physical strength in which swords and spears had the final say. The knights wore heavy armored suits, weighing fifty pounds or more, and it took much muscle power just to move about in such suits. In an attempt to extend the reach of physical power, the Macedonian soldiers were armed with spears as long as twenty-one feet. To elicit the best results, one Western military principle said that force must be applied continuously and constantly during the whole process of a battle. The best example is the phalanx, a tactical formation developed by the ancient Greeks and perfected by Alexander the Great, which dominated the Western battlefield until the Middle Ages. It was an unbroken linear array of heavily armed infantry standing shoulder to shoulder in files that were normally eight-men deep and sometimes deeper. Archers and slingers were placed in front lines followed by spearmen with shields and swordsmen. The phalanx was to be kept both in battle and in march so that immediate and constant application of force could be ensured.

In calculating military balance, the West tends to take numerical advantage in troops and superiority in weapons as preconditions for winning a war. Armies that have more troops and better equipment enjoy greater strength than those that have not. And, as a rule, the stronger armies will defeat weaker ones by pure force. This has marked a very different approach from that of the Chinese. Western wars have rarely been taken as a competition of wisdom but rather a contest in strength. This tradition has left its impact even on modern Western military thinking.

Likewise, different reflections on what wins in war have their roots in philosophical thinking. First, as far as subject matter is concerned, Chinese philosophy has a tradition of treating "man" as the focus of studies, whose aim is to seek "the meaning of life" or "the way to lead a fulfilling life." Traditional philosophy advocates self-cultivation and teaches people how to improve themselves and become wiser. In a certain sense, Chinese philosophy is a philosophy of man, of human life, and of the human mind. Its introspective nature leads to dependence on human intellect to overcome obstacles and also leans toward a tendency of turning inward to find solutions to any problem. Chinese philosophy might as well be termed a "philosophy of wisdom" as it teaches such dialectical military propositions as "destroying hardness with softness" (*yirou kegang*), "defeating a stronger enemy with a weaker force" (*yiruo shengqiang*), and "advancing by retreating" (*yitui weijin*) to cope with the inherent uncertainty of war. Given the nutrition of such a philosophical culture, there ought to be no question why "stratagem" instead of "strength" is taken as the most determining variable in deciding the outcome of war.

Second, as far as methods of philosophical studies are concerned, the Chinese tradition emphasizes direct observation, perception by senses, introspection, and personal experience. Such a way of thinking tends to be ambiguous and equivocal in defining concepts and establishing categories. However, it is the ambiguity and equivocality that make philosophical concepts flexible and all-inclusive. The "soft thinking" provides fertile soil for the art of military stratagem to flourish. For example, *ying* and *yang*, the negative and the positive, are two opposing philosophical concepts that might be contradictory, complementary, interchangeable, and so on. *Ying* and *yang* may emerge in many intriguing forms according to the ways they are combined with each other or separated from each other. A clever manipulation of *ying* and *yang* may mislead the enemy, for example by making them think you are weak when you are actually strong or vice·versa and to make them expect your defensive actions when you are actually going to take offensive actions or vice versa. Further, *ying* can transform into *yang* and *yang* into *ying*, as expressed in the following well-known military maxims: Courage comes from fright and superiority ends up with inferiority; only when one is pushed into a desperate position can he experience the thrill of survival and only when one is in caught in a death trap can he find a way to life. In short, the all-inclusive *ying* and *yang* may be presented as a series of relationships between orthodox and unorthodox (*qi* and *zheng*), toughness and softness (*gang* and *rou*), superiority and inferiority (*qiang* and *ruo*), offensive and defensive (*gong* and *shou*), and emptiness and fullness (*xiu* and *shi*).

The ambiguity featured in Chinese philosophical conceptualization provides comprehensiveness and flexibility in a philosophical tradition with which Chinese military thinkers enjoy ample freedom to direct their thoughts anywhere and everywhere when pondering upon military stratagem.

The West adopts the method of abstract logical analysis and systematic deduction in its philosophical thinking. In their thinking process, Western philosophers try to define every concept, category, and proposition precisely and explicitly. Every hypothesis has to be exactly defined, logically analyzed, and convincingly proved. Frequently, means of quantitative and qualitative analysis are resorted to in the thinking process. Such a king of "hard thinking" who seeks accuracy, hard truth, and no ambiguity could be tolerated. It is not difficult to trace the link between this way of philosophical thinking to the consideration of "strength" as the most dynamic factor in war, because "strength" is much easier to define and to quantify than "wisdom."

Turning back to today's PLA, we can see it continues to depend very much on stratagem as a means to achieve victory (*moulue zhisheng*). While a part of this dependence can be accounted for by the fact that the PLA had no weapon advantages to rely on in most of its war experience, much of it has to be explained by the dominance of traditional thinking.

Westerners frequently complain about the ambiguity and secrecy surrounding the Chinese PLA—its military doctrine, its deployment, its budget, its weapon research and development program, and its current order of battle. They attribute lack of transparency to the nature of China's social system. It is easy to overlook the role played by deep-rooted philosophical thinking, in which ambiguity is not only a means to achieve an end but also an art to be explored with imagination. Concealment, deception, and secrecy, all salient ingredients of traditional military stratagem, have much more to do with China's traditional culture than with its current social system.

It is not difficult to see that different philosophies lead to different approaches in military thought as well as military practice. Chinese military tradition illustrates its focus on "man" by stressing ethical and moral dimensions in war, valuing the "human factor" more than the "weapon factor" and preferring victory won by wisdom to victory won by brutal force. Western military tradition shows its emphasis on the "material" by focusing on national interest, the weapons factor, and strength in military calculations. It is fair to say that both traditions have their merits and shortcomings, and both are still exerting influence on today's military thinking. Hence cross-cultural studies in military traditions are more than necessary to enhance mutual understanding and to minimize misperceptions.

Notes

1. *Zhou Yi: Xugua* (Divination Book of the Zhou Dynasty: Initial Divination).
2. *Wu Zi: Tuguo* (Quotations of Wu Zi: On State Governing).
3. *Xun Zi: Yibing* (Quotations of Xun Zi: On Force).
4. Mao Zedong, *Mao Zhuwi Yulu* (Chairman Mao's Quotations) (Beijing: 1968), p. 234.
5. *Dangdai Zhongguo* (Contemporary China). The Chinese People's Liberation Army, vol. 1, p. 581.
6. Deng Xiaoping's Meeting with Former U.S. President Richard Nixon in October 1989, *Selected Works of Deng Xiaoping*, vol. III (Beijing: Remnin Chubanshe, 1993), p. 330.
7. *Meng Zi: Gongsun Chou Xia* (Quotations of Mencius: Talk with Gongsun Chou), part III.

8. *Shangjun Shu: Huace* (Quotations of Shang Yang: Giving Advice).
9. *Guan Zi: Ba Yan* (Quotations of Guan Zi: On Gaining the Position of an Overlord).
10. *Hu Ling Jing: Shengbai* (Hu Ling Jing: Victory and Defeat).
11. *Wu Zi: Zhibing* (Quotations of Wu Zi: On Army Management).
12. *Zhou Yi: Shi* (Divination Book of the Zhou Dynasty: The Launching of War).
13. *Dengtan Bijiu: Jian Quan* (Dengtan Bijiu: The Responsibilities of a General).
14. *Guan Zi: Zhongling* (Quotations of Guan Zi: Emphasis on Discipline).
15. *Lushi Chunqui: Lunwei (Lu's Spring and Autumn Annals: On Authority)*.
16. Ibid.
17. *Shiyijia Zhu Sun Zi* (Art of War with Notes and Comments from Eleven Schools).

8

Beijing's Perception of the International System and Foreign Policy Adjustment after the Tiananmen Incident

Suisheng Zhao

In explaining China's foreign policy behavior, scholars have employed a range of analytical and theoretical approaches. With some risk of oversimplification, two broad categories of such approaches may be identified: international system-centered approaches and domestic state-centered approaches. Domestic state-centered approaches emphasize the authoritarian structure of the Communist state and focus on the values, preferences, and objectives of key decision-makers and their factional conflicts or bureaucratic cleavages, explaining foreign policy essentially as an extension of domestic politics. International system-centered approaches explain Chinese foreign policy as a function of the attributes or capability of China relative to other nation-states. In this view, the Chinese government is perceived as responding to the particular set of opportunities and constraints that China's position in the international system creates at any moment in time.

Many Chinese foreign policy specialists have focused on a variety of domestic factors in the search for a fitting explanatory model.[1] The rationale for the domestic state-centered approaches is provided partly by the fact that "the Chinese state is an extremely strong one, and it is also highly articulated,"[2] and partly by the obvious shortcomings of the international system-centered approaches in which the policy-making process is treated as a "black box" and foreign policy is reduced to a predictable reaction based solely on the national interest or on *realpolitik* considerations. Nevertheless, the rationale for domestic state-centered approaches by no means overrules the merits of the international system-centered approaches. Even those scholars who favor the domestic state-centered approaches recognize that, by abstracting from domestic politics and focusing on the relative attributes of China relative to other countries, the international system-centered approaches can be a necessary "first cut" in the analysis of Chinese foreign policy.[3] Given the progress of reform and of a more open-door policy in recent decades, China has increasingly become a part of a larger world environment that provides opportunities for, as well as constraints on, its policy options. Policy-makers in Beijing have become more and more sensitive to China's position in the changing international environment, which Beijing has had only a limited role in shaping. Thus, identifying Beijing's perceptions of the international system and explaining China's policy responses to the "invisible hand" of the international system become increasingly important not only because it is the necessary "first cut" but also because it provides a serious alternative to the analysis of Chinese foreign policy.[4]

This chapter, applying the international system-centered approach, represents a modest attempt to analyze Beijing's perception of the change in the international system from the Cold War bipolar confrontation between the United States and the Soviet Union to a new post–Cold War multipolar world and to examine China's foreign policy adjustment after the Tiananmen incident in the early 1990s. It finds that Beijing's perception of a multipolar system is a matter of normative truth rather than an empirical or analytical assessment, which should surprise no one familiar with the history of Chinese analysis of the world situation. Because the multipolar system is its goal, Beijing "perceives" it. While working hard to encourage a multipolarity, Beijing in fact finds a unipolar reality in the post–Cold War era and has accommodated to it through pragmatic foreign policy adjustment.

Beijing's Perception of the Changing International System

From a neorealist perspective, the international system is "composed of a structure and interacting parts."[5] The structure is an arrangement of its parts, namely nation-states. One of the most important attributes of an international system is its distribution of power across nation-states.[6] Whether or not Beijing's leaders have ever read any neorealist works, they have perceived, since the 1970s, changes in the international system in a neorealist sense of power politics.

Power occupies a central place in Chinese thinking on international politics, although it was covered by ideological language in the Mao Zedong era. In post-Mao China, the power consideration becomes crystal clear. A Xinhua (China's official news agency) article states that, in foreign activities, "China does not define its stand according to the ideologies and social systems of other countries, but entirely on the basis of the merits of the matters themselves."[7] That is, China's stand is defined solely on the basis of enhancing its national interest and international influence.

In light of the power consideration, the dynamics of international politics have been perceived, in Beijing's mind, as a change of power distribution across the world that is structured around several great powers or a few poles. What constitutes a pole? Major General Wang Pufeng, deputy director of the Department of Research on Strategy at the Academy of Military Science of China, explains that "the term 'pole' represents the interests of one party which has the capacity to exert influence on international affairs and has certain control over other world forces." According to the major general, "each country deals with international affairs in accordance with its own interests and exerts its influence."[8]

To act in accordance with its own interests and to exert its influence on international affairs, Beijing has worked hard, since the early 1970s, to find or shape an international system that is in its favor or at least not to its disadvantage. Beijing put forth a hierarchical structure of three worlds in the 1970s. According to this perception, Beijing believed that, in cooperation with developing countries (the Third World), a developed Japan and Western Europe, which were the second world, could be a force to counter the alleged hegemonism of the two superpowers that constituted the first world.[9] China thus could act as a leader of the Third World. After the United States extended diplomatic recognition to Beijing in 1979, while Beijing admitted a Washington-Moscow bipolar system, it was also looking forward very much to a strategic triangle in which China played a global role by maneuvering between the United States and the Soviet Union. Since 1983, a new concept of multipolarity (*duojihua*) has occurred among Beijing's foreign policy makers and schol-ars.[10] This concept, however, did not become a dominant and official perception until the end of the Cold War, which was marked by the reunification of Germany and the disinte-gration of the Soviet Union during 1990–1991.

A review of Beijing's foreign relations literature indicates that it was not until 1990 that Chinese scholars, such as Chen Qimao in Shanghai and Song Yimin in Beijing, became convinced that the Cold War and bipolar system had come to an end and the world had entered a new period of transition toward multipolarization.[11] Although it is not clear when the political leaders in post-Tiananmen China accepted the scholars' view and, whether or not a consensus among these leaders was reached, a review of the speeches of senior Chinese government officials indicates that the official view about the change in the inter-national system was very cautious at first. In an interview in December 1990, Qian Qichen, the foreign minister at the time, said that the world was in a transitional phase. The old order had dissolved but no new one had emerged to take its place.[12] A similar view was expressed by Li Peng in March 1991 when he told the National People's Congress that "the old world structure, which lasted for over four decades, disintegrated and a new one has yet to take shape."[13] After the end of the Gulf War and the breakup of the Soviet Union in 1991, Beijing's view became clear-cut and it asserted that the world was evolving to-ward a multipolar system. In the 1991 year-end assessment of the international situation, Qian Qichen stated, "although the world is in the transitional period and a new pattern has not yet taken shape, there is a rough structure in international relations, in which one superpower and several powers depend on and struggle against each other." To make his point clear, Qian indicated that "this is the initial stage of the evolution towards multipolarization."[14] In a press conference on March 23, 1992, Qian said once again, "The breakup of the old world pattern means the end of the post-war bipolar system character-ized by the hostility between the two superpowers. A new world . . . is likely to be a multipolar pattern."[15]

What would the multipolar world look like in Beijing's mind? In his book, Xi Shuguang, a Chinese policy consultant in the Security Market Research Center of China, systemati-cally discussed the emerging world after the collapse of the Soviet Union. He presented a new structure of one system (the Western capitalist system) and five "geopolitical plates" (European plate, African plate, Middle East and Central Asian plate, Asia-Pacific plate, and North-South American plate). He believed that "unless the life-death interests of the West are in danger, the Western group will not be likely to interfere in various regional issues."[16] Another Chinese scholar perceived two types of multipolarization. One is a three-polar world, in which the European community constitutes the European pole; the United States, Canada, and Mexico may form the North American pole; and Japan is working toward an Asia-Pacific economic rim, leading to the formation of the Asia-Pacific pole. Another is a five-polar structure, which consists of the United Sates, United Germany, Russia, China, and Japan.

Beijing's perception of multipolarity apparently stood against the speculation of West-ern commentators that a unipolar world characterized by U.S. hegemony was emerging from the ashes of the Cold War. Beijing's leaders reacted with great alarm to President George Bush's "New World Order" proclamations during the Gulf War in the early 1990s. A Chinese analyst declared that this was a ruse for extending U.S. hegemony throughout the globe.[17] Underlying Beijing's condemnation of a unipolar world lies a belief that the U.S. global power was declining absolutely because of its troublesome economy at the time. In Qian Qichen's words, "although the U.S. influence in the world had been growing since the Gulf War, its ability was not equal to its ambition because it was meeting many economic difficulties at home and restrained by many factors."[18] Major General Wang Pufeng took the example of the Gulf War and suggested that "Although the United States appears to have won the Gulf War, in fact, it demonstrates a weakening capability of controlling the world. In the past, the United States financially supported foreign troops in fighting wars. This time, however, it sent its troops to the Gulf, backed financially by other countries. This reflects its weak economy."[19] A Chinese analyst summarized that "the United States dreams of a unipolar hegemony, but it cannot afford it economically."[20]

Implications of Multipolarity

What opportunities and constraints was Beijing's leadership responding to when it perceived a multipolar international system in the early 1990s? The constraints imposed by a multipolar world on Beijing's strategic position in the world are transparent. A multipolar world is a world bereft of the much-coveted balancing third force in global triangular geopolitics. After all, it was bipolarity that served as the lodestar of Chinese foreign policy over the Cold War years. It was the bipolar world that enabled Beijing to exploit superpower rivalry as a fulcrum to gain strategic leverage, economic and trade benefits, and global influence. The structural reality of the bipolar system to a great extent explains the puzzle as to how China, as a developing country, managed to exert global influence and how it acted as if it were a global power and was treated as such by the rest of the world, including the two superpowers. China played the Cold War triangle game well and often to its advantage. The end of the Cold War bipolar system, therefore, "left China's leaders without a definition of their place in the world."[21] With the collapse of the old order, Chinese leaders began to feel vulnerable and marginalized in world affairs. These feelings were compounded by the deep paranoia that took hold in the wake of the anti-Communist revolution across Eastern Europe. Beijing's leadership found it difficult to identify quickly a satisfactory niche in a transforming international system. The end of the Cold War, especially the collapse of the Soviet Union and East European Communist regimes, together with China's own immense economic and political problems after the Tiananmen massacre, profoundly frightened Beijing's leadership.[22]

However, multipolarization also provides opportunities for Beijing. A multipolarizing world may bring about a change of world order from superpower contention toward more egalitarian international relations. It may be a world of multiple opportunities for China to assert itself forcefully on multiple chessboards by increasing its trade, peddling arms, attracting foreign aid and investment, and exporting construction workers, all designed to enhance China's domestic modernization and to improve its international relations.

Beijing perceived these opportunities to be in accordance with its prediction that in the process of multipolarization, a new West–West contradiction would arise and the existing North–South as well as South–South contradictions would be intensified. Beijing's leadership believes that U.S. relations with its allies in Europe and Asia would be more difficult to manage in a world where the "Soviet threat" disappeared. According to Chen Xiaogong, a

Chinese analyst, under the previous bipolar system of U.S.-Soviet confrontation, even though the situation was tense, it remained stable. He called it a "strained stability." With the disintegration of the Soviet Union, the West lost its common enemy and other conflicts of interest began to damage the Western alliance. Some problems, held down by the two superpowers' strife for hegemony and by the two blocs' confrontation, surfaced, making the situation tumultuous and capricious. He described this situation as one of "turbulent détente." In Chen's view, West–West contradictions involved economic frictions, political control and anticontrol, and division in defense matters. He provided statistics showing that, in 1990, the European Community's GNP (Gross National Product) already surpassed that of the United States. Japan's GNP was about three-fifths that of the United States. The narrowing gap of strength between the United States and Europe and Japan would sharpen their economic frictions and political differences. While Washington was still able to use defense as its trump card to control Europe and Japan, the latter were becoming more independent in this field. France and Germany had proposed to establish a "European corps" and Japan was transgressing the line that forbade it to send troops abroad. Chen concluded that it was no longer an easy task for the United States, Europe, and Japan to coordinate their relations.[23]

Therefore, a multipolarizing world, from Beijing's vantage point, meant an imbalance of power in Europe and other parts of the world that would produce new conflicts between the United States and its allies, preventing unipolarity from occurring. After all, a U.S.-dominated unipolar world would be the worst of all possible world systems for the Chinese leadership, because the troublesome issues such as human rights and nuclear proliferation had become real problems with which the post-Tiananmen Chinese regime had to deal. As such, a Western scholar indicated that "China would prefer to find itself in a multipolar world in which U.S. global power declines absolutely and regional powers, such as China, are able to resist external interference in their respective region."[24]

In terms of North–South contradictions, Chen Xiaogong held that because the West turned its attention and assistance to Russia and other republics of the former Soviet Union, the South obtained fewer funds and bore a heavier debt burden. North–South contradictions had hence become acute. In addition, South–South contradictions were developing as well in the multipolarization process, since the gap between rich and poor Third World countries was widening. Of the developing countries' GNP, the newly industrialized countries and oil-exporting states made up about 80 percent. Many low-income South Asian and African countries with a heavy debt burden were confronted with the threat of hunger and poverty. Racial, religious, territorial, and resource conflicts between Third World countries were thus on the increase.[25]

In contrast to the contradictions and turbulence that had taken place in the world transiting toward a multipolar system, Beijing believed that China was on a stable and upward trajectory. As Xi Shuguang put it, in the 1990s, major problems would still be concentrated in Europe, with the turmoil in the former Soviet Union and Eastern Europe becoming the focus of Western worries. He suggested that "Since the early 1980s, China has developed good relations with its neighbors. Although the strategic triangle no longer exists, China has already become a power with enormous economic and military potential. It is generally agreed that many world issues cannot be solved without the participation of China."[26] In a similar fashion, Qian Qichen also stated that, as a balancing and stable factor in a turbulent world, China could overcome constraints and explore opportunities.[27]

Foreign Policy Adjustment

How much of the above is "official optimism" and "whistling in the graveyard" for public and elite consumption and how much is the real view of the leadership? It is difficult for anyone to judge without access to highly classified documents circulated among the top leaders in China. However, one way to find an answer is to examine Beijing's foreign policy behavior in the early 1990s. Beijing made a major foreign policy adjustment after the Tiananmen crackdown. At least the following three aspects of the adjustment are relevant to China's perception of changes in the international system. First, Beijing temporarily played down its pretense to being a global power and acted more as a regional power, focusing its policy objectives primarily on seeking better relations with, and greater influence on, neighboring Asia-Pacific countries. Second, Beijing's policy toward the United States and other Western powers became characterized by a combination of ever-ready concessions with an official anti-West sentiment and rhetorical toughness. Third, Beijing took more and more cooperative actions in multilateral initiatives within and outside the United Nations, striving to establish an image as an independent and responsible partner in the community of nations.

Policy toward the Asia-Pacific Region

Prior to Tiananmen, Beijing seemed on the verge of an era of unprecedented tranquillity as a global power; as such, Gorbachev's summit visit to Beijing in 1989 attracted the attention of the whole world. After the Tiananmen crackdown and the disintegration of the Soviet Union, however, Beijing's international status declined drastically and the elderly Chinese leaders perceived China as under siege, the potential victims of a massive external effort to subvert their rule and to bring about the "peaceful evolution" (*heping yianbian*) of their system. These international situations forced Beijing to modify its foreign policy pretensions of being a global power and attend to its primary interests as a regional power.[28]

This adjustment came when Beijing found its vital interests increasingly tied to the Asia-Pacific region. In Beijing's calculation, China, with a population of 1.1 billion and as a permanent member of the UN Security Council, would eventually carry greater weight in future international relations. Nevertheless, in a global environment that was considered by Chinese officials as hostile to China as the only remaining major power holding to Marxist socialism, China had to build a regional base to maintain its leverage. The Asia-Pacific region fitted this need particularly well. Beijing well understood that China's influence and impact were more appreciated in this region than anywhere else. While some countries in this region might not trust China, neither did they want to see a power vacuum open to possible Japanese hegemony.[29] In addition, as Qian Qichen indicated, "The Asian-Pacific region is comparatively stable and East Asia is the most dynamic region in the world economy."[30] In response to the worldwide trend toward forming regional economic and trade blocs, if Beijing exerted great influence on major players in this region and skillfully maneuvered the regional cooperative efforts such as the Asian-Pacific Economic Cooperation (APEC), China, it is believed, could gain the cooperation of Japan and other countries and eventually become the leader of this dynamic region.[31]

In light of these calculations, "to establish good neighborly relations with neighboring countries" became a basic objective of China's foreign policy in the 1990s.[32] Beijing made some important diplomatic breakthroughs in developing relations with its neighbors in the Asia-Pacific region at both multilateral and bilateral levels. At the multilateral level, relations between China and the Association of the Southeast Asian Nations (ASEAN) made great progress. While China had no diplomatic relations with three of the six ASEAN nations before Tiananmen, it had established formal diplomatic relations with all the members of ASEAN by 1991 and received invitations to attend the opening ceremonies of ASEAN foreign ministers' meetings and to hold talks with foreign ministers of the six countries in 1992.[33] In addition, trying not to be left out or left behind, Beijing developed a more flexible policy toward Asia-Pacific multilateral cooperation. Beijing accepted the proposal that China as a sovereign state, and Taiwan and Hong Kong as separate economic regions, could participate individually in APEC and chose to join at its ministerial conference in Seoul in 1991. Beijing also took initiative in subregional cooperation, including what it called a "Northeast Asian Economic Zone," composed of China (its three northeast provinces, Inner Mongolia, Hebei, and Shangdong), Russia (Far Eastern part), the Korean Peninsula, and Japan; a "South China Economic Zone," involving China's southern provinces, Hong Kong, Macao, and Taiwan; and a "Southeast Asian Economic Zone," comprising ASEAN and Indochina, with cooperation from China and Japan.[34]

At the bilateral level, immediately following the breakup of the Soviet Union, a Chinese government delegation was sent to visit Russia, with which China shared a 4,000 kilometer border. The two sides signed a summary of their talks, confirming that the principles laid out in the two Sino-Soviet joint communiqués would guide Sino-Russian relations. In the meantime, China quickly recognized the other eleven former Soviet republics when they declared independence. Japanese Prime Minister Toshiki Kaifu's visit to China in 1991 made Japan the first among industrial countries to restore normal relations with China after Tiananmen. In April 1992, with the twentieth anniversary of the normalization of Sino-Japanese relations, the visit of General Secretary Jiang Zemin of the Chinese Communist Party to Japan further strengthened Beijing-Tokyo cooperation. Moreover, in 1991–1992, Chinese President Yang Shangkun visited the People's Republic of Mongolia, becoming the first Chinese president to visit that country. China and South Korea, soon after the exchange of trade offices, established formal diplomatic relations. The signing of a boundary agreement with Laos put an end to the long unsettled border issue. General Secretary Do Muoi of the Vietnamese Communist Party and Chairman Vo Van Kiet of the Council of Ministers of Viet Nam visited China, signaling that the ten-year-old hostilities between the two countries had faded away and that state-to-state relations were normalized. And Beijing's premier Li Peng visited India, reinforcing the relationship between the two largest Asian nations.

Policy toward the West

Immediately after Tiananmen, while the triangle relationship had already changed in such a way that China's room to maneuver between the two superpowers significantly diminished, some leverage remained. A warmer relationship with the Soviet Union, for example, would to a certain extent have offset pressure from the United States. The visits of Premier Li Peng and General Secretary Jiang Zemin to the Soviet Union respectively in April 1990 and May 1991 certainly implied this strategic consideration. Nevertheless, with the breakup of the Soviet Union at the end of 1991, this leverage finally vanished. For the United States, Beijing was no longer a counterweight to the Soviet Union, and to this extent, China became strategically less important to the West. Nevertheless, maintaining and developing a good relationship with the West, especially with the United States, was still crucially important to China's long-term interests. China had to rely upon U.S. help for its economic and technological modernization drive. In addition, the relationship with the United States was not only critical for China's national reunification plan but also important to offer Beijing leverage in its relationship with Japan in the Asia-Pacific region. In light of these calculations, Beijing made a number of compromises with the United States for its long-term interests after Tiananmen, even though officially it reasserted firm rhetoric in order for the old leaders to save "face."

The discrepancy between Beijing's policy pronouncements and its policy performance on many issues concerning Sino-West relations was striking. On the official level, China never admitted that it yielded to any Western pressures after the Tiananmen crackdown. It accused America of intervening in its domestic affairs and pretended to be a victim of Western sanctions. Deng Xiaoping told his Japanese visitors that, "China is not afraid of sanctions, which in the long run will backfire at those imposing them."[35] On the policy level, however, Beijing took pragmatic steps to ameliorate America's anger and made every effort to maintain or rebuild its relations with the West although very much in its own way. Beijing made concessions on all three issues that the United States raised: human rights, nuclear proliferation and conventional arms sale, and the trade deficit.

For example, Beijing made concessions in reaction to Western, especially to United States, criticism of its human rights record.[36] At the time of the Tiananmen massacre, Beijing stubbornly insisted that the incident was nobody else's business; however, it eventually became willing to listen to complaints from the West and ultimately made concessions. In June 1990, it allowed prominent dissident Fang Lizhi, who had taken refuge in the U.S. embassy in Beijing after Tiananmen, to leave China. Most prominent political prisoners in Beijing jailed after Tiananmen were released as well. In October 1991 and August 1992, China for the first time issued a White Paper on human rights and a White Paper on prison conditions, which presented a systematic defense of the regime's human rights policy instead of simply refuting Western criticisms.

Nuclear proliferation was another issue on which Beijing made concessions. Prior to 1991, Western intelligence had repeatedly reported that Beijing had transferred nuclear technology to such countries as Algeria, Iran, Pakistan, and North Korea. Beijing's refusal to participate in a multinational accord to prevent nuclear proliferation seemed to confirm this speculation.[37] In order to avoid upsetting the United States and other Western countries, in 1991, Beijing assured James Baker, the U.S. secretary of state, that China would abide by international agreements banning nuclear proliferation and the export of missile technology. At the end of 1991, the National People's Congress (NPC) standing committee adopted a resolution on signing the nuclear Nonproliferation Treaty, and Beijing finally joined the list of 140 other signatories.

On the trade issue, China wanted to placate the U.S. Congress's anger over its huge $11 billion trade deficit in 1991 and the export of prison labor products. China also wanted U.S. support of its effort at entering the General Agreement on Tariffs and Trade (GATT) and to maintain Most Favored Nation (MFN) status-vital to the future of China's as well as Hong Kong's economy. To achieve these objectives, China sent a large delegation of more than a hundred people in 1991 to the United States to purchase more than $700 million in American goods and again, in July 1992, a delegation to make deals with all three automakers in Detroit. In 1991 and once again in June 1992, China also sent delegations to Western Europe and purchased more than $1.6 billion in European goods.[38]

Policy toward UN Activities

After Tiananmen, Beijing's participation in UN activities became progressively expansive and its orientation cooperative. China acted more and more within the UN charter system rather than trying to reform or transform the system. Beijing understood well that the United Nations, as a multinational organization facilitating cooperative interaction among sovereign states, would play a more important role in a multipolar world than in the old bipolar one.

As a result, Beijing made some important breakthroughs to accommodate UN activities. For example, Beijing cooperated with the United States in UN-Paris negotiations on the settlement of the Cambodian issue, and on April 16, 1992, China for the first time sent forty-seven military observers and four hundred military engineers to join the UN peace-keeping forces in Cambodia.[39] Beijing, also for the first time, hosted the Forty-eighth Session of the UN Economic and Social Commission for the Asian-Pacific on April 14–16, 1992, forty-five years after its establishment in Shanghai in 1947.[40]

Through participating in UN activities, Beijing attempted to establish an image as an independent but responsible member of the community of nations rather than an oppositional force. For instance, during the Gulf crisis in 1990, Beijing, as a member of the UN Security Council, skillfully preserved its independent status and also satisfied the United States and its allies by abstaining from rather than opposing the UN resolution on Iraq. When the Gulf War broke out, Beijing criticized both Iraq's invasion of Kuwait and the war to expel Iraq from Kuwait. After the war, Beijing announced that, as the Iraqi people were innocent, it was necessary to lift restrictions on Iraq's import of food and daily necessities out of humanitarian consideration and, at the same time, Beijing also wanted to see continued Iraq cooperation with the UN Security Council in implementing its resolutions.[41] Another example involved the Libya issue. Beijing voted for UN Resolution 731 in favor of making a thorough, fair, and objective investigation into the Lockerbie aircraft explosion in accordance with the UN charter and international law. In the meantime, Beijing also made it clear that it was against the UN Security Council applying sanctions against Libya, because, as Beijing's foreign minister said, "it will not contribute to the settlement of the problem, but will only intensify the situation in this region and entail serious consequences."[42] In addition, Beijing endorsed the call for the establishment of "a new global partnership" and signed "the Rio Declaration on Environment and Development" which was signed by most UN members (although not by the United States) at the UN Conference on Environment and Development in June 1992.[43]

Conclusion

Judging by China's cooperative stance on the issues discussed here, it is apparent that Beijing's foreign policy adjustment in the early 1990s was not only seemingly contradictory but also did not necessarily follow from the official perception of a multipolar system. While Asian-focused, Beijing sought global power status by its participation in the Middle East Peace Talks, and its role in the United Nations, where with its veto power China could extract strategic concessions from the United States and other major powers. The fact that Beijing adjusted its policy priority to the Asia-Pacific region and took cooperative actions in international activities may be interpreted as China's understanding of the need for a regional base and the importance of their role as an independent but responsible partner in a multipolar world. Its concessions to the United States reflected Beijing's perception of a unipolar reality and its lack of confidence in the emerging multipolarization. From this perspective, it may be argued that because the multipolar system was its goal, Beijing "perceived" it. Beijing in fact perceived a unipolar reality after the Tiananmen incident and accommodated to it while also working hard to keep open all options and to encourage multipolarity.

Notes

1. For one example, see David Bachman, "Domestic Sources of Chinese Foreign Policy," in Samuel S. Kim, ed., *China and the World* (Boulder, CO: Westview Press, 1989), pp. 31–54; also see Kenneth Lieberthal, "Domestic Politics and Foreign Policy," in Harry Harding, ed., *China's Foreign Relations in the 1980s* (New Haven, CT: Yale University Press, 1984), pp. 43–79.

2. Bachman, "Domestic Sources of Chinese Foreign Policy," p. 33.

3. In comparative foreign policy studies, some scholars even go further, arguing that systemic theories deserve a certain primary status over other levels of analysis. See, for example, Robert Keohane "The World Political Economy and the Crisis of Embodied Liberalism," in John H. Goldthorpe, ed., *Order and Conflict in Contemporary Capitalism* (Oxford, UK: Clarendon Press, 1984), p. 16.

4. Pat McGowan and Charles W. Kegley, Jr., eds., "Foreign Policy and the Modern World-System," vol. 8, *Sage International Yearbook of Foreign Policy Studies* (Beverly Hills, CA: Sage, 1988), p. 9.

5. Kenneth N. Waltz, "Political Structure," in Robert Keohane, ed., *Neorealism and Its Critics* (New York: Columbia University Press, 1986), p. 71.

6. Ibid., p. 92.

7. *Beijing Review* 35, no. 19, May 11–17, 1992, p. 16.

8. *Conmilit*, September 1991. Reprinted in *Beijing Review*, 34, no. 47, November 25–December 1, 1991, p. 11.

9. "Chairman Mao's Theory of the Differentiation of the Three Worlds is a Major Contribution to Marxism-Leninism," in *Beijing Review*, 45, November 4, 1977, pp. 29–33.

10. Samuel Kim, "Mainland China and a New World Order," in *Issues and Studies* 27, no. 11 (November 1991): 29.

11. See, for example, Chen Qimao, "A Study of the Development of the World System from Bipolarity to Multipolarity," in *Guojiwenti Yanjiu* (The Studies of International Issues), 1990, no. 4: 1–6; Song Yimin, "The New Problems Associated with the Changing World Situation," in *Liaowang Zhoukan* (Outlook Weekly), 1990, no. 36, pp. 39–42; Wang Guang, "What New World Order Does the West Want?" in *Liaowang Zhoukan*, 1990, no. 37: 38–43; and Huang Dingwi and Wang Yulin, "The Rapidly Changing World Situation," *Xiandai Guoji Guanxi* (Contemporary International Relations), 1990, no. 3: 1–8.

12. *Renmin Ribao, Haiwaiban* (People's Daily, Overseas Edition), December 17, 1990.

13. *Beijing Review* 34, no. 15, April 15–21, 1991, Document XX.

14. "Adhering to Independent Foreign Policy," in *Beijing Review* 34, no. 52, December 30, 1991–January 5, 1992, pp. 7, 10.

15. *Beijing Review* 35, no. 14, April 6–12, 1992, p. 15.

16. Xi Shuguang, *Shijie Xinjiegou* (The New Structure of the World), Chendu: Sichuan Renmin Chuban She, 1992. Quoted from Min Chen's book review, in *The Journal of Contemporary China* 2, no. 1 (Winter/Spring 1993): 97.

17. Ba Benwang, "Haiwan zhanzheng dui shijie zhanlue xingshi de yingxiang" (Impact of the Gulf War on the World Military Strategy and Situation), *Guoji Zhanlue Yanjiu* (Research on International Strategy), no. 1 (January 1991): 17.

18. *Beijing Review* 34, no. 52, December 30, 1991–January 5, 1992, p. 7.

19. *Beijing Review* 34, no. 47, November 25–December 1, 1991, p. 11.

20. *Beijing Review* 35, no. 5–6, February 3–16, 1992, p. 15.

21. Michel Oksenberg, "The China Problem," in *Foreign Affairs* 70, no. 3 (1991): 9.

22. For one discussion of the collapse of communism in east Europe and the Soviet Union, see Michael B. Yahuda, "Chinese Foreign Policy and the Collapse of Communism," in *SAIS Review* 12, no. 1 (Winter/Summer 1992): 125–137.

23. Chen Xiaogong, "The World in Transition," in *Beijing Review* 35, no. 5–6, February 3–16. 1992, pp. 13–14.

24. David Shambaugh, China's Security Policy in the Post–Cold War Era," in *Survival* (Summer 1992): 92.

25. Chen Xiaogong, "The World in Transition," p. 14.

26. Xi Shuguang, *Shijie Xinjiegou.* Quoted from *The Journal of Contemporary China* 2, no. 1 (Winter/Spring 1993): 98–99.

27. *Beijing Review* 34, no. 52, December 30, 1991–January 5, 1992, p. 8.

28. Robert Scalapino, "The United States and Asia: Future Prospects," in *Foreign Affairs* 70, no. 5 (Winter 1991/92): 34.

29. Lin Zimin "'Walking on Two Legs': The Domestic Context of China's Policy toward the Asian-Pacific Region," in *China Report* 3, no. 2 (April 1992): 11.

30. *Beijing Review* 35, no. 14, April 6–12, 1992, p. 14.

31. This impression is from the author's interviews with a number of Chinese foreign policy scholars in July 1992.

32. Xin Hua, "China's Successful Diplomacy," in *Beijing Review* 35, no. 19, May 11–17, 1992, p. 12.

33. *Beijing Review* 35, no. 14, April 6–12, 1992, p. 15.

34. Liu Jiangyong, "Sino-Japanese Cooperation in a Changed Situation," in *Beijing Review* 35, no. 14, April 6–12, 1992, p. 17.

35. *Beijing Review* 32, no. 40, October 2–8, 1989, p. 5.

36. Pi Ying-hsien, "Beijing's Foreign Relations in the New International Situation," in *Issues and Studies* 28, no. 5 (May 1992).

37. *The Christian Science Monitor*, November 22, 1991, p. 18.

38. *Renmin Ribao*, June 23, 1992.

39. *Beijing Review* 35, no. 17, April 27–May 3, 1993, p. 11.

40. Ibid., p. 12.

41. *Beijing Review* 35, no. 14, April 6–12, 1992, p. 14.

42. Ibid.

43. *Beijing Review* 35, no. 24, June 15–21, 1992, p. 13.

9

China's Pragmatic Approach to Nonproliferation Policies in the Post–Cold War Era

Jing-dong Yuan

China's nonproliferation policies have undergone significant changes since the 1990s. A clear indication of these changes is reflected in Beijing's accession to major international arms control and nonproliferation treaties, bilateral nonproliferation commitments given to the United States pledging adherence to the guidelines of the Missile Technology Control Regime (MTCR), and new domestic regulations governing exports of nuclear, chemical, and dual-use materials and technologies. During the 1990s, China acceded to the Treaty on the Nonproliferation of Nuclear Weapons (NPT), signed and ratified the Chemical Weapons Convention (CWC), and signed the Comprehensive Test Ban Treaty (CTBT). In short, there was a significant transformation of China's perspectives and policy, changing from accusation and suspicion in the 1970s and 1980s to more active participation and guarded endorsement of the international norms in arms control, disarmament, and nonproliferation in the 1990s. These developments have been prompted by Beijing's growing recognition of proliferation threats, an acute concern over its international image, its assessment of how progress in nonproliferation could promote better Sino-U.S. bilateral relations, and by U.S. nonproliferation initiatives aimed at influencing Chinese behavior.

Needless to say, significant problems remain and continue to haunt Sino-U.S. relations. Beijing has different perspectives on arms control and nonproliferation and tends to interpret its commitments narrowly. There are continuing controversies over Chinese transfers of nuclear, chemical, and missile components and technologies to countries of proliferation concern. Beijing is also increasingly linking fulfillment of its nonproliferation commitments to changes in U.S. policies in arms sales to Taiwan and missile defense. This gap between Beijing's policy declarations and its actual practices has presented successive U.S. administrations with serious challenges.

This article discusses the evolution of Chinese nonproliferation policies over the last ten years. It documents China's changing perspectives on nonproliferation during this period and discusses the rationales behind the policy changes. It examines continuing controversies over Chinese transfers of nuclear, chemical, and missile components and technologies to countries of proliferation concern and explains the gap between Beijing's policy declarations and its actual practices. Finally, the article discusses the U.S. role in seeking to influence Chinese policy and the effects of unresolved disputes on bilateral relations.

China and Nonproliferation: Positive Developments

In the 1980s, China emerged as one of the leading suppliers of arms and dual-use technologies. Toward the end of the decade, revelations of Chinese nuclear and missile transfers to countries in the Middle East, the Persian Gulf, and South Asia raised serious proliferation concerns and were a contributing factor in the "China threat" debate in the United States.[1] Among the controversial Chinese arms transfers were the sale of the Dong Feng 3 (CSS-2) intermediate-range ballistic missiles to Saudi Arabia, HY-2 ("Silkworm") antiship missiles to Iran, the nuclear reactor deal with Algeria, and missile transfers to Pakistan.[2]

Since the end of the Cold War, Beijing has made gradual yet significant progress in its nonproliferation policies, specifically in three key areas:

- accession to major international arms control and nonproliferation treaties and conventions;[3]
- bilateral arrangements with the United States pledging Chinese commitment to missile nonproliferation; and
- promulgation of domestic export control regulations.[4]

An important indicator of China's acceptance of international nonproliferation norms can be found in its participation in major international treaties and conventions (see Table 9.1). Since the early 1990s China has joined the NPT (1992), signed (1993) and ratified (1997) the CWC, and signed the CTBT (1996). Beijing has on various occasions enunciated in clear terms the three principles governing its nuclear exports: (1) acceptance of IAEA safeguards; (2) peaceful use; and (3) no retransfers to a third country without China's prior consent. In May 1996, the Chinese government further pledged not to provide assistance to unsafeguarded nuclear facilities.[5] In October 1997, China formally joined the Zangger Committee.[6]

China has also undertaken specific steps to dispel concerns over its past proliferation activities and to promote its image as a responsible member of the international community. China had been known for years as a supplier of nuclear items and technology to Third World countries without proper safeguard requirements. For instance, China reportedly sold unsafeguarded nuclear reactors to Algeria, Pakistan, and Iran.[7] Since joining the NPT in 1992, China has supported the treaty's indefinite extension (1995) and actively participated in succeeding NPT Review Conferences. Beijing also played a positive role in defusing the North Korean nuclear crisis and facilitating the conclusion of the October 1994 Agreed Framework between the Democratic People's Republic of Korea (DPRK) and the United States.[8] After the South Asian nuclear tests of May 1998, China, exercising its authority as the rotating chairman of the UN Security Council at the time, actively sought and coordinated the P-5 consultation leading to the adoption of the United Nations Security Council Resolution (UNSCR) 1172 that condemned the tests and demanded that India and Pakistan unconditionally sign the CTBT and NPT and refrain from weaponization. During President Bill Clinton's visit to China in June 1998, China and the United States issued a joint communiqué on South Asia that called on India and Pakistan to "stop all further nuclear tests and adhere immediately and unconditionally to the CTBT, . . . and to enter into firm commitment not to weaponize or deploy nuclear weapons or the missiles capable of deliv-

Table 9.1

China and International Nonproliferation Regimes

International Treaties and Negotiations	Multilateral Export Control Regimes
Acceded to the Non-Proliferation Treaty (NPT), March 1992	Pledged to abide by the original 1987 Missile Technology Control Regime (MTCR) guidelines in February 1992.
Signed the Chemical Weapons Convention (CWC), January 1993; ratified CWC and joined the Organization for the Prohibition of Chemical Weapons (OPCW) as a founding member, April 1997	Agreed in the October 1994 U.S.-China joint statement to adhere to the MTCR and agreed to apply the concept of "inherent capability" to its missile exports.
Participated in the UN Register of Conventional Arms from 1993 to 1997	Officially joined the Zangger Committee, October 1997.
Indicated in the U.S.-China joint statement of October 1994 support of the negotiation and "earliest possible achievement" of a Fissile Material Cut-Off Treaty (FMCT)	Promulgated the Regulations on Nuclear Export Control in September 1997, and the Regulations on Export Control of Dual-Use Nuclear Goods and Related Technologies in June 1998.
Supported the indefinite extension of the NPT, May 1995	Announced a series of decrees and circulars governing chemical exports: Circular on Strengthened Chemical Export Controls (August 1997), Decree No.1 of the State Petroleum and Chemical Industry Administration (June 1998).
Signed the Comprehensive Test Ban Treaty (CTBT), September 1996	Issued the Regulations on Export Control of Military Items in October 1997.
Signed the Additional Protocol to its Safeguards Agreement with International Atomic Energy Agency (IAEA) ('93+2') in 1998; domestic legal procedures for entry of force of the Protocol completed in 2002 (however, China has yet to adopt IAEA full-scope safeguards)	U.S.-China official talks during 1997–1998 on China's possible membership in the MTCR; issued statement in November 2000 pledging not to transfer missile-related technologies.

Sources: Adapted from database compiled by the East Asia Nonproliferation Program, Center for Nonproliferation Studies, http://www.nti.org/db/china/index.html.

ering them." The posttest situation thus provided a unique opportunity for China to present itself as a major, responsible power on the world stage, protecting the integrity of the international nonproliferation regime.

Beijing's endorsement of global nonproliferation treaties and conventions has been paralleled by bilateral agreements and understanding with the United States to address the latter's proliferation concerns. While remaining outside of the MTCR, Beijing has made a number of pledges committing it to abide by the regime's original 1987 guidelines in missile transfers. In November 1991, China gave its verbal pledge to adhere to the MTCR Guidelines and Parameters, which was reaffirmed in a classified February 1992 letter to the Bush administration. In October 1994 China announced in a joint statement with the United States that it would "not export ground-to-ground missiles featuring the primary parameters of the Missile Technology Control Regime (MTCR) . . . inherently capable of reaching a range of at least 300 km with a payload of at least 500 kg."[9] Prior to the May 1999 NATO bombing of the Chinese embassy, Beijing reportedly had been actively considering MTCR membership. In addition, China promised in a statement in November 2000 that it would not assist states in developing "ballistic missiles that can be used to deliver nuclear weapons" and that it would issue "at an early date" a "comprehensive" list of missile-related and dual-use items that would require government licenses for export.[10]

A third significant development in China's evolution toward international nonproliferation norms over the last decade has been the introduction of domestic export control regulations (see Table 9.2).[11] Beginning with the May 1994 Foreign Trade Law, the Chinese government has issued a series of regulations, decrees, and circulars. In August 2002, China finally issued the long-awaited Regulations on Export Control of Missiles and Missile-related Items and Technologies and the Control List. In October, Beijing further issued regulations and controls lists governing the exports of chemical and biological materials that could be used in weapons of mass destruction (WMD) development. In addition, the Chinese government amended in the same month the regulations governing military product exports. Taken together, these regulations constitute a nascent export control system. In addition, arms control and nonproliferation is increasingly assuming a higher profile in the making of China's national security policy. In April 1997, a new Department of Arms Control and Disarmament was established within the Ministry of Foreign Affairs (MFA), and there has been increasing coordination among MFA, MOFTEC (Ministry of Foreign Trade and Economic Cooperation), and CAEA (China Atomic Energy Agency) officials in implementing export control regulations. In conventional arms trade, the Commission on Science, Technology, and Industry for National Defense (COSTIND) and the People's Liberation Army General Armament Department take a leading role.[12]

The evolution of China's nonproliferation policies over the past decade is significant in several respects. First is a clear change of China's position on nonproliferation from the earlier one of dismissal to one of selected support.[13] It should be noted that in the 1950s and 1960s, China was pursuing a nuclear weapons program of its own and was in effect calling for the proliferation of nuclear weapons to other socialist and anti-imperialist/anticolonial countries.[14] However, once China secured a place in the "nuclear club" and as its confidence grew with increasing nuclear capability, its attitude regarding the proliferation of nuclear weapons began to change. China's endorsement of the NPT's indefinite extension, its positive contribution to the defusing of the North Korean nuclear crisis, and its response to the May 1998 nuclear tests in South Asia are testimony to this change. While there remains a significant gap between Chinese nonproliferation policy pronouncements and its proliferation activities (see below), in overall terms, China may be seen as moving along the nonproliferation "learning curve" toward adopting certain norms shared by the international community.[15]

Another important indicator can be found in the serious efforts Beijing has undertaken to integrate itself into the formal international nonproliferation regime through accession to key

Table 9.2

Evolution of China's Export Control System

Sectors	Laws and Regulations
General	• Foreign Trade Law, 1994
Chemical & Dual-Use	• Regulations on Chemical Export Controls, December 1995 • Supplement to the December 1995 regulations, March 1997 • A ministerial circular (executive decree) on strengthening chemical export controls, August 1997 • Decree No.1 of the State Petroleum and Chemical Industry Administration (regarding chemical export controls), June 1998 (Note: These regulations have expanded the coverage of China's chemical export controls to include dual-use chemicals covered by the Australia Group.) • Measures on Export Control of Certain Chemicals and Related Equipment and Technologies and Certain Chemicals and Related Equipment and Technologies Export Control List, issued on October 19, 2002
Biological & Dual-Use	• Regulations of the People's Republic of China on Export Control of Dual-Use Biological Agents and Related Equipment and Technologies and Dual-Use Biological Agents and Related Equipment and Technologies Export Control List, issued October 14, 2002
Nuclear & Dual-Use	• Circular on Strict Implementation of China's Nuclear Export Policy, May 1997 • Regulations on Nuclear Export Control, September 1997 (Note: The control list included in the 1997 regulations is identical to that used by the Nuclear Suppliers Group, to which China is not a member) • Regulations on Export Control of Dual-Use Nuclear Goods and Related Technologies, June 1998 • Nuclear export control list as amended, June 28, 2001
Military & Dual-Use	• Regulations on Export Control of Military Items, October 1997 • The Procedures for the Management of Restricted Technology Export, November 1998 (Note: The new regulations cover 183 dual-use technologies, including some on the Wassenaar Arrangement's "core list" of dual-use technologies) • China's Ministry of Foreign Trade and Economics Cooperation (MOFTEC) released a Catalogue of Technologies Which Are Restricted or Banned in China, presumably also in late 1998 • Decision of the State Council and the Central Military Commission on Amending the PRC Regulations on Control of Military Products Export, issued on October 15, 2002
Missile Systems & Components	• Chinese government's verbal assurance of its intention to adhere to MTCR, November 1991, followed by written commitment, February 1992 • U.S. and Chinese governments' joint statement on missile proliferation, October 1994; Beijing's agreement to ban all MTCR-class missiles and to the "inherent capability" principle in defining MTCR-class missile systems • The Chinese government's statement in November 2000 promising for the first time to promulgate missile export control regulations and to issue a control list • China's announcement of the promulgation of the Regulations on Export Control of Missiles and Missile-related Items and Technologies and the Control List, August 2002

Sources: Adapted from database compiled by the East Asia Nonproliferation Program, Center for Nonproliferation Studies, http://www.nti.org/db/china/excon.htm.

treaties and conventions and by active participation in multilateral negotiations, in particular in such form as the Conference on Disarmament in Geneva. By adhering to international treaties and conventions, Beijing not only demonstrates its commitment to nonproliferation principles, it has also placed itself, to some extent, under international legal constraints. This is clearly reflected in the signing of the CTBT, which limits its nuclear weapons modernization programs, and the ratification of the CWC, which introduces intrusive verification provisions. Finally, China's promulgation of domestic export control regulations further indicates a conscientious effort to adapt to internationally accepted standards and practices. While a functional domestic export control system still remains on the drawing board, the fact that such steps have been undertaken reflects both attitudinal change and commitments in resources and personnel on the one hand, and coordination within government, on the other.

There is no single explanation, but several factors have influenced the evolution of Chinese nonproliferation policies. There are security, image, and bilateral considerations. China has gradually begun to realize that proliferation of WMD and delivery systems can affect its own security interests negatively. In this regard, the existing international nonproliferation regimes such as the NPT offers tangible benefits for China, not the least of which would be the prohibition of Japan, the Koreas, and Taiwan to acquire nuclear weapons. At the same time, while the CTBT imposes constraints on China's own nuclear weapons modernization programs, Beijing is willing to pay the price if such mechanisms would prevent countries such as India from joining the nuclear club. China's response to North Korea's nuclear program provides an illustration. While Beijing insisted on alternative measures other than sanctions as the more practical means of dealing with the issue, it understandably shared similar concerns with the other regional powers—South Korea, Japan, and the United States—and assisted in averting a nuclear crisis. A nuclear North Korea and the potential fallout, which could include the nuclearization of Northeast Asia with South Korea and Japan following suit, are definitely not in China's interest. Similarly, a North Korea that continues to develop its ballistic missiles could also cause instability in the region, leading to reactions such as theater missile defense and Japanese participation in its development and deployment. These security concerns may explain Beijing's role in defusing the nuclear crisis and its quiet efforts to urge Pyongyang to halt its missile test.[16]

China's international image is another factor. This must be understood in the broader context of China's own perception of its place in the international community. All considered, Beijing does not want to be seen as an outcast or impediment to international nonproliferation efforts. Events in the late 1980s and early 1990s created an environment under which China felt obliged to move closer to the international nuclear nonproliferation norms. The revelations of Iraq's secret nuclear weapons program, the disclosure of China's export of a nuclear reactor to Algeria, and France's announcement to accede to the NPT helped push China into announcing its own accession to the NPT.[17] China's endorsement of the NPT extension and abandonment of delaying tactics in the final days of the CTBT negotiations also provide evidence of its concern with its image as a responsible power.

Third, China's policy changes to a certain degree have also been influenced by its bilateral relationship with the United States. Several elements can be identified. The first reflects an asymmetrical economic interdependence where Beijing's need for advanced U.S. technologies obliges it to undertake the necessary policy adjustments required by Washington. One example is Sino-U.S. nuclear cooperation during both the negotiation and implementation of the bilateral NCA (Nuclear Cooperation Agreement). Beijing applied for membership and later joined the International Atomic Energy Agency (IAEA) in early 1984. Subsequently, it declared that it would apply IAEA safeguards to all of its nuclear exports and announced its so-called three principles. As Michael Brenner has suggested, "The negotiation with the United States of a nuclear cooperation agreement proved to be the vehicle through which the PRC came to terms with the wider implications of its growing, and seemingly unrestrained, program of nuclear commerce. China had to move up a steep learning curve on proliferation matters."[18] It was only after the Clinton administration was able to certify that China had complied with U.S. nonproliferation legislation that the 1985 U.S.-China agreement on nuclear cooperation finally went into effect in March 1998.

Another element is the growing tendency on the part of the Chinese to view its arms control and nonproliferation cooperation with the United States purely through the prism of bilateral relations.[19] For example, important progress in Chinese nonproliferation policy, such as the formal pledge to not provide assistance to Pakistan's missile programs and the announcement of official, new export control regulations on dual-use nuclear transfers, was achieved during the Clinton-Jiang summits in 1997–1998 when bilateral relations were relatively stable and improving. However, when bilateral relations sour, as happened in early 1999 with the release of the Cox Report, U.S. allegations of Chinese campaign contributions, NATO/U.S. bombing of the Chinese embassy in Belgrade, and the controversial Wen-ho Lee case, Beijing has been less cooperative in arms control and nonproliferation. Indeed, China immediately cancelled its security and arms control dialogues with the United States after the May 1999 embassy bombing.

Continuing Concerns and Controversies

Over the past decade, in particular since the mid-1990s, Chinese proliferation activities have narrowed in terms of both their scope and character. In the 1980s and early 1990s, China was involved in numerous controversial arms transfers to countries in the Middle East, the Persian Gulf, and South Asia. These included nuclear weapons designs, exports and assistance to unsafeguarded nuclear facilities, and sales of complete missile systems, including HY-2 ("Silkworm") surface-to-ship missiles to Iran, Dong Feng 3 (CSS-2) intermediate-range ballistic missiles to Saudi Arabia, and 34 M-11 short-range ballistic missiles to Pakistan. In the latter half of the 1990s, Chinese transfers have moved away from complete missile systems to exports of largely dual-use nuclear, chemical, and missile technologies. At the same time, the number of recipient countries has also declined significantly.[20]

Despite these generally positive developments, serious concerns remain over China's proliferation policies and activities. These refer to both Beijing's perspectives on nonproliferation and its continuing transfers of dual-use nuclear, chemical, and missile items and technologies. China has shown different attitudes toward the existing international nonproliferation regimes. It has acceded to most international treaties and conventions that are broadly based with universal membership (e.g., NPT, CWC), and has by and large complied with their norms and rules. On the other hand, it has been less than forthcoming, and occasionally quite critical, with regard to the largely Western-initiated, supply-side, multilateral export-control regimes. China has declined to join such arrangements as the Nuclear Suppliers

Group (NSG), the Australia Group (AG), the Wassenaar Arrangement, and the MTCR. There are a number of reasons. First of all, China is highly critical of these cartel-like regimes, considering them to be discriminatory, selective, imbalanced, unequal, and arbitrary.[21]

Second, while Beijing supports the principles of nuclear nonproliferation, it has also emphasized the importance of promoting peaceful use of nuclear energy. China expresses strong concern over the imbalance between nuclear nonproliferation and legitimate needs of developing countries for nuclear assistance and technology transfers.[22] Indeed, it is perhaps for this reason and also for practical purposes, in particular its ongoing nuclear cooperation programs with Pakistan that China does not require full-scope safeguards (FSS) as a condition of supply for its nuclear exports.

Third, China has its own security concerns when considering the various nonproliferation regimes and their impact. This is particularly revealing in its positions on MTCR compliance and missile transfer issues. China has pledged to comply with the MTCR guidelines but only with the original 1987 version and not the revised 1993 version, nor has China ever agreed to adhere to the MTCR annex. Chinese concerns revolve around a number of issues. One is what Beijing regards as the regime's discriminatory approach regarding the controlled items and its failure to address the demand side of missile proliferation. Another key Chinese argument has been that ballistic missiles per se are not weapons of mass destruction but rather delivery vehicles just as high-performance fighter aircraft, which are also capable of delivering nuclear, biological, and chemical weapons. Indeed, the Chinese do not consider missiles with conventional warheads as inherently destabilizing. While acknowledging that the MTCR has played some positive role in delaying missile proliferation, Chinese officials have pointed out the inherent limitation of the regime and have suggested that its potential merge with the ABM Treaty with stricter export controls could better address missile proliferation challenges.[23] A more effective control mechanism should be comprehensive, fair, and reasonable, and should also be linked to other disarmament measures and to efforts to settle regional conflicts. Such a regime should restrict both missiles and other offensive weapons, including fighter aircraft.

China is also critical of the "lack of objective criteria and the double standard applied by certain MTCR members in implementing requirements of the regime."[24] Beijing argues that the MTCR does not prohibit missile proliferation between member states. Indeed, the recent U.S.-led initiatives in research and development of theatre missile defense only reinforce Beijing's views in this regard. In particular, Beijing views Washington's intention to incorporate Japan and Taiwan into Theatre Missile Defense (TMD) as an act of missile proliferation since it is hardly distinguishable between defensive and offensive applications of missile technology. Within such a context, continuing U.S. arms sales to Taiwan, along with the development and possible deployment of TMD systems that could extend to Japan and Taiwan, convince Beijing's leadership that Washington only cares about its own absolute security without consideration of others' interests. China has made enough concessions already by, for example, suspending nuclear cooperation with and missile transfers to Iran but has received nothing from the United States in return, in particular with regard to arms sales to Taiwan. As a result, Beijing is now more determined to hedge its future nonproliferation cooperation with the United States on the latter's attitude toward issues of greater salience to China.[25]

Indeed, since early 1999, China has intensified its objection to the research, development and deployment of regional TMD systems.[26] Beijing has five areas of concerns.

First, the Chinese see TMD as yet another deliberate step by the United States to strengthen the U.S.-Japan military alliance and to contain China. Second, China contends that TMD research and development will encourage and facilitate Japanese remilitarization. The Chinese argue that TMD technology could help Japan to build long-range missiles. An even greater concern is that once Japan develops "the shield" then it would be encouraged to develop offensive military technologies or "the sword." Third, China contends that U.S. TMD sales to Taiwan are a violation of the 1982 U.S.-China communiqué on arms sales. Selling TMD to Taiwan would also represent a gross violation of China's territorial integrity, a blatant act of interference in China's domestic affairs. Fourth, Beijing is very concerned that TMD will give a false sense of security, hence encouraging the independence elements in Taiwan. Fifth, and perhaps China's greatest concern, is that TMD sales to Taiwan will constitute a de facto recreation of the 1954 U.S.-Taiwan military alliance. Many Chinese argue that TMD will require a high degree of technical and policy coordination between Washington and Taipei, which could lead both sides into closer military cooperation.

Finally, one should also note that ballistic missiles have increasingly become a critical element in China's defense modernization programs, and a "niche" or "comparative advantage" in its arms exports.[27] For Beijing, development and deployment of ballistic missiles serve the purposes of deterring independence elements in Taiwan and discouraging potential U.S. intervention in a military operation in the Taiwan Strait. At the same time, with China's conventional arms exports suffering precipitous decline since the early 1990s, and given MTCR member states' more restrictive policies, the importance of missile transfers and assistance has risen for Beijing. Given these overall strategic and commercial considerations, China has been reluctant to part with what little it retains both as bargaining leverage and as a source of income. This by and large explains Beijing's resistance to fully embrace missile nonproliferation norms that could negatively affect its interests.

The record of Chinese proliferation activities over the past decade remains mixed and contentious.[28] These include ring magnets export and the transfer of complete M-11 systems to Pakistan; sales of cruise missiles, missile technology, and chemical weapons–related items to Iran; and continued missile assistance to Iran, Pakistan, and North Korea.[29] These controversies draw attention to the gap between Beijing's public pronouncement on nonproliferation and its reported proliferation activities, raising questions about China's commitment and intentions. Recent reports by the National Intelligence Council and the Central Intelligence Agency continue to identify China as one of the key suppliers of materials and technologies that contribute to the proliferation of weapons of mass destruction and their delivery systems.[30] Table 9.3 provides a summary of reported Chinese proliferation activities over the last decade.

There can be a number of explanations regarding this apparent word-deed gap. These can be summarized as different perspectives and narrow interpretations, geostrategic and commercial interests, a nascent domestic export control system, deliberate lapse in enforcement, and issue linkage.

Different Perspectives, Narrow Interpretation

While supporting the general principles of nonproliferation, China has often emphasized that there should be a proper balance between nonproliferation obligations and the need

Table 9.3

Western Media Reports of Suspected Chinese Proliferation Activities

Recipient States	Missiles, Missile Components, & Technologies	Nuclear Materials & Technologies	Chemical Agents Used in CWs and Missile Systems
Algeria		- Nuclear reactor (*WT* 11 Apr. 91)	
Iran	- Silkworm (*LAT* 14 Feb. 89); - M-1B (*WP* 29 March 90); - M-11/Tondar-68 (*JDW* 1 Feb. 92); - Oghab (Jacob & McCarthy); - 150 C-802 (*ACT* Feb. 96; *WT* 29 Jan. 97; [China has promised to halt future transfers] - Guidance systems & propellant ingredients (*DN* 19–25 June 95 & *LAT* 3 Apr. 92); - Telemetry equipment used in flight tests for Shahab-3 and Shahab-4 MRBMs (*WT* 10 Sept. 97, 7 Dec. 98) - Suspected sales of specialty steel, telemetry equipment, and training on inertial guidance (*WT* 15 Apr. 99)	- Nuclear reactors (*WT* 16 Oct. 91 & *ACT* May 95); [suspended Sept. 95] - Calutron, or electromagnetic isotope separation system (*WT* 25 Sept. 95)	- Facilities and chemicals suited for making chemical weapons (*WP* 8 March 96 & *WT* 21 Nov. 96); - 400 metric tons of chemicals, including carbon sulfide (*JDW* 8 Jan. 97; *WP* 8 March 96); - Delivery of two tons of calcium hypochlorate, a chemical used for decontamination (Cordesman 1999); - Dual-use equipment and vaccines with both civilian medical applications and biological weapons applications (*WT* 24 Jan. 97); - Supply of 500 tons of phosphorus pentsulphide (*ST* 27 May 98)
Iraq		- Uranium enrichment high-speed centrifuges (*WT* 14 Dec. 89)	- Lithium hydride (*WP* 1 Oct. 90)
Libya	- Planned to build a hypersonic wind tunnel for missile design (*WT* 21 Jul 00) - Training Libyan missile experts at the Beijing University of Aeronautics and Astronautics (*WT* 30 June 00)		- Chemical agents (*WP* 7 June 90)
North Korea	- Reportedly transferred accelerometers, gyroscopes, and precision-grinding machinery (*WT* 20 July 99)		- Reportedly transferred 20 tons of tributyl phosphate (TBP), a chemical used to extract fissile material from spent nuclear fuel (*WT* 17 Dec. 02)

Pakistan	- M-9/M-11 (*WP* 6 April 91/*NYT* 6 May 93); - M-11 components (*IHT* 23 June 95); - 34 M-11s but missiles remain inside crates at Sagodha Air Base (*WSJ* 15 Dec. 98) - Blueprints and equipment to build missile-producing plant (*WP* 25 Aug. 96) - Specialized metal-working presses and special furnace (*WT* 15 April 99) - Reported assistance re development of the Shaheen MRBM (*JDW* 13 Dec. 00) - Specialty steel, guidance systems, and technical aid for a second missile plant (*FEER* 22 June 00; *NYT* 2 July 00) - Shipped "substantial amounts" of missile components, technology, and know-how for Shaheen-I and Shaheen-II (*WT* 6 Aug. 01)	- Nuclear reactor (*FEER* 23 June 92); - Ring magnets (*NYT* 8 Feb. 96); - Furnace and diagnostic equipment (*WT* 9 Oct. 96; *WP* 10 Oct. 96) - May have provided equipment for heavy water production plant (*NW* 23 March 00)	
Saudi Arabia	- CSS-2 IRBM (*WP* 29 March 88)		
Syria	- Negotiated M-9 deal (Jacob & McCarthy) - Chinese firms contributed equipment and technology to liquid fuel missile program (*WT* 30 June 00)	- Nuclear reactor (Shuey & Kan)	- Chemical agents (*NYT* 31 Jan. 92)

Sources: ACT (*Arms Control Today*); Cordesman (Anthony Cordesman, *Iraq and the War of Sanctions*, Westport, CT: Praeger, 1999); DN (*Defense News*); FEER (*Far Eastern Economic Review*); IHT (*International Herald Tribune*); JDW (*Jane's Defence Weekly*); LAT (*Los Angeles Times*); WP (*Washington Post*); WSJ (*Wall Street Journal*); WT (*Washington Times*); NW (*Nucleonics Week*); NYT (*New York Times*); Jacob & McCarthy (Gordon Jacob and Tim McCarthy, "China's Missile Sales—Few Changes for the Future," *Jane's Intelligence Review*, December 1992, 559–563); Shuey & Kan (Robert Shuey and Shirley Kan, *Chinese Missile and Nuclear Proliferation: Issues for Congress* CRS Issue Brief, April 1994); ST (*Sunday Telegraph*).

for legitimate peaceful use of nuclear, chemical, and space technologies. One plausible explanation, therefore, could be that Beijing simply views many of the controversial transfers, such as its nuclear reactor sales to Iran and Pakistan, as legitimate commercial transactions allowed by international treaties and under IAEA safeguards (even though not necessarily in compliance with full-scope safeguards). At the same time, economic reform and opening up also encourage domestic defense industrial sectors to seek overseas markets for their products to compensate for the difficult defense conversion process and declining military procurement.[31] Commercial interests and a different perspective on nonproliferation, therefore, provide for China's strict interpretation of its treaty obligations.

Geostrategic and Commercial Interests

Geostrategic considerations and the drive for commercial gains have been important factors behind Chinese transfer decisions. One is to expand its influence to regions of increasing importance such as the Middle East and the Persian Gulf. China's sale of the CSS-2 to Saudi Arabia gained the latter's diplomatic recognition. China's resilient defense cooperation with Pakistan is manifestation of Beijing's commitment to its loyal ally. Meanwhile, with China's conventional arms exports suffering precipitous decline since the early 1990s, sales of ballistic and cruise missiles became a "niche" or "comparative advantage" for Beijing, given MTCR member states' more restrictive export policy. These factors explain China's reluctance to fully embrace missile nonproliferation norms, which could deprive it of both the geostrategic and commercial benefits.[32]

A Nascent Domestic Export Control System

Another reason may be the inability of the central government to monitor, much less control, the activities of various companies due to the nascent nature of the domestic export control system and ambivalence in interagency coordination of policy from license review to approval, to customs inspections.[33] Meanwhile, decentralization and institutional pursuit of parochial interests encourage companies to dodge regulations and even openly defy rules.[34] The controversial sale of 5,000 ring magnets to Pakistan has often been cited as such an example of inadequate government oversight and effective control. At the same time, the magnitude of the chemical industry and the growing number of dual-use items make control efforts exceedingly difficult if not entirely futile.

Deliberate Lapse in Enforcement

China may deliberately choose not to enforce its nonproliferation commitments as a way to retain its bargaining leverage with the United States on issues such as NMD and TMD, to seek expansion of its influence to regions of increasing importance such as the Persian Gulf and the Middle East,[35] or merely as a retaliatory response to what it considers as an affront to its own national security interests by others. One area where this linkage operates is with U.S. arms sales to Taiwan, where China sees continuing arms sales as a violation of the U.S. commitment in the 1982 communiqué. In addition, when bilateral relations experience a downturn, Beijing has been less cooperative in arms control and nonproliferation. Such instances would include the release of the Cox Report charging the Chinese with nuclear espionage, U.S. allegations of Chinese campaign contributions, the accidental bombing of the Chinese embassy in Belgrade, and the controversial Wen-ho Lee case.

Issue Linkage

Finally, Beijing increasingly links further progress on proliferation issues to U.S. actions on its security concerns. This is clearly reflected in China's missile transfer activities. Beijing seeks to obtain tangible gains (e.g., satellite launches) in its negotiations with Washington and occasionally offers limited concessions. However, China never ignores the larger picture and has increasingly conditioned (although implicitly) its interpretation and implementation of missile nonproliferation commitment on U.S. policy in areas of direct concern to itself, namely, arms sales to Taiwan and developments in missile defenses. For instance, China promised in November 2000 to promulgate the laws governing missile technology exports, but it was not until August 2002 that China finally issued such a control list.[36]

Between Carrot and Stick: The U.S. Role

U.S.-Chinese disputes over nonproliferation issues remain a serious problem in bilateral relations. Over the years, successive U.S. administrations have sought to influence Chinese policies through a combination of inducements and sanctions. These range from suspension of technology transfers and imposition of economic sanctions against selected Chinese companies implicated in violation of U.S. laws, to incentives in the forms of technology transfers to and commercial space launch contracts with China.[37] In May 1991, responding to Chinese transfers of M-11 components to Pakistan, the Bush administration denied licenses for selling twenty high-speed computers to China.[38] The administration also turned down licenses for exporting satellite parts and other advanced technologies after revelation that two Chinese companies—the China Precision Machinery Import-Export Corporation and the China Great Wall Industry Corporation—had engaged in missile technology proliferation activities.[39] It was only after securing both verbal and written pledges from China that the U.S. government lifted these sanctions in February 1992.[40]

The Clinton administration continued the policy of pressure and sanctions to force change in China's proliferation behavior. On August 25, 1993, the U.S. government, after determining that China in 1992 had transferred M-11 missiles and related technologies to Pakistan in violation of the MTCR guidelines, imposed sanctions on eleven Chinese arms-exporting companies and one Pakistani entity in accordance with both the Arms Export Control Act (AECA) and the Export Administration Act (EAA).[41] U.S. companies would be banned from selling various satellite- and rocket-related items and equipment worth an estimated $1 billion to China's Ministry of Aerospace Industry, including the China Precision Machinery Import-Export Corporation, and Pakistan over the next two years.[42] In May 1997, and again in June 2001, the U.S. government imposed sanctions against Chinese individuals, state-run companies, and one Hong Kong company accused of exporting dual-use chemical precursors and/or related production equipment and technology to Iran.[43] In September 2001, the Bush administration imposed sanctions on the China Metallurgical Equipment Corporation for allegedly transferring MTCR Category-II technology and components to Pakistan. U.S. officials suggested that China failed to live up to its November 2000 pledge to desist from missile-related exports and to issue a missile export control list.[44] Subsequent bilateral talks to resolve the dispute prior to President George W. Bush's October 2001 trip to the APEC summit meeting in Shanghai failed to yield any result.[45] Additional sanctions were imposed on China in 2002. Table 9.4 lists present U.S. sanctions against those over the past decade.

The United States has also resorted to other mechanisms to enforce its nonproliferation policies with regard to China. A highly controversial case involved the 1993 forced inspection of the Chinese cargo ship, the *Yinhe*, suspected of carrying chemical precursors en route to Iran. The incident, which ended with U.S. failure to detect the suspected materials on board, incurred strong Chinese protests and deeply felt rancor.

Despite U.S. pressure, Beijing reportedly has continued to transfer missile components and to provide assistance to countries like Pakistan and Iran. Whether or not U.S. sanctions have been effective in affecting Chinese behavior remains inconclusive at this point. What can be said is that a mixture of U.S. sanctions, imposed or threatened, and economic benefits, withheld or offered, have had some impact on Chinese policy and behavior. The October 1994 agreement by the United States and China is a good example. China had long maintained that the M-11, with a range of 280 kilometers and a payload of 800 kilograms, fell within the original guidelines of MTCR for acceptable transfers. In the end, Beijing did accept the U.S. argument that missile systems with inherent capability for modification are to be restricted if the modified model then falls within MTCR guidelines. Beijing also agreed to hold in-depth discussions with the United States on the MTCR and possible membership. In exchange, the Clinton administration lifted the sanction imposed in August 1993.[46]

The recognition that sanctions may not "bite" without support from other countries and concerns that America's competitors may gain commercial advantages as a result of unilateral U.S. sanctions probably explain why the United States has turned to alternative policy initiatives to influence Chinese behavior.[47] Given that an important motivation behind Chinese weapons transfers is the pursuit of commercial interests, economic incentives in the forms of technology transfers and trade benefits, and the lifting of existing sanctions can, under the right conditions, induce Beijing to change its arms transfer policy.[48] Both the Bush and Clinton administrations have either offered to allow China greater access to U.S. technology or waived sanctions in return for Beijing's pledges and demonstrated actions to halt selling items and technologies of proliferation concern. Since 1989, Presidents Bush and Clinton have granted twenty waivers for U.S. satellites to be sent into orbit by Chinese launch vehicles.[49] This practice has been used to encourage positive Chinese nonproliferation behavior by providing tangible economic benefits. For instance, a 1995 Sino-U.S. agreement allowed China to launch fifteen American geostationary satellites through 2001.[50] Indeed, the Clinton administration specifically offered the prospect of expanding the space launch program, including waiving the post-Tiananmen sanctions on satellite launches on Chinese boosters to induce China to join the MTCR.[51]

Another example of economic incentives at work was the 1998 certification by the Clinton administration that paved the way to implementing the 1985 Sino-U.S. agreement on peaceful use of nuclear energy.[52] This allowed the U.S. nuclear industry to tap into China's potential billion-dollar nuclear market, in addition to encouraging more responsible Chinese nuclear export controls.[53] Over the years since the conclusion of the U.S.-China NCA, successive U.S. administrations have indicated that implementation of the agreement required China to make specific nonproliferation commitments. Persistent U.S. efforts gradually brought about noticeable change in Chinese nonproliferation policies. In May 1996, China made a formal pledge not to provide nuclear and dual-use assistance to unsafeguarded foreign facilities. In addition, China phased out its nuclear cooperation programs with Iran by suspending the sale of two 300–megawatt Qinshan-type nuclear power reactors, canceling the transfer of a uranium conversion facility, and turning down Iranian requests for other sensitive equipment and technology.[54] In October 1997, China formally joined the Zangger Committee.

However, the strategy of economic incentives, in particular in the form of technology transfers, has its limitations and is not without its controversies. For instance, the Clinton administration's effort to get China to join the MTCR in exchange for greater access to American commercial space technology has been declined by Beijing.[55] At the same time, U.S. technology transfers risk diversion to Chinese military end-use or, more worrying still, reexports to third countries. There already have been a number of such cases where U.S. machine tools and computers supposedly designated for civilian end-use have found their way in factories manufacturing Chinese cruise missiles and new-generation fighter aircraft.[56] Another prominent case involves two U.S. satellite makers, Loral and Hughes, which allegedly provided sensitive information to China. In 1995–1996, the two companies conducted investigations into the causes of the failed launches of the Apstar 2 and Intelsat 708 by Chinese Long March rockets, but they disseminated the results of the findings to China without obtaining the necessary export control license. The sensitive information transmitted could potentially help China improve its ballistic missile guidance systems.[57] Critics charge that not only have satellite sales failed to bring any tangible gains in U.S. nonproliferation goals, they have actually contributed to China's missile modernization programs that eventually would pose threats to U.S. security interests.[58] This controversy, coupled with the release of the Cox Committee report alleging systematic Chinese nuclear espionage, prompted Congress to impose a moratorium on U.S. satellite sales to China. Under such pressure the Clinton administration decided to deny Hughes Electronics Corp. a $450 million sale of a U.S. communications satellite to China.[59]

In sum, U.S. attempts to pressure China into accepting Western arms-transfer guidelines through the use of releasing/withholding advanced technologies have so far produced mixed results. Although one cannot deny that from time to time China has exercised restraint and has made good on its pledges, this is likely a reflection of Beijing's assessment of its national interests after weighing expected rewards (Western technologies) against forsaken commercial opportunities (missile/nuclear transfers). One important factor that may have influenced China's nonproliferation policies is its perception of how progress in these policy areas could contribute to the overall bilateral relationship with the United States. This may have influenced China's decision to discontinue sales of antiship missiles (C-802, C-801) to Iran.[60] It may also provide the rationale for China's decision to issue its key nuclear and dual-use export control regulations in 1997–1998: to facilitate the development of a "strategic partnership" between China and the United States, as well as to secure the Clinton administration certification for implementation of the 1985 NCA. This linkage suggests that a serious deterioration in Sino-U.S. relations could cause China to increase its proliferation activities.

Table 9.4

U.S. Nonproliferation Sanctions against China, 1989–2002

Date	Sanctions	Description	Status
9 July 2002	- Imposed against the Jiangsu Yongli Chemicals and Technology Import and Export Corporation (China); Q. C. Chen (China); China Machinery and Equipment Import Export Corporation (China); China National Machinery and Equipment Import Export Corporation (China); CMEC Machinery and Electric Equipment Import and Export Company Ltd. (China); CMEC Machinery and Electrical Import Export Company, Ltd. (China); China Machinery and Electric Equipment Import and Export Company (China); Wha Cheong Tai Company Ltd. (China); China Shipbuilding Trading Company (China); Hans Raj Shiv (India)	- Imposed pursuant to the Iran-Iraq Nonproliferation Act of 1992	Duration of a minimum of two years
9 May 2002	- Imposed against the Liyang Chemical Equipment Company (Liyang Yunlong of China); the Zibo Chemical Equipment Plant (Chemet Global Ltd of China); the China National Machinery and Electric Equipment Import and Export Company; the Wha Cheong Tai Company of China; the China Shipbuilding Trading Company; the China Precision Machinery Import/Export Corporation; the China National Aero-Technology Import and Export Corporation, and O. C. Chen, a Chinese businessman.	- Imposed pursuant to the Iran Nonproliferation Act of 2000	Duration of a minimum of two years
16 Jan 2002	- Imposed against Liyang Chemical Equipment, China Machinery and Electric Equipment Import and Export Company, and a Chinese individual	- Imposed pursuant to the Iran Nonproliferation Act of 2002	Duration of a minimum of two years
1 Sept 2001	- Imposed against China Metallurgical Equipment Corporation and its subunits and successors	- Imposed pursuant to the Arms Export Control Act and the Export Administration Act of 1979, as amended	Duration of a minimum of two years

Date	Action	Legal basis	Duration/Status
18 June 2001	- Imposed against Jiangsu Yongli Chemicals and Technology Import and Export Corporation	- Imposed pursuant to Section 3 of the Iran Nonproliferation Act of 2000	Duration of a minimum of one year
21 May 1997	- Imposed against five Chinese individuals, two Chinese companies, and one Hong Kong company for knowing and materially contribution to Iran's chemical weapons program	- Imposed pursuant to the Chemical and Biological Weapons Control and Warfare Elimination Act of 1991	Waived 1 November 1994; sanctions against Pakistani Ministry of Defense expired August 1995
24 August 1993	- Imposed against China's Ministry of Aerospace Industry that had engaged in missile technology proliferation activities, and Chinese government organizations involved in development or production of electronics, space systems, or equipment and military aircraft and Pakistan's Ministry of Defense	- Imposed pursuant to the 1990 Missile Technology Control Act	
25 May 1991	- Prohibition of the export of missile-related computer technology and satellites	- Imposed pursuant to the 1990 Missile Technology Control Act - Restricting the export of missile technology, missile-related computers, and satellites - No waivers on satellite export licenses	Waived 23 March 1992; sanctions against Pakistan's SUPAR-CO expired

Sources: Adapted from database compiled by the East Asia Nonproliferation Program, Center for Nonproliferation Studies, http://www.nti.org/db/china/sanclist.htm.

On the whole, China has continued to resist overt U.S. pressures on proliferation issues. While the United States has noted progress in Chinese nuclear export controls, it still sees problems in the areas of missile and chemical transfers. At the same time, the United States has moved its focus on bilateral cooperation on export control to enforcement mechanisms.[61] Consequently, from the U.S. perspective, a critical issue will be identifying the factors underlining Chinese progress in arms control and nonproliferation policies while designing effective policy instruments and using incentive/sanction packages to engender positive change. This remains a serious challenge in the years ahead.[62]

The difficulty in securing China's full compliance with U.S. nonproliferation policies lies in differences in perceptions, interests, and policy goals. While the United States has introduced broad-ranging nonproliferation measures and targeted particular states in implementing its policies, China has only committed to the universally accepted *global* nonproliferation norms as embodied in the NPT and the CWC. It is therefore not difficult to understand why Beijing resisted U.S. pressure to suspend nuclear exports to Iran, since the latter complies with IAEA safeguard provisions, including full-scope safeguards.

There are also differences in interests. Washington seeks to stem proliferation of WMD and their delivery systems to the Middle East, the Persian Gulf, and South Asia out of its interests for the protection of U.S. troops deployed in these regions, secure supplies of oil, the security of Israel, and stability in Indo-Pak relations. Beijing, on the other hand, regards its nuclear and missile exports as an important source of foreign exchange as well as a way to gain influence in these regions.[63] Indeed, China's refusal to adopt IAEA full-scope safeguards may be due to concerns that such measures would deprive it of potential markets for nuclear technology. With regard to its continued missile technology transfers and assistance to Pakistan, Beijing's motive may be more strategic than commercial. Islamabad has remained an important factor in Beijing's strategic calculation regarding South Asia and useful in its competition with India.[64]

The Taiwan issue features prominently in Beijing's arms control and nonproliferation calculus. China views U.S. arms sales to Taiwan as a blatant violation of the August 17, 1982, joint communiqué and an infringement of China's sovereignty and territorial integrity. Continued U.S. arms sales to Taiwan in both quantitative and qualitative terms have convinced Beijing that nonproliferation cooperation with Washington may turn out to be a one-way street and, therefore, not worth pursuing. Chinese officials have publicly warned that any U.S. transfer of TMD systems to Taiwan would be the "last straw" in the Sino-U.S. relationship and would result in serious consequences.[65]

Finally, China is increasingly concerned with the ultimate goal of U.S. nonproliferation policies—what it views as the U.S. drive for absolute security. Consequently, Beijing wants to retain flexibility and bargaining leverage with Washington. The latter has become more relevant given the developments since early 1999—the bombing of the Chinese embassy in Belgrade, the release of the Cox Report, and the U.S. decisions to develop and deploy both national and theater missile defense systems. Beijing is especially concerned with the last development, which it considers as the most potent threat to its national security interests.[66] China's predictable response will be to build up its missile forces and develop countermeasures; Beijing will also hold any progress in global arms control hostage to U.S. missile defense decisions. China is already pushing for setting up an ad hoc committee at the Conference on Disarmament to negotiate an outer space nonweaponization treaty and has held up work on a fissile material cutoff treaty.[67]

Missile defenses and Taiwan have emerged as the key issues likely to divide Beijing and Washington over the priorities of the arms control and nonproliferation agenda. Unless serious efforts are made to address some of China's core security concerns, Beijing can be expected to be less concerned about issues of greater significance to the United States, such as weapons proliferation, when it perceives that its own interests are either being ignored or even harmed by U.S. actions. One way to register unhappiness and to avenge its grievance is to make military transfers to regions/countries of U.S. concern, or to be less responsive to U.S. calls to tighten up China's own export control and international nonproliferation commitments. Other retaliation measures have been cancellation of high-level visits and bilateral talks on nonproliferation issues such as missile transfers and fissile material cutoff.[68]

Given that Sino-U.S. disputes over proliferation issues reflect differences in threat perceptions and derive from lack of mutual understanding of each other's positions and security concerns, extended high-level talks are particularly important and can result in substantive progress in the area of nonproliferation.[69] Indeed, constructive dialogue and better understanding between China and the United States on various weapons transfer-related issues may increase the chance of their eventual solution. Clearly, efforts must be made to encourage Beijing to comply with, in spirit as well as in letter, the norms and practices of nonproliferation. In this regard, the United States can and should play an important role given its concern over the proliferation of WMD and its leadership role in various multilateral nonproliferation export control regimes. However, the U.S. failure to ratify the CTBT, its rejection of a verification protocol to the Biological and Toxic Weapons Convention (BTWC), its pursuit of missile defenses, its withdrawal from the ABM Treaty, and the announcement to deploy national missile defenses by 2004 have in China's eyes greatly weakened American credibility in global nonproliferation leadership.

Conclusion

China has made significant progress in its nonproliferation policies over the last decade. This is reflected in its gradual acceptance of the core elements of the international nonproliferation norms, rules, and code of conduct. China has also pledged adherence to the MTCR's original guidelines governing missile transfers, and introduced elements of a domestic export control system. The factors that have contributed to these positive developments include China's concern over its international image, a growing awareness of the danger that WMD proliferation can pose to its own security, and its interest in maintaining a stable U.S.-China relationship. U.S. policy initiatives to engage, induce, and punish have also had some impact on Chinese proliferation behavior. However, the pace and future direction of Chinese nonproliferation policy will be closely linked to Beijing's overall assessment of its security interests, threats, and policy priorities. Given recent developments in missile defenses and the growing salience of the Taiwan issue, continued Chinese support of global arms control, and nonproliferation cannot be taken for granted.

The Bush administration has both opportunities to seize and major obstacles to overcome in its efforts to enlist continued Chinese cooperation in arms control and nonproliferation. It is clear from the above discussion that engagement must remain a key element of U.S. China policy, but the choice of appropriate policy tools remains a challenge. The September 11 terrorist attacks provide the opportunity for a fresh start for Sino-U.S. relations.[70] Chinese leaders hope that increased cooperation in the area of antiterrorism can help rebuild a stable bilateral relationship. Despite major differences between the United States and China over human rights, humanitarian intervention, and regional security issues, the two countries have also pursued common interests in combating narcotics trafficking, international organized crime, and terrorism. Many Chinese analysts also hope this attack could mark a turning point for enhanced consultation and expanded cooperation in fighting terrorism and pursuing other areas of mutual interest. This could help restore Sino-U.S. relations following a series of setbacks, including 1998–1999 allegations of Chinese nuclear espionage, the accidental U.S. bombing of the Chinese embassy in Belgrade in May 1999, and the collision between a U.S. reconnaissance plane and a Chinese fighter in April 2001.

Several observations can be made here. One is the clear need to take Chinese concerns over missile defenses and its objection to arms sale to Taiwan more seriously, even as Washington must continue to take issue with Beijing's proliferation activities. This does not mean that the United States should compromise its fundamental security interests and foreign policy objectives. What is required is a clear sense of balance and priorities in managing U.S.-China relations, promoting global nonproliferation agendas, protecting America against ballistic missile threats, and honoring its commitment to supporting a peaceful settlement of the Taiwan issue. In this context, continued high-level official dialogues on security, arms control, and nonproliferation between the United States and China must be maintained and regularized.

Past experience has indicated that intensive and patient bilateral consultation and negotiations have yielded noticeable results. Dialogue should not merely focus on U.S. concerns over specific Chinese proliferation activities but also on the potential threats that WMD proliferation can pose to China's own security. An important strategy, therefore, should be to engage China in expressing concerns, exchanging ideas, and negotiating nonproliferation arrangements in which China has a stake in following through on its norms, codes of behavior, and specific rules. China is more willing to follow through on pledges it has either negotiated with the United States or voluntarily made rather than being seen as forced to accept a certain code of behavior imposed by others.[71] Behind-the-door diplomacy has been more effective than open confrontation in getting controversial issues clarified, settled, or resolved.

The September 11, 2001, terrorist attacks on the World Trade Center and the Pentagon could also prompt important changes in Chinese arms transfer and nonproliferation policies. Beijing must now reconsider its relations with key countries in the Middle East. Chinese ties to countries like Iran, Iraq, and Libya have been driven by a host of geostrategic, commercial, and foreign policy considerations, often straining relations with Washington in the process. China's alleged transfer of nuclear and chemical-related items and technology to Iran and Pakistan and the role of Chinese companies in rebuilding the Iraqi air defense system have triggered U.S. complaints and sanctions. One motive for Chinese efforts to develop diplomatic and economic relations with countries such as Iran and Iraq was to prevent them from supporting separatist groups in Xinjiang. The terrorist attacks could again bring the issue of China's ties to these countries under the U.S. microscope and increase pressure on China to curtail relations and end arms sales to countries harboring or sponsoring terrorist groups.[72] In the context of an international war against terrorism, Beijing may have to make some difficult choices.

The United States must also recognize the tremendous changes that have taken place in China since economic reform and opening up began more than two decades ago. One of the consequences of this transformation and decentralization is the decreasing capability of the central government to oversee and control economic activities, some of which can cause proliferation concerns. While China has issued various export control regulations and is moving closer to harmonize its policy with the existing multilateral nonproliferation export control practice, significant challenges lie ahead. These include regime adherence and participation; legal framework development and improvement, and in particular the promulgation of open and comprehensive control lists; bureaucratic processes and division of labor, including interagency review procedures; export application, review, and approval procedures; enforcement, customs inspections, and punitive measures; prelicense checks and postshipment verification systems; and infrastructure development, personnel training, and education. The United States can and should encourage and assist Chinese efforts in strengthening its domestic export control system by offering training and institutional development support. In addition, the United States also can help China in developing technical means and know-how in the area of safeguarding nuclear facilities and supporting the emergence of an arms control community in that country.[73] It requires foresight and courage to maintain key programs and contacts even when bilateral relations are undergoing difficult times.

A third observation is the utility of and resort to economic sanctions to punish Chinese proliferation behavior. A rush to impose sanctions without giving time for clarification, checking evidence, and negotiation can generate a lot of animosity but not necessarily produce the desired outcome. At the same time, judicious and selective use of sanctions may continue to serve useful purposes, especially when there are undeniable Chinese violations of its nonproliferation commitments and when such activities are clearly sanctioned by the government. Whenever possible, broad allied support should be sought; otherwise sanctions cannot be effective either as an instrumental (forcing policy change in Beijing) or a punitive (denying Beijing what it wants) tool. At the same time, sanctions (which impose high costs on certain U.S. industries) could become increasingly difficult to sustain and could incur growing opposition from American business communities.

Finally, there must be greater coordination between the executive and legislative branches to achieve greater credibility on U.S. nonproliferation policies. The implementation of the China policy must remain the purview of the executive branch, with congressional and bipartisan consultation and support. In other words, there should be only one China policy and consistency in its interpretation and implementation. Rather than seeking to introduce additional China specific legislation, Congress should work with the administration and focus on oversight issues so as to ensure that existing laws that are in line with global nonproliferation norms and principles are enforced.

Notes

1. On this point, see Evan S. Medeiros, "China, WMD Proliferation, and the 'China Threat' Debate," *Issues and Studies* 36 no. 1 (January/February 2000): 19–48.
2. Gordon Jacobs and Timothy McCarthy, "China's Missiles Sales—Few Changes for the Future," *Jane's Intelligence Review* (December 1992): 559–563.
3. Wendy Frieman, "New Members of the Club: Chinese Participation in Arms Control Regimes, 1980–1995," *The Nonproliferation Review* 3, no. 3 (Spring-Summer 1996): 15–30.
4. Jing-dong Yuan, Phillip C. Saunders, and Stephanie Lieggi, "New Developments in

China's Export Controls: New Regulations and New Challenges," *The Nonproliferation Review* 9, no. 3 (Fall–Winter 2002): 153–167.

5. Tim Weiner, "U.S. Says China Isn't Helping Others Build Nuclear Bombs," *New York Times*, December 11, 1997, p. A5; Evan S. Medeiros, "China Offers U.S. New Pledge on Nuclear Exports, Avoids Sanctions," *Arms Control Today* 26, no. 4 (May/June 1996): 19.

6. Statement of Robert J. Einhorn, Deputy Assistant Secretary of State for Nonproliferation, before the Committee on International Relations, U.S. House of Representatives, February 4, 1998.

7. R. Jeffrey Smith, "Algeria to Allow Eventual Inspection of Reactor, Envoy Says," *Washington Post*, May 2, 1991, p. 36; James L. Tyson, "Chinese Nuclear Sales Flout Western Embargoes," *The Christian Science Monitor*, March 10, 1992, pp. 1, 3; "China Sales to Iran Raise Nuclear Concern," *Arms Control Today* 21, no. 10 (December 1991): 21, 26.

8. See Chapter 3, "Dealing with China," in Ashton B. Carter and William J. Perry, *Preventive Defense: A New Security Strategy for America* (Washington, DC: Brookings Institution Press, 1999), pp. 92–122.

9. "Joint United States-People's Republic of China Statement on Missile Proliferation," October 4, 1994.

10. Associated Press, "China Pledges It Will Not Aid Foreign Missile Development," November 21, 2000.

11. Yuan, Saunders, and Lieggi, "New Developments in China's Export Controls."

12. Bates Gill and Evan S. Medeiros, "Foreign and Domestic Influences on China's Arms Control and Nonproliferation Policies," *The China Quarterly*, no. 161 (March 2000): 66–94; Wen L. Hsu, "The Impact of Government Restructuring on Chinese Nuclear Arms Control and Nonproliferation Policymaking," *The Nonproliferation Review* 6, no. 4 (Fall 1999): 152–167. However, the ultimate overall impact on China's decision-making processes remains to be seen.

13. Wu Yun, "China's policies towards arms control and disarmament: From passive responding to active leading," *The Pacific Review* 9, no. 4 (1996): 577–606; Frieman, "New Members of the Club."

14. See, for example, Mingquan Zhu, "The Evolution of China's Nuclear Nonproliferation Policy," *The Nonproliferation Review* 4, no. 2 (Winter 1997): 40–48.

15. Weixing Hu, "Nuclear Nonproliferation," in Yong Deng and Fei-Ling Wang, eds., *In the Eyes of the Dragon: China Views the World* (Lanham, MD: Rowman and Littlefield, 1999), pp. 119–140.

16. Ashton B. Carter and William J. Perry, *Preventive Defense: A New Security Strategy for America* (Washington, DC: Brookings Institution Press, 1999), pp. 92–122; "PRC Played 'Crucial Role' in Halting DPRK Missile Launch," *The Korean Times* (Internet version, http:// times.hankooki.com/), September 20, 1999.

17. Zachary S. Davis, "China's Nonproliferation and Export Control Policies: Boom or Bust for the NPT Regime?" *Asian Survey* 35, no. 6 (June 1995): 591.

18. Michael Brenner, "The People's Republic of China," in William C. Potter, ed., *International Nuclear Trade and Nonproliferation: The Challenge of the Emerging Suppliers* (Lexington, MA: Lexington Books, 1990), p. 254.

19. Robert J. Einhorn, "Nonproliferation Challenges in Asia," speech given before the Asia Society in Hong Kong, June 7, 2000.

20. Evan S. Medeiros, "The Changing Character of China's WMD proliferation activities" in Robert Sutter, ed., *China and Weapons of Mass Destruction* (Washington, DC: Congressional Research Service, Library of Congress, Spring 2000), pp. 111–148.

21. Statement by Mr. Sha Zukang, ambassador of the People's Republic of China for Disarmament Affairs at the First Committee of the 52nd Session of United Nations General Assembly, New York, October 14, 1997.

22. Ambassador Sha Zukang, Director-General, Department of Arms Control and Disarmament, Ministry of Foreign Affairs of China, "Some Thoughts on Non-Proliferation," speech at

the 7th Annual Carnegie International Non-Proliferation Conference on Repairing the Regime, January 11–12, 1999, Washington, DC.

23. Barbara Opall-Rome, "Chinese Official Urges Broader, Revised MTCR," *Defense News*, January 25, 1999, p. 1.

24. Ibid.

25. Evan S. Medeiros, *Missiles, Theater Missile Defense and Regional Security*, Conference Report, the 2nd U.S.-China Conference on Arms Control, Disarmament and Nonproliferation (Monterey, CA: Center for Nonproliferation Studies, July 1999). On U.S. arms sales to Taiwan, see John P. McClaran, "U.S. Arms Sales to Taiwan: Implications for the Future of the Sino-U.S. Relationship," *Asian Survey* 40, no. 4 (July/August 2000): 622–640; Wei-chin Lee, "US Arms Transfer Policy to Taiwan: from Carter to Clinton," *Journal of Contemporary China* 9, no. 23 (March 2000): 53–75.

26. Sha Zukang, "Some Thoughts on Non-Proliferation," address given at the 7th Carnegie International Non-Proliferation Conference, January 11–12, 1999, Washington, DC, http://www.ceip.org/programs/npp/sha/html; Howard Diamond, "China Warns U.S. on East Asian Missile Defense Cooperation," *Arms Control Today* 29, no. 1 (January/February 1999): 27.

27. On this point, see Bates Gill, James Mulvenon, and Mark Stokes, "China's Strategic Rocket Forces: Transition to Credible Deterrence," in James Mulvenon and Richard Yang, eds., *The People's Liberation Army as Organization* (Santa Monica, CA: RAND, in 2001); Evan S. Medeiros and Bates Gill, *Chinese Arms Exports: Policy, Players, and Process* (Carlisle, PA: Strategic Studies Institute, U.S. Army War College, August 2000).

28. Media coverage in this area is extensive. See also, the Majority Report of the Subcommittee on International Security, Proliferation, and Federal Services of the Committee on Governmental Affairs, U.S. Senate, *The Proliferation Primer* (January 1998); and Shirley A. Kan, *Chinese Proliferation of Weapons of Mass Destruction: Current Policy Issues*, CRS Issue Brief (Washington, DC: Congressional Research Service, updated March 12, 2001).

29. Testimony by Robert J. Einhorn, Deputy Assistant Secretary of State for Nonproliferation, Before the Subcommittee on International Security, Proliferation, and Federal Services, Senate Committee on Governmental Affairs, April 10, 1997; Shirley A. Kan, *China's Proliferation of Weapons of Mass Destruction and Missiles: Current Policy Issues*, CRS Issue Brief for Congress, updated March 12, 2001.

30. National Intelligence Council, Foreign Missile Developments and the Ballistic Missile Threat to the United States through 2015, September 1999. Director of Central Intelligence, Unclassified Report to Congress on the Acquisition of Technology Relating to Weapons of Mass Destruction and Advanced Conventional Munitions, July 1 through December 31, 1999 (August 2000).

31. See John Frankenstein and Bates Gill, "Current and Future Challenges Facing Chinese Defence Industries," *The China Quarterly* 146 (June 1996): 394–427.

32. Evan S. Medeiros and Bates Gill, *Chinese Arms Exports: Policy, Players, and Process* (Carlisle, PA: Strategic Studies Institute, U.S. Army War College, August 2000); Yitzhak Shichor, "Mountains out of Molehills: Arms Transfers in Sino-Middle Eastern Relations," *Middle East Review of International Affairs* 4, no. 3 (Fall 2000): 68–79.

33. See Cupitt and Murayama, *Export Controls in the People's Republic of China*; Hsu, "The Impact of Government Restructuring."

34. John W. Lewis, Hua Di, and Xue Litai, "Beijing's Defense Establishment: Solving the Arms-Export Enigma," *International Security* 15, no. 4 (Spring 1991): 87–109; Kevin F. Donovan, "Chinese Proliferation Bureaucracies," in Barry R. Schneider and William L. Dowdy, eds., *Pulling Back from the Nuclear Brink: Reducing and Countering Nuclear Threats* (London: Frank Cass, 1998), pp. 218–234.

35. See Bates Gill, "Chinese Arms Exports to Iran," and David Dewitt, "The Middle Kingdom Meets the Middle East: Challenges and Opportunities," *China Report* 34, nos. 3 and 4 (July–December 1998): 355–379, and 441–455, respectively.

36. Jing-dong Yuan, "Missile Export Controls Significant Step for Beijing," *South China Morning Post*, August 29, 2002, p. 12.

37. See "U.S. Nonproliferation Sanctions Against China" (Monterey, CA: East Asia Nonproliferation Program Database, Center for Nonproliferation Studies, 1999); Duncan L. Clarke and Robert J. Johnston, "U.S. Dual-Use Exports to China, Chinese Behavior, and the Israel Factor: Effective Control?" *Asian Survey* 39, no. 2 (March/April 1999): 193–213; Victor Zaborsky, "Economics vs. Nonproliferation: U.S. Launch Quota Policy Toward Russia, Ukraine, and China," *The Nonproliferation Review* 7, no. 3 (Fall-Winter 2000): 152–161.

38. "U.S. Bars Export of High-Speed Computers, Other Items to China, Citing Arms Concerns," *International Trade Reporter*, May 29, 1991, p. 805.

39. U.S. President George Bush, 1989, "Statement by Press Secretary Fitzwater on Restrictions on U.S. Satellite Component Exports to China, April 30, 1991," *Weekly Compilation of Presidential Documents* 27, May 3, 1991, p. 531; Clyde H. Farnsworth, "Bush Denies Satellite Parts to China," *New York Times*, May 1, 1991, p. A15; "Bush Defends China Policy, Implements Export Controls," *International Trade Reporter*, June 19, 1991, p. 940; "Bush Renewing Trade Privileges for China, but Adds Missile Curbs," *New York Times*, May 28, 1991, pp. A1, A8.

40. Jon B. Wolfsthal, "China Promises to Join NPT by March, Will Follow Missile Export Guidelines," *Arms Control Today* 21, no. 10 (December 1991): 22; Elaine Sciolino, "U.S. Lifts Its Sanctions on China Over High-Technology Transfers," *New York Times*, February 22, 1992.

41. The nonproliferation requirements for China can be found in *Omnibus Export Amendments Act of 1991* (H.R.3489). House of Representatives, 102nd Cong., 1st Sess., October 23, 1991.

42. "China Missile Sanctions to Block U.S. High-Tech Exports," *Export Control News* 7, no. 8. August 26, 1993, pp. 2–3; "Diplomacy Hit by Missile," *The Economist*, August 28, 1993, p. 32; Daniel Williams, "U.S. Weighs Trade Curbs against China," *Washington Post*, August 25, 1993, p. A1, and "US Punishes China Over Missile Sales," *Washington Post*, August 26, 1993, p. A1; Kerry Dumbaugh, *China-US Relations*, CRS Issue Brief, IB94002 (Updated June 9, 1994), p. 5.

43. "US Arms Control/Nonproliferation Sanctions Against China," January 1998. Center for Nonproliferation Studies Database; Jonathan S. Landay, "Clinton's Curveball on China," *The Christian Science Monitor*, May 23, 1997, pp. 1, 9; Department of State, Bureau of Nonproliferation, "Imposition of Nonproliferation Measures against a Chinese Entity, Including Ban on US Government Procurement, Public Notice 3707," June 18, 2001, in the Federal Register, vol. 66, no. 123, June 23, 2001.

44. State Department. Bureau of Nonproliferation, "Imposition of Missile Proliferation Sanctions Against a Chinese Entity and a Pakistani Entity, Public Notice 3774," in the Federal Register, vol. 66, no. 176, September 11, 2001; Rob Wright, "U.S. to Sanction Chinese Firm," *Los Angeles Times*, September 1, 2001.

45. Reuter, "US, China Fail to Resolve Pakistan Missile Dispute," October 12, 2001.

46. "U.S., China Reach New Accords on MTCR, Fissile Cutoff Issues," *Arms Control Today* 24, no. 9 (November 1994): 28.

47. The economic rationale for not using sanctions as a policy instrument is captured in David M. Lampton, "America's China Policy in the Age of the Finance Ministers: Clinton Ends Linkage," *The China Quarterly* 139 (September 1994): 597–621; Lampton, *Same Bed, Different Dreams: Managing U.S.-China Relations, 1989–2000* (Berkeley: University of California Press, 2001), chap. 3. It has been estimated in a recent U.S. government study that billions of dollars in potential sales to China could be lost as a result of unilateral U.S. sanctions. See United States General Accounting Office, *US Government Policy Issues Affecting US Business Activities in China* (Washington, DC), May 1994.

48. William J. Long, "Trade and Technology Incentives and Bilateral Cupertino," *International Studies Quarterly* 40, no. 1 (March 1996): 77–106.

49. Warren Ferster, "Sanctions Legislation Frustrates Industry," *Space News*, May 25–31, 1998, p. 20.

50. Victor Zaborsky, "U.S. Missile Nonproliferation Strategy toward the NIS and China: How Effective?" *The Nonproliferation Review* 5, no. 1 (Fall 1997): 88.

51. Howard Diamond, "U.S. Renews Effort to Bring China into Missile Control Regime," *Arms Control Today* 28, no. 2 (March 1998): 22.

52. "Text: President Certifies China under U.S.-China Nuclear Agreement," United States Information Agency, January 16, 1998; Howard Diamond, "Clinton Moves to Implement Sino-U.S. Nuclear Agreement," *Arms Control Today* 28, no. 1 (January/February 1998): 30.

53. Jennifer Weeks, "Sino-U.S. Nuclear Cooperation at a Crossroads," *Arms Control Today* 27, no. 5 (June/July 1997): 7–13.

54. R. Jeffrey Smith, "China's Pledge to End Iran Nuclear Aid Yields U.S. Help," *Washington Post*, October 30, 1997, p. 1.

55. Howard Diamond, "U.S. Renews Effort to Bring China into Missile Control Regime," *Arms Control Today* 28, no. 2 (March 1998): 22; Jim Mann, "China Rejects Joining Missile-Control Group, U.S. Officials Say," *Los Angeles Times*, April 17, 1998, http://www.latimes.com/HOME/NEWS/NATIONS/t000036404.html.

56. Nigel Holloway, "Cruise Control," *Far Eastern Economic Review*, August 14, 1997, pp. 14–16; Jonathan S. Landay, "Is China Diverting High Technology to US Foes?" *The Christian Science Monitor*, July 11, 1997, pp. 1, 8.

57. "Hughes and Loral: Too Eager to Help China?" *BusinessWeek*, September 13, 1999; Juliet Eilperin, "GOP Leaders Demand Satellite Export Data," *Washington Post*, May 12, 1998, p. A5.

58. Richard D. Fisher, Jr., *Commercial Space Cooperation Should Not Harm National Security*, The Heritage Foundation Backgrounder no. 1198, June 26, 1998; Bill Gertz, "Eased Export Controls Aided Beijing's Missile Technology," *Washington Times*, May 7, 1999, http://www.washtimes.com/news/news3.html.

59. Howard Diamond, "House Seeks to Limit Space Cooperation with China," *Arms Control Today* 28, no. 4 (May 1998); Jeff Gerth and David E. Sanger, "Citing Security, U.S. Spurns China on Satellite Deal," *New York Times*, February 23, 1999, http://www.nytimes.com/library/world/asia/022399satellite-sale.html.

60. Bill Gertz, "China to Halt Missile Sales to Iran," *Washington Times*, January 20, 1998.

61. "China and Non-Proliferation: Interview with Senior US Official," *US Foreign Policy Agenda* (January 1998): 30–31; Press Conference by John D. Holum, Acting Under Secretary of State for Arms Control and International Security Affairs and the Director of the Arms Control and Disarmament Agency, Beijing, China, March 26, 1998.

62. Bates Gill, "U.S., China and Nonproliferation: Potential Steps Forward," *The Monitor* 3/4, no. 4/1 (Fall 1997/Winter 1998): 27–32.

63. See John Calabrese, "China and the Persian Gulf: Energy and Security," *The Middle East Journal* 52, no. 3 (Summer 1998): 351–366.

64. Mushahid Hussain, "Pakistan-China Defense Co-operation: An Enduring Relationship," *International Defense Review* 2/1993, pp. 108–111; Cameron Binkley, "Pakistan's Ballistic Missile Development: The Sword of Islam?" in William C. Potter and Harlan W. Jencks, eds., *The International Missile Bazaar: The New Suppliers' Network* (Boulder, CO: Westview Press, 1984), pp. 75–97.

65. John Pomfret, "Chinese Warn U.S. Not to Arm Taiwan; Official Says Transfer of Missile Defenses Would Be 'Last Straw,'" *Washington Post*, March 6, 1999, p. A1.

66. Paul H.B. Godwin and Evan S. Medeiros, "China, America, and Missile Defense: Conflicting National Interests," *Current History* (September 2000): 285–289.

67. See, for example, statement by Mr. Hu Xiaodi, ambassador for Disarmament Affairs of China at the Plenary of the Conference on Disarmament, Geneva, June 15, 2000, http://www.fmprc.gov.cn/eng/c464.html.

68. Barbara Opall, "U.S. Queries China on Iran," *Defense News*, June 19–25, 1995, pp. 1, 50.

69. Bates Gill and Matthew Stephenson, "Search for Common Ground: Breaking the Sino-U.S. Non-Proliferation Stalemate," *Arms Control Today* 26, no. 7 (September 1996): 15–20.

70. Bonnie S. Glaser, "Terrorist Strikes Give U.S.-China Ties a Boost," in Brad Glosserman and Eun Jung Cahill Che, eds., *Comparative Connections* 3, no. 3 (October 2001): 35–45, http://www.csis.org/pacfor/cc/0103Q.pdf; Michael Swaine, "Reserve Course? The Fragile Turnaround in U.S.-China Relations," *Policy Brief* 22 (Washington, DC: Carnegie Endowment for International Peace, February 2003), http://www.ceip.org/files/pdf/Policybrief22.pdf.

71. Rodny W. Jones, "China and the Nonproliferation Regime: Renegade or Communicant?" in *China & Nuclear Nonproliferation: Two Perspectives*, PPNN Occasional Paper No.3 (Southampton, UK: Centre for International Policy Studies, University of Southampton, 1989), p. 26.

72. Jennifer Lee, "Complaints That Chinese Companies Supply Rogue Nations," *New York Times*, November 12, 2001.

73. One such program is the currently dormant U.S.-China lab-to-lab technical exchange initiative. See Nancy Prindle, "The U.S.-China Lab-to-Lab Technical Exchange Program," *The Nonproliferation Review* 5, no. 3 (Spring–Summer 1998): 111–118. See also, Yuan, Saunders, and Lieggi, "Recent Developments in China's Export Controls."

Part Three

Strategic Relations with the Major Powers
and Asian-Pacific Neighbors

10

Patterns and Dynamics of China's International Strategic Behavior

Joseph Y.S. Cheng and Zhang Wankun

International strategy normally refers to a long-term, macro plan to maximize a country's national interests.[1] The formulation and implementation of an international strategy usually depends on three variables: (1) the international system, (2) domestic conditions, and (3) national leadership. The three variables interact organically through the national leaders' subjective assessments of the objective international and domestic environments. The inhibiting effects of the international system are probably more conspicuous when a country's capability is limited and when the international power transfiguration is in a stage of transition.

Some scholars divide strategic behavior into three categories: conquering territory, initiating war, and power balancing.[2] The international strategic behavior patterns of China in this chapter, however, refer to the behavior characteristics in China's handling of its relations with major powers. Not every country's foreign policy has a substantial strategic content, but in the case of the People's Republic of China (PRC), its foreign policy consistently possesses a predominant strategic content based on geopolitical considerations. Since 1949, the foreign policy of the PRC has gone through the Cold War and post–Cold War stages. In the Cold War era, China's international strategy evolved more or less in ten-year cycles: the Sino-Soviet alliance in the context of "leaning to one side" (1949–1960); a two-fronts strategy of anti-imperialism and antirevisionism (1960–1972); a pseudo-alliance with the United States against the Soviet Union (1972–1982); and the adoption of an independent, peaceful foreign policy line (1982–1989).[3] Correspondingly, four international strategic behavior patterns emerged: the alliance model, the united front model, the pseudo-alliance model, and the balance of power model.[4]

In the early 1990s, the disintegration of the Soviet Union witnessed the end of bipolarity and the emergence of a new post–Cold War international system in which the danger of a world war has been much reduced. In this transition, major powers and groups of major powers are engaged in a competition of comprehensive national power, in attempts to secure more advantageous positions in the new global configuration of power. In view of its spectacular economic growth since 1978, by the end of the twentieth century, China's potential to become a major power was fully appreciated.[5] In the post–Cold War era, such potential has become a significant factor in the strategic adjustments on the part of all major powers.

China's active push for a multipolar world has signified a new stage in China's international strategy in the post–Cold War era. This has been most conspicuous in the new network of partnerships in China's relations with the major powers of the world since the mid-1990s. Since April 1996, China has established different types of partnerships with Russia, France, the United States, the United Kingdom, the Association of Southeast Asian Nations (ASEAN), the European Union (EU), and Japan, practically all the major powers and regional organizations (see Table 10.1).[6]

The acceptance and promotion of partnerships among major countries reflect an attempt on the part of China to redefine its position in the new international strategic pattern. The partnerships also reveal a strategic idea offered by China as an emerging major power in the post–Cold War era. The key questions regarding this idea are: (1) What is the content of such partnerships among major powers? (2) What are the differences between this partnership model among major powers and the Chinese foreign policy of independence and peace in the 1980s? (3) What are the motivations behind China's promotion of this model of partnerships among major powers? (4) Does the new model of partnerships among major powers provide an important innovation in the management of relations among major powers in the coming century? This chapter attempts to answer these questions so as to better understand China's international strategic behavior patterns in the post–Cold War era, the inner logic of partnership relationships, and the trends of China's international strategy in the coming century.

What Is Partnership?

"Partnership" is a new concept in Chinese diplomacy in the post–Cold War era. According to the *Xiandai Hanyu Cidian* (Modern Chinese Dictionary), the phrase *huoban* (partnership) comes from an ancient military system. "Ten men formed one *huo*, the head of the *huo* was also in charge of the meals. Those who belonged to the same *huo* were *huoban*. Today, it refers to those who have joined the same organization or engaged in the same activities."[7] *The Webster's Encyclopedic Unabridged Dictionary of the English Language* defines "partner" as "one who shares or is associated with another in some action or endeavor; sharer; associate," and a "partnership" as "the state or condition of being a partner; participation; association; joint interest."[8] Countries in partnership, therefore, have common interests and are ready to be in association for joint action or endeavor.

On the basis of the speeches and statements from Chinese leaders and spokesmen of the Chinese foreign ministry, the kind of partnership relationships that China promotes is one of equality, friendly cooperation, and lack of confrontation. The Chinese leadership has been emphasizing the "three no's" principle in the Sino-Russian strategic, cooperative partnership, meaning that the relationship involves no alliance, no confrontation, and is not directed against or at the expense of a third party.[9] It also reiterates, in the building of a constructive strategic partnership between China and the United States, that "China is not a potential adversary of the United States, much less an enemy of the United States, instead it is a trustworthy partner for cooperation."[10]

An analysis of Chinese foreign policy documents reveals that the partnership relationships promoted by the Chinese leadership share a number of characteristics. In the first place, as indicated by the former Chinese foreign minister, Qian Qichen, at the end of 1997, states of the world should respect each other in a multipolar power configuration. They should treat each other as equals, promote mutual benefit, seek to establish consen-

Table 10.1

Partnerships between China and the Major Powers/Regional Organizations

April	1996	China and Russia established a strategic, cooperative partnership of equality and mutual trust facing the twenty-first century.
May	1997	China and France established a comprehensive partnership.
October	1997	China and the United States declared to build a constructive strategic partnership in the twenty-first century.
November	1997	China and Canada agreed on a comprehensive partnership straddling over the present and the next centuries.
December	1997	China and Mexico agreed on a comprehensive partnership straddling over the present and the next centuries.
December	1997	Chinese and ASEAN leaders agreed to establish a partnership facing the twenty-first century based on good-neighborliness and mutual trust.
December	1997	Chinese and Indian leaders reached a consensus on the establishment of a partnership of constructive cooperation facing the twenty-first century.
February	1998	China and Pakistan decided to establish a partnership of comprehensive cooperation facing the twenty-first century.
April	1998	Chinese and European Union leaders agreed to establish a constructive partnership of long-term stability facing the twenty-first century.
October	1998	China and the United Kingdom decided to establish a comprehensive partnership.
November	1998	China and Japan declared to establish a partnership of friendship and cooperation facing the twenty-first century, working for peace and development.
February	1999	China and South Africa decided to establish a constructive partnership facing the twenty-first century.
April	1999	China and Egypt confirmed to establish a relationship of strategic cooperation facing the twenty-first century.
October	1999	China and Saudi Arabia confirmed to establish a relationship of strategic cooperation facing the twenty-first century.

sus, and tolerate differences among them. They should engage in friendly cooperation, not engage in confrontation, not form alliances, and not act against the interests of a third state. Chinese leaders believe that the above constitutes the foundation of the new relationships among states in the twenty-first century. [11]

In the Sino-Russian joint communiqué released during Boris Yeltsin's visit to China in April 1996, it was stated:

> The trend towards global multipolarity is developing. The quest for peace, stability, cooperation and development has already become the mainstream of contemporary international life. . . . Both sides reaffirm that mutual respect and equality are the important principles of maintaining and developing normal, healthy state relations. Countries irrespective of being large or small developed or developing, or whose economies are in transition, are equal members of the international community. People of every country have the right to choose independently their social system, developmental path and model, based on their respective national conditions, in the absence of foreign interferences.[12]

Chinese leaders appeal to their counterparts of major countries to actively identify common interests. They hope that differences and contradictions in social systems and values will not affect the healthy development of state-to-state relations. The use of phrases such as "facing the twenty-first century" and "straddling over the present and the next centuries" in the partnership relationships between China and other major powers signifies that Chinese leaders hope to maintain such relationships among major powers on a long-term, stable basis. Chinese leaders apparently consider state-to-state relations in terms of nonzero-sum games, and they appeal for the abandonment of the Cold War mentality.

Mutual cooperation should cover all sectors, especially economic, trade, scientific, and technological exchanges. This is in line with the Chinese leadership's position that peace and development constitute the main themes of the post–Cold War era, and that China would concentrate on its objective of modernization. Strengthening of communications is another important aspect of partnership relationships. Exchanges of visits at the heads of state level, scheduled meetings between premiers, joint commissions involving senior officials, hot lines, and various channels of communications are typical provisions in China's partnerships with other major powers. [13]

These high-sounding principles are basically sincere, but they are obviously in accord with China's strategic objectives: (1) to maintain a peaceful international environment to concentrate on economic development; (2) to improve China's dialogue with all major powers and regional organizations to ensure recognition of China's major power status and enhance China's influence in international affairs; (3) to push for multipolarity and to prevent the United States from dominating international affairs; (4) to facilitate progress in China's economic diplomacy to open up markets and attract investment, advanced technology, and management expertise from diverse sources; and (5) to improve China's image in the international community.

At the same time, China's relations with the major powers are not without problems, particularly with regard to the United States and Japan. China has yet to convince its neighbors that it does not constitute a threat to them and will not do so even when its national power has considerably improved in twenty years' time. Chinese leaders have to exercise restraint to prevent China's territorial disputes with its neighbors and the Taiwan question from escalating into military conflicts. They also have to show that their advocacy for a new international political and economic order remains constructive, and that China's experts have meaningful proposals to offer. All these are formidable challenges for Chinese strategic planners in the twenty-first century.

Partnership in the Context of Principles and Traditions of Chinese Foreign Policy

In the Maoist era, China formed an alliance with the Soviet Union in the 1950s and engaged in a "pseudo-alliance" with the United States in the 1970s. In the era of Deng Xiaoping, however, China adopted a foreign policy of independence and peace, taking into consideration the new international strategic situation and the lessons of the past decades. This was enunciated in the report presented to the Twelfth National Congress of the Chinese Communist Party (CCP) and written into the 1982 constitution. In March 1986, the fourth session of the Sixth National People's Congress endorsed the report on the Seventh Five-Year Plan (1986–1990).

Following the Chinese leadership's usual practice, Premier Zhao Ziyang in the fourth part of the report provided a detailed account of the ten principles guiding China's foreign policy. Besides reaffirming the general principles of defending world peace, opposing hegemonism, observing the Five Principles of Peaceful Coexistence, [14] and support for the Third World, China also reiterated its position on arms control and disarmament and its open-door policy and support for the United Nations. As regards China's position of never establishing an alliance or a strategic relationship with any big power, the Chinese premier further stated that China's relations with various countries would not be determined by their social systems and ideologies, and that China's position on various international issues would be guided by the criteria of defending world peace, developing relationships of friendship and cooperation among various countries, and promoting international prosperity. Premier Zhao Ziyang also stressed China's emphasis on various exchanges at the people-to-people level. These ten principles constituted an action program for China's foreign policy of independence and peace.

China's foreign policy in the Deng Xiaoping era may be described as modernization diplomacy, as it mainly serves the national goal of modernization. [15] It still inherited the antihegemony tradition of the Maoist era, but it had lost the distinct ideological connotation in opposing hegemonism. Enemy states were no longer identified along ideological lines, while China would oppose specific hegemony policies of any country. [16] Chinese leaders in the 1980s and 1990s considered that hegemonism and power politics were the greatest threats to international peace and stability. In his report to the Twelfth National Congress of the CCP in September 1982, Hu Yaobang, general secretary of the party, stated: "At present, the major forces threatening the peaceful coexistence of various countries in the world are imperialism, hegemonism and colonialism." [17] His successor, Jiang Zemin, in his report to the Fifteenth National Congress of the CCP in September 1997, reiterated that, "The mentality of Cold War still exists, hegemonism and power politics remain the major sources of the threats to international peace and stability." Jiang then made the appeal: "To oppose hegemonism, and protect world peace." [18] When the vice-president of the People's Republic of China, Hu Jintao, presented a televised address on

May 9, 1999, the day following the NATO bombing of the Chinese embassy in Belgrade, he also reaffirmed China's opposition to hegemonism and power politics. [19]

The concept of partnership is also related to the Chinese diplomatic principle of "not establishing an alliance or strategic relationship with any major power." In his opening address to the Twelfth Party Congress in 1982, Deng stated: "China's affairs have to be handled in accordance with China's conditions, their handling has to depend on Chinese people's strength. Independence and self-reliance, in the past, present and future, remain our standpoint."[20] In May 1984, in his meeting with the Brazilian president, Joâo Baptista de Oliveira Figueiredo, Deng's statements were more straightforward: "China's foreign policy is a policy of independence, of genuine non-alignment. China has no alliance relationship with any country, it adopts a completely independent policy. China doesn't play the United States card, nor does it play the Soviet Union card. China does not allow others to play the China card too."[21] China's position of never establishing an alliance or a strategic relationship with any big power was one of the ten principles guiding Chinese foreign policy as presented in Zhao Ziyang's report to the National People's Congress in March 1986. [22] Indeed this was a core principle of China's foreign policy of independence and peace in the 1980s, and an important criterion to assess whether China had actually implemented a foreign policy of independence and peace. Undeniably, China established a formal alliance with the Soviet Union in the 1950s, and a "pseudo-alliance" with the United States and its Western allies in the 1970s. But, according to the Chinese scholars interviewed by the authors, these strategic choices had been made under specific historical conditions and when China was in a relatively weak position. It may be argued that these decisions were made against the idealistic strategic principle of Chinese leaders, which included the emergence of China as a strong and independent international force. Besides being in line with the above position, Chinese leaders believed that in view of the improving national strength of China, its formation of an alliance or strategic cooperation with other countries targeted against either the United States or the Soviet Union would be unfavorable to the international balance of power and to global peace and stability.[23] Even in the mid-1990s, some members of the Chinese diplomatic community still regarded the pursuit of a "strategic relationship" among countries as the reflection of a Cold War mentality.[24]

The Chinese foreign policy research workers interviewed by the authors in recent years defended China's strategic partnerships with Russia and the United States by denying that they implied a return to an alliance policy or a significant change in China's international strategy. As a principle guiding China's foreign policy of independence and peace, "not establishing an alliance or a strategic relationship with any major power" specifically referred to the Sino-Soviet alliance in the 1950s and the Sino-American strategic cooperation in the 1970s. The Chinese leadership considers that both models no longer serve China's national interests and international stability. Second, the Chinese government has been emphasizing the "three no's" principle in the Sino-Russian strategic partnership, thereby setting limits to the strategic partnership. Moreover, despite the common Sino-Russian interests in promoting multipolarity and in the establishment of a new international order, China has not deviated from its nonalignment, nonconfrontation strategic position. This was reaffirmed in the Chinese government's rejection of the Russian proposal of a strategic triangle among Russia, China, and India, as presented by the Russian prime minister, Yevgeny Primakov, at the end of 1998. In the Persian Gulf crisis in February 1998, when an American air war against Iraq seemed imminent, and in the bombing of Kosovo by NATO, which started in March 1999, China did not engage in close coordination or joint action with Russia.[25] China's stress on the independence of its foreign policy has preserved the flexibility of its implementation. Apparently, China's partnerships with various major powers since 1996 have not affected its position on rejecting alliances.

This emphasis on independence, however, may have an undesirable impact. In the 1970s, China largely behaved like a bystander in international politics, and it was not active in international organizations. A strong concern for independence may discourage joint action with other countries and limit China's influence in the establishment of a new international political and economic order. At the same time, China has accepted the international financial and trade system dominated by the West and has been seen by its neighboring Third World countries as a serious competitor. Emphasizing independence may exacerbate China's aloof position in the Third World. After all, the domestic authoritarian regime already discourages the Track-2 processes of international interaction.

It has also been argued that the strategic partnership concept is an extension of the Five Principles of Peaceful Coexistence. China, India, and Burma initiated those principles in the mid-1950s. The principles were supposed to guide relations among countries with different social systems, and they were later extended to cover relations among countries with the same social system as well. On October 31, 1984, in a talk with President U San Yu of Myanmar, Deng Xiaoping explained that the Five Principles of Peaceful Coexistence would have a potentially wide application: "The Five Principles of Peaceful Coexistence provide the best way to handle the relations between nations. Other ways—thinking in terms of 'the socialist community,' 'bloc politics' or 'spheres of influence,' for example—lead to conflict, heightening international tensions. Looking at the history of international relations, we find that the Five Principles of Peaceful Co-existence have a potentially wide application." [26]

Further, even before the end of the Cold War, Chinese leaders already proposed that the Five Principles of Peaceful Coexistence could serve as the principles for the establishment of a new international order. On December 21, 1988, Deng, in his meeting with the Indian prime minister, Rajiv Gandhi, stated:

The general world situation is changing, and every country is thinking about appropriate new policies to establish a new international order. Hegemonism, blocs politics and treaty organizations no longer work. Then what principle should we apply to guide the new international relations? . . . Two things have to be done at the same time. One is to establish a new international political order; the other is to establish a new international economic order. . . . As for a new international political order, I think the Five Principles of Peaceful Co-existence, initiated by China and India, can withstand all tests. . . . We should take them as norms for international relations. If we want to recommend these principles as a guide to the international community, first of all, we should follow them in our relations with each other and with our other neighbors. [27]

After the Tiananmen incident, as well as the dramatic changes in Eastern Europe and the former Soviet Union in the late 1980s and early 1990s, Deng Xiaoping believed that China should "avoid the limelight and maintain a low profile," but he also considered that

China had to make its contribution. Obviously, the Chinese leadership hoped that China would be able to make a contribution in the establishment of a new international political and economic order, especially through the promotion of multipolarity. [28] The Chinese official line on the above is best summarized in the Sino-Russian joint declaration on global multipolarization and the establishment of a new world order issued in April 1997 during Jiang Zemin's visit to Moscow, as well as the document "Sino-Russian Relations at the Turn of the Century" released by the two governments in November 1998.

There are differences between the concept of strategic partnership in the post–Cold War era and the tradition of the Five Principles of Peaceful Coexistence. The former reflects the importance attached to relations with major powers on the part of the Chinese leadership; it also implies that China looks upon itself as a major power and pursues the legitimate interests of one in a multipolar world. Chinese leaders and foreign policy researchers believe that China's legitimate interests have not been sufficiently respected in the existing international order dominated by the United States and the Western powers. The more significant the neglect of China's interests, the greater its eagerness to push for a new international order. Yet China has not been able to mobilize other countries toward this end; it also appears to lack the expertise to formulate proposals of a broad appeal. Vague principles continuously articulated by the Chinese leadership may help to improve China's image to a small extent, but they are far from promoting a new international order.[29]

The Hierarchy of Partnerships

From Table 10.1, one may detect a certain hierarchy in China's partnership network with the major powers and regional organizations established in recent years: (1) strategic partnership (the strategic, cooperative partnership between China and Russia and the Sino-American declaration to build toward a constructive strategic partnership); (2) comprehensive partnership (between China and France, China and Canada, China and Mexico, China and Pakistan, as well as China and the United Kingdom); (3) constructive partnership (the China-EU constructive partnership of long-term stability); (4) partnership based on good neighborliness and mutual trust (China and ASEAN); and (5) partnership of cooperation (Sino-Indian partnership of constructive cooperation and the Sino-Japanese partnership of friendship and cooperation). Chinese leaders have been very careful in their use of diplomatic language; different types of partnerships may well imply different mutual expectations and different levels of significance in China's international strategy.

China's strategic partnerships with Russia and the United States represent a redefinition of the two important bilateral relationships in the post–Cold War era, as well as an extension and reshaping of the Cold War strategic triangle in the new international environment. On one hand, China attempts to stabilize the two bilateral relationships through the framework of strategic partnerships so as to consolidate a peaceful and stable international environment to realize its modernization program. On the other hand, China also intends to engage in strategic cooperation with Russia to a certain extent in order to push for global multipolarization and to deter the United States from dominating international affairs as the sole superpower in the world.

Comprehensive partnerships have already been established between China and the other two permanent members of the UN Security Council, as well as major regional powers such as Canada, Mexico, Pakistan, South Africa, Egypt, and Saudi Arabia. Apparently, countries enjoying comprehensive partnerships with China are major powers or regional major powers. They share parallel interests with China in that they desire a more active role in international affairs through global multipolarization. They also do not have any significant conflicts of interests with China.

Chinese leaders consider that China's relations with the European Union have been making progress in recent years. China is the fourth largest trade partner of the European Union, while the European Union is the third largest trade partner of China. At the same time, the European Union is China's fifth largest source of foreign investment, as well as the most important source of foreign technology and equipment.[30] The average scale of an EU investment project is considerably larger than one funded by the United States and Japan. Many EU investment projects are in the automobile, telecommunications, pharmaceutical, and other industries with significant advanced technology inputs. Chinese experts believe that vast potential for cooperation also exists in the information technology, insurance, and financial services sectors. They consider that in view of the escalating competition among the United States, Japan, and the European Union in the global market, the latter would enhance the priority given to the development of the China market.

Exchange of visits between Chinese leaders and those from the European Union and its member countries have been expanding. In February 1998, Sir Leon Brittan, vice-president of the European Commission, visited China. He revealed to the Chinese foreign minister, Qian Qichen, that the European Union was deliberating a new policy initiative to establish closer ties with China. He also reaffirmed the EU's backing for accelerated negotiations in support of Chinese membership of the World Trade Organization. In the following March, the European Union announced a proposal to hold annual summit meetings with China as part of its efforts to upgrade relations with China. The New China News Agency described the initiative as an attempt by the European Union to raise its political relations with China to the same level as those with the United States, Japan, and Russia.[31] There have been held three annual China-EU summits since 1998. Similarly, China appreciates the initiating of the Asia-Europe Meeting, first held in Bangkok in March 1996, then in London in 1998, and Seoul in 2000. Chinese leaders believe that cooperation with the European Union has been strengthening in a comprehensive manner.

In line with the Chinese leadership's push for multipolarization, it certainly wants to see China and the European Union emerging as strong poles in the global power transfiguration in the next century. They share the common concern regarding the American ambition of securing a predominant role in international affairs and the future direction of Russia. Their common interests are expected to grow in areas ranging from nuclear proliferation and weapons proliferation, energy development, environmental protection, and food programs to the combat of international organized crime. Chinese leaders appreciate that the European Union, relative to the United States, is more sympathetic to development issues in the Third World and the North-South question. They therefore endorse the integration of the European Union and value a constructive partnership of long-term stability with it.

China obviously wants to maintain a friendly relationship with its neighbors. The ASEAN is significant in that Chinese leaders believe it is an important actor, a pole, in the regional multipolar power configuration.[32] They consider that though China has some territorial disputes with Malaysia, Brunei, the Philippines, and Vietnam over the Spratlys, as well as with Vietnam over the Paracels, there are no fundamental conflicts of strategic interests between China and the ASEAN countries. In fact, ASEAN enjoys the tacit support of the major powers in the region in its role as a balancer, as the major powers all understand that each of them cannot secure a predominant position in the regional balance of power nor would they accept any one major power assuming that position. Hence, China believes that it can establish a partnership with the ASEAN countries based on good-neighborliness and mutual trust. If cooperation between the two improves and the problems between them are resolved, then the relationship may well become a comprehensive partnership.

Among China's neighbors, Japan and India are potential threats, and they too perceive China as a potential threat. The Chinese leadership is very concerned with the strengthening of a security cooperation between Japan and the United States and the possibility of their development of a theatre missile defense program involving South Korea and Taiwan. Chinese leaders are suspicious that "the state of emergency" provisions in the U.S.-Japan defense cooperation guidelines released in 1997 cover Taiwan, thus constituting an infringement of Chinese sovereignty over Taiwan and a serious interference in China's domestic affairs. Though the Japanese government has formally stated that "the state of emergency" is not a geographical concept, a number of Japanese politicians and officials have openly declared that Taiwan is covered by such provisions in the U.S.-Japan defense cooperation guidelines. So far the Chinese government and the Chinese people remain dissatisfied with the fact that Japan has not made a formal apology to China for the atrocities committed in the country in the Second World War, especially after a formal apology had been delivered to South Korea during President Kim Dae Jung's visit to Japan in the autumn of 1998. Moreover, the territorial dispute concerning the Diaoyutai (Senkaku) Islands has not been settled, and the prospects for a satisfactory settlement are far from bright. In view of the rising nationalism in the two countries, the bilateral relationship is not free from potential conflicts of fundamental strategic interests.

The Chinese leadership, however, is still interested in a friendly relationship with Japan, because Japan is a major power and an important neighbor. Chinese leaders appreciate the need to reassure Japan so that it will continue to avoid the option of massive rearmament. They have exercised considerable restraint in dealing with the Diaoyutai Islands dispute, so much so that they have been criticized as being weak in overseas Chinese communities. Both China and Japan intend to play a more significant role in the Asia-Pacific region and in the international community, especially in the United Nations. Both, however, have been exercising considerable self-restraint so that their pursuits are a non-zero-sum game and can be mutually supportive. The Asia-Pacific region needs effective representatives in negotiations with Western countries and in various international organizations. China and Japan, on the basis of close bilateral and regional consultations, can perform that role together. After all, they have no intention of isolating each other in regional affairs, and they do not have the ambition of assuming a predominant leadership role in the region. From the Korean Peninsula to Cambodia, it is obvious that a settlement can hardly be reached without some agreement and cooperation between Beijing and Tokyo.

Many Japan experts in China see Japan adopting a strategy of "escaping America to return to Asia" in the 1990s, in contrast to its "escaping Asia to enter Europe" strategy at the end of the previous century, and they welcome Japan's new strategy.[33] They consider that Japan needs a reliable base in order to become a political power, and they perceive Japan playing a leading role in Asia through its superiority in economic power and technology. They see no problem of China fitting into the economic integration among Japan, the NIEs in East Asia, and the ASEAN countries. This is understandable now that China continues to receive aid from Japan, enjoys a substantial trade surplus in the bilateral trade, and attracts increasing Japanese investments in the manufacturing industries whose products are intended for the Japanese market. Moreover, China has other sources for importing technology and management expertise and does not have to be overly dependent on Japan. In the eyes of the Chinese leadership, this is an acceptable foundation for a partnership of friendship and cooperation, though the problems in the bilateral relationship are not to be underestimated.

In November 1998, Jiang Zemin, president of the PRC, visited Japan. The Chinese side did not secure a formal apology from Japan during the visit comparable to that just offered to South Korea. Similarly, it had also hoped that the Obuchi government would follow the Clinton administration's example in opposing Taiwan memberships in international organizations whose members are sovereign states. The disappointment of the Chinese side resulted in the Sino-Japanese joint statement issued without the leaders of the two states signing it, which was a strong, unmistakable signal of the difficulties in the bilateral relationship. Under such circumstances, frictions such as Chinese warship operations in what Japan regarded as its own exclusive economic zone and the visit to the Yasukuni shrine by Japanese cabinet ministers further exacerbated the mutual distrust between the two countries. Leaderships in both countries are under increasing domestic pressures to stand firm and are, therefore, tempted to use minor sanctions to indicate their respective displeasures and to exert pressure on their counterparts. The visit of the Japanese foreign minister, Kono Yohei, to China in August 2000 and Premier Zhu Rongji's visit to Japan two months afterwards represented efforts by both governments to arrest the deterioration in the bilateral relationship. The Chinese side is especially keen to generate new momentum to improve relations with Japan so as to fulfill its objectives of forging a partnership with its important neighbor and winning over Japan to support multipolarity as defined by the Chinese leadership.

In contrast to Sino-ASEAN relations, Chinese leaders have neglected Sino-Indian relations. Indians resent this neglect, and it has adversely affected Beijing's objective of maintaining friendly relations with its neighbors. In 1988, the Indian prime minister, Rajiv Gandhi, visited China, and he was followed by his successor P.V.N. Rao in September 1993. In the latter visit, the two countries reached an agreement on maintaining peace and tranquility along the line of actual control in the border area. In November 1996, President Jiang Zemin visited India, and the two governments agreed on troop withdrawals from the border area and the non-use of force in the settlement of differences. Jiang also expressed the wish that China and the South Asian countries would build a relationship of strategic partnership facing the twenty-first century.[34] In December 1997, Chinese and Indian leaders reached a consensus on the establishment of a partnership of constructive cooperation facing the twenty-first century during the Indian visit of Wei Jianxing, member of the CCP Politburo Standing Committee.

The consensus apparently did not bring about much improvement in the bilateral relationship. When the H.D.D. Gowda government was eventually replaced by the A.B. Vajpayee

Figure 10.1   **Levels of Partnership Relationships Established by China with Major Powers and Regional Organizations Facing the Twenty-first Century**

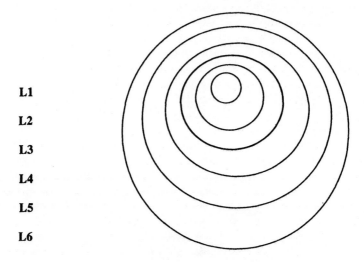

L1

L2

L3

L4

L5

L6

L1 = Sino-Russian and Sino-American Strategic Partnerships
L2 = Comprehensive Partnerships Between China on One Hand and France, Canada, Mexico, Pakistan, and the United Kingdom Respectively on the Other.
L3 = Sino-EU Constructive Partnership of Long-term Stability
L4 = Sino-ASEAN Partnership Based on Good-neighborliness and Mutual Trust
L5 = Sino-Japanese Partnership of Friendship and Cooperation and Sino-Indian Partnership of Constructive Cooperation
L6 = Sino-South African Constructive Partnership, Sino-Egyptian Relationship of Strategic Cooperation and Sino-Saudi Arabian Relationship of Strategic Cooperation

---

government, Sino-Indian relations sharply deteriorated. In April 1998, George Fernandes, defense minister of the Vajpayee government, blamed China for providing Pakistan with missile technology and openly stated that China was the "potential threat No.1."[35] Then came the Indian nuclear tests in the following May. An opinion survey in China showed that 80 percent of the respondents were worried about the impact on China's national security.[36] There was also a conspiracy theory that India's nuclear tests had the tacit approval of the United States, which wanted to enlist India as a "frontline" state against China.[37] At the same time, there were assessments of India's potential in the development of nuclear weapons, indicating that it might surpass that of the United Kingdom and would be comparable to that of France and China. There were also reports that India might have its first nuclear submarine in 2004 or 2007. Worse still, Chinese foreign policy and military experts speculated that India might, in response to the needs of domestic politics and foreign policy, create border incidents in the coming year or so, and these incidents along the Sino-Indian border might even led to a limited war. The Chengdu Military Region, responsible for Tibet's defense, was asked to prepare for a limited defensive war that could last one and a half months.[38]

In the foreseeable future, Chinese leaders have no choice but to exercise self-restraint. They seem to have avoided severe criticism of India's nuclear tests and have joined other governments to encourage the Indian government to sign the Comprehensive Test Ban Treaty. They appear to have demonstrated greater understanding of India's nuclear policy and the similarities between the respective positions of Beijing and New Delhi on nuclear tests and nuclear disarmament. It is hoped that with friendly exchanges of visits between leaders of the two countries, military buildup along the Sino-Indian border can be avoided and, despite the limited prospects for progress, border negotiations can also continue. The maintenance of the partnership of constructive cooperation would certainly be an important component of China's policy of preventing a sharp deterioration in the bilateral relationship. When the Indian president visited China in May–June 2000, both governments managed to discover more common ground on global multipolarization and other international issues.[39]

In recent years, Chinese leaders have been keen to demonstrate that they have not neglected the Third World. They are certainly attaching more importance to regional powers such as Mexico, Pakistan, Egypt, South Africa, and Saudi Arabia. These countries not only have considerable influence within their respective regions, they can contribute much to the promotion and maintenance of a multipolar balance within their respective regions too.

The Sino-Russian and Sino-American Strategic Partnerships

The Sino-Russian strategic, cooperative partnership was the first partnership established between China and another country. At present, this is undoubtedly the most developed partnership with well-established operational mechanisms.

The foundation for this strategic partnership was laid in the final years of the Cold War. The Sino-Soviet summit in May 1989 symbolized the common attitude of "let us put the past behind us and open up a new era."[40] The Chinese position, as defined by Deng Xiaoping, was that the two countries "would not be engaged in confrontation, nor would they form an alliance, but would simply normalize their relations." Deng emphasized that the Chinese position was based on the lessons drawn from the history of the international Communist movement and that the Five Principles of Peaceful Coexistence should constitute the foundation of the bilateral relationship.[41] The joint communiqué released also indicated that the normalization of Sino-Soviet relations was not directed against any third country nor would it damage the interest of any third country.[42] At the end of 1991, when the Soviet Union broke up and Russia emerged as an independent sovereign state, the Chinese government immediately dispatched a delegation headed by the minister of foreign economic relations and trade, Li Lanqing, to visit Russia and declared the recognition of Russia as well as its replacement of the Soviet Union as a permanent member of the UN Security Council. Both sides also signed the "Notes on Sino-Russian Talks." In January 1992, Chinese Premier Li Peng met Russia President Boris Yeltsin at the United Nations; both sides affirmed that they would develop the bilateral relationship on the basis of the joint communiqué in 1989 and the notes in 1991.

Sino-Russian relations then entered a new stage and began progressing rapidly. Close contacts between the top leaders on both sides have been providing a significant direct push. Since the China visit of Boris Yeltsin in December 1992, there have already been seven more summit meetings between Chinese and Russian leaders, and they now enjoy a "strategic, co-operative partnership of equality and trust facing the twenty-first century."[43]

While bilateral trade remains limited and major energy joint ventures take time to materialize, Sino-Russian military cooperation proceeded smoothly in the 1990s. Boris Yeltsin indicated that China bought US$1.8 billion worth of Russian arms in 1992. The Russians have encouraged the People's Liberation Army to buy much more, and Chinese military delegations have often toured Russian military plants and research facilities to see what is on offer. In addition, Russian scientists are working with their Chinese counterparts to develop new defense technology.[44]

From 1992 to 1995, China acquired 72 Su-27 fighter aircraft, 4 Kilo-class submarines, 10 Ilyushin transport aircraft, and 100 S-300 air-defense missile systems from Russia.[45] In 1996, Beijing reportedly applied for a license from Rosvooruzheniye, Russia's arms-exporting monopoly, to manufacture up to 200 Su-27 fighters in Shenyang.[46] The Kilo-class submarines are a formidable addition to the Chinese submarine fleet, which had been largely equipped with obsolete 1950s Whiskey and Romeo models from the Soviet Union. The Kilos have a range of 9,650 kilometers and can stay out at sea for up to forty-five days. The submarines would be able to provide protection for Chinese surface vessels operating at increasingly longer distances from their home ports. In 1992, China reportedly agreed to buy around 50 T-72 tanks and 70 BMP vehicles from Russia at a cost of around US$250 million, probably for equipping a rapid reaction force.[47] China also showed interest in securing from Russia the A-50 airborne warning and control aircraft and long-range, early-warning radar systems.[48] At the end of 1997, China had spent an estimated US$6 billion on Russian weapons since 1991. More contracts are expected. Chinese orders have become vital for many Russian defense plants that are fighting for survival because of drastic cuts in government procurements.[49] After acquiring two Sovremenny-class destroyers for about US$1 billion, the Chinese navy bought two more in 2000. These destroyers could also be armed with the latest Yakhont missiles designed to destroy aircraft carriers and other warships.[50]

The Sino-Russian strategic, cooperative partnership may be perceived as the result of over forty years of tumultuous relations between the two countries. Chinese leaders consider it an extension of the normalization of the bilateral relationship, as advocated by Deng Xiaoping.[51] It is also a product of the strategic adjustments on the part of the Chinese and Russian leaderships in response to the post–Cold War situation to push for global multipolarization and to limit the predominance of the United States in international affairs. Chinese leaders and the Chinese official media highly evaluate this strategic, cooperative partnership. President Jiang Zemin and Premier Li Peng both stated that this partnership was not a measure of expediency but a historical choice facing the twenty-first century.[52]

Sino-American relations are probably the most complicated and most troublesome among China's relations with the major powers in the post–Cold War era. The Tiananmen incident and the significant changes in Eastern Europe and the Soviet Union made a great impact on Sino-American relations. With the break up of the Soviet Union, the strategic foundation of Sino-American relations in the Cold War era—coordinated deterrence of the Soviet threat—disappeared; the Tiananmen crackdown also dramatically altered the China policy agenda in American domestic politics. At the same time, China's spectacular economic growth supported its emergence as a major power with superpower potential in the post–Cold War era. Its unique international position as the largest socialist country in the post–Cold War era, its promotion of a new international political and economic order, its arms sales to countries like Pakistan and Iran, as well as the Taiwan question have also introduced new elements of geopolitical confrontation into the bilateral relationship, besides the ideological clashes that continue into the post–Cold War era.[53] Due in part to the above-mentioned factors, Sino-American relations have experienced significant ups and downs in the last decade.[54]

Chinese leaders want to avoid such fluctuations in Sino-American relations. They hope that the bilateral relationship can remain on a course of stable, healthy development. Despite the two relatively successful summits in 1997–1998 and the commitment to build toward a constructive strategic partnership in the twenty-first century, however, there is little room for optimism. Chinese leaders believe that the most serious challenge to the bilateral relationship is the United States' strategy to maintain its sole superpower status and its hegemonic behavior. Toward this end, the Clinton administration has been trying to strengthen the North Atlantic Treaty Organization (NATO) and the U.S.-Japan security treaty system in an attempt to mould the new international order for the next century according to its own will. Recently, the United States' inclination to involve Taiwan in its Theatre Missile Defense System, its military actions against Iraq without the endorsement of the UN Security Council, its intervention in Yugoslavia under the umbrella of NATO, and the bombing of the Chinese embassy in Belgrade are all considered significant threats to the political foundation of the Sino-American constructive strategic partnership.[55] There is a view in China that without the settlement of the Taiwan question, there cannot be genuine normalization of Sino-American relations, not to speak of a constructive strategic partnership. After the NATO bombing of the Chinese embassy in Belgrade, many considered that "the behaviour [sic] of the United States reflects that the so-called strategic partnership is only Beijing's wishful thinking."[56] At the same time, there is much doubt in the United States as to the substance of a Sino-American constructive strategic partnership.

Chinese leaders have been advocating partnerships among major powers as the new model for the twenty-first century. From an idealistic point of view, Chinese leaders hope to secure a multipolar world in which the major powers can establish relationships based on equality, mutual trust and respect, and mutual checks and balances. Stable, balanced relations among the major powers will provide a better guarantee of regional and global balance and stability. This is in line with Chinese leaders' objectives of peace and development and their pursuit of a peaceful international environment to concentrate on economic development. Chinese leaders want to avoid a unipolar world or a chaotic, multipolar situation with crises such as that generated by the Indo-Pakistani nuclear arms race. Chinese Vice-President Hu Jintao appeared on television after the NATO bombing of the Chinese embassy in Belgrade and appealed to Chinese people to act in accordance with China's fundamental interests, so as not to damage social stability and solidarity and thus adversely affect the construction of socialism with Chinese characteristics. *Renmin Ribao* followed with five editorials emphasizing the upholding of the party's basic line of economic construction as the central task, persisting in economic reforms and opening to the

Table 10.2

Sino-Russian Summit Meetings in the 1990s and Their Major Achievements

	Date	Place	Major achievements
First	December 17–19, 1992	Beijing	Signed "Joint Statement on the Foundation of Sino-Russian Mutual Relations" and another 24 documents on cooperation in various areas; both sides would perceive each other as friendly countries.
Second	September 2–6, 1994	Moscow	Signed "Sino-Russian Joint Statement," a joint statement on the non-first-use of nuclear weapons against each other and not to aim strategic nuclear weapons at each other and an agreement on the western section of the Sino-Russian border, both heads of state announced that the two countries have established a long-term, stable "constructive partnership facing the twenty-first century."
Third	April 24–26, 1996	Beijing	Signed the third joint statement, announcing the establishment of a strategic, cooperative partnership of equality and trust facing the twenty-first century; both sides agreed on the establishment of a scheduled meetings mechanism between leaders of both sides; formed a Sino-Russian Friendship, Peace and Development Committee; on April 25, heads of state of China, Russia, Kazakhstan, Kyrgyzstan, and Tajikistan concluded an agreement on the mutual reduction of military forces in the border area.
Fourth	April 24–26, 1997	Moscow	Signed the fourth joint statement and a number of agreements on bilateral cooperation; both sides also announced the completion of the demarcation of the eastern section of the Sino-Russian border.

Fifth	November 9–11, 1997	Beijing	Signed the fifth joint statement and a number of agreements on bilateral cooperation; both sides also announced the completion of the demarcation of the eastern section of the Sino-Russian border.
Sixth	November 23–24, 1998	Moscow	Released a joint statement on Sino-Russian relations at the turn of the century.
Seventh	December 9–10, 1999	Beijing	Both heads of state attended the signing ceremony regarding the boundary, released a joint communiqué, and a joint statement. Both sides pointed out that the contents of their strategic cooperative partnership relationship affirmed in 1996 were being substantiated; Russia supported Beijing's position on Taiwan; China also supported Russian action in Chechnya. During the meeting, both sides signed the protocol on the eastern and western sections of the boundary and the agreement on the joint economic exploitation of the islands and their neighboring waters in the boundary rivers.
Eighth	July 17–18, 2000	Beijing	Both heads of state concluded a joint declaration and a joint statement on the question of antiballistic missiles.

Source: Renmin Ribao (People's Daily), Beijing.

outside world, maintaining social stability, and implementing independent, peaceful foreign policy.[57] These messages implied that China would continue to "avoid the limelight" and "refrain from confrontation" in its foreign policy line.

Motivations for China to Promote Partnership among Major Powers

On the basis of the authors' interviews with Chinese foreign policy experts, it was Russia who first proposed the concept of a strategic, cooperative partnership in their bilateral relationship with China.[58] Similarly, it was also President Bill Clinton who first raised the idea of a strategic partnership with China when he visited Australia in November 1996.[59] Though the Chinese leadership was presented with a brand-new concept only when China and Russia agreed to establish a strategic, cooperative partnership facing the twenty-first century in April 1996, one year later, Chinese leaders already showed considerable initiative in promoting further development of the strategic, cooperative partnership. At the April 1997 summit, the Chinese side drafted the joint declaration on global multipolarization and the establishment of a new international order and secured the Russians' endorsement. The document provided a comprehensive elaboration of the theoretical content of the strategic partnership relationship. At the same time, China established various types of partnerships with other major powers.

It appears that China's motivation included the following elements. In the first place, China wanted to secure a peaceful, stable international environment to pursue its modernization. After the Tiananmen incident, Deng Xiaoping still stressed the need to "adhere to the basic line for one hundred years, with no vacillation,"[60] and "quietly immerse ourselves in practical work to accomplish something—something of China."[61] On August 28, 1998, Jiang Zemin, in his speech at the ninth ambassadorial conference held by the Ministry of Foreign Affairs, indicated that Chinese foreign policy straddling across the next century "should work hard to generate a better environment along China's periphery and internationally to realize the strategic objective of China's socialist modernization construction."[62] A stable international environment includes the global environment, the regional environment, and the environment along one's borders. In the case of China, Chinese leaders believe that geopolitical factors dictate that the stabilization of China's relations with the major powers is crucial to its securing a stable international environment.

Second, Chinese leaders wanted to promote global multipolarization and seek due recognition of China's status as an important pole in a mulitpolar world. In March 1990, Deng Xiaoping made the following observation in his discussions with some leading cadres: "The situation in which the United States and the Soviet Union dominated all international affairs is changing. Nevertheless, in future when the world becomes three-polar, four-polar or five-polar, the Soviet Union, no matter how weakened it may be and even if some of its republics withdraw from it, it will still be one pole. In the so-called multi-polar world, China will be counted as a pole."[63]

In August 1998, in his speech to the ninth ambassadorial conference organized by the Chinese Ministry of Foreign Affairs, Jiang Zemin indicated that the world structure would be developing in the direction of mulitpolarity. This trend occurred at the end of the Cold War, when the international situation became more relaxed, and when world peace forces continued to expand. It reflected the deep changes in international relations and the progress of the times. Further development of this trend of multipolarization at various levels and in various areas would help to weaken and restrain hegemonism and power politics, promote the establishment of a just and rational international political and economic new order, and bring a peaceful, stable and prosperous world into the next century. Chinese leaders believe that China's pursuit of multipolarity in the international political and economic order is only natural in view of its position as a major power with considerable influence, as the largest developing country, and as a member of the Third World.[64]

Finally, the Chinese leadership has been engaging in developing new strategic thinking and ideological positions in support of China's emergence as a major power in the coming century. The emergence of a major power relies not only on "hardware" such as economic power and military force, it also depends on "software," including an international strategy, values, and so on.[65] As argued by Paul Kennedy, the wisdom of the governments concerned will be a crucial factor determining the outcome of the competition among the United States, Russia, China, Japan, and the European Union.[66] Some Chinese scholars consider that the correctness of the international strategy is a fundamental factor contributing to the rise and fall of a country.[67]

In contrast to the two reestablishments of an international order after the two world wars,[68] China for the first time in this century feels that it has the opportunity to take part in the rebuilding of an international order under favorable conditions, particularly in view of its being recognized as a major power in the post–Cold War era. Chinese leaders consider that they have a genuine opportunity to exert some positive influence in shaping the new world order. During the Cold War, China was probably the largest and strongest country outside the two camps, but this position meant that basically China could only attempt to block other major powers' designs without any chance of realizing its own blueprint.

Chinese leaders anticipate that China's emergence as a major power may attract a strong negative response, especially from other major powers. The "China threat" theory is obviously one type of response. Hence, Chinese leaders consider the handling of China's relations with its neighbors and other major powers a serious challenge. The partnerships that China has developed with various major powers since the mid-1990s reflect the strategic thinking of the Chinese leadership in preparing the country for the next century. By now, China's network of partnerships is more or less in shape. Within this network, China hopes to forge a pattern of relations among major powers based on equality, cooperation, mutual respect, and mutual checks and balances. This, to some extent, also reflects the changing values of Chinese foreign policy from classical realism to legalism.[69] There is a recognition among Chinese scholars that international law is not a strong point among Chinese diplomats, and China must work hard to enhance its influence in international organizations.

Toward a New Model in Managing Relations among Major Powers in the Twenty-first Century

In international politics, relations among major powers tend to be the focus of attention. The balance among major powers has a direct impact on the evolution of patterns of international relations.[70] The rise and decline of major powers are the fundamental forces shaping the changes of international orders and the establishment of new ones.[71] After the conclusion of the Cold War, whether or not the major powers can surpass the confrontational alliances of the previous era and establish constructive, cooperative partnerships will be the crucial question determining the new world order in the next century.

Apparently, the above two models are the fundamental choices. Alliances are common forms of relations among states. Through formal treaties and agreements, states bind themselves together against another state or group of states, leading to confrontation, conflicts or even wars.[72] Military alliances were the earliest type of alliances, and it spread to include economic, political, and diplomatic alliances. The most significant recent example was the Cold War in which NATO and the Warsaw Pact engaged in decades of political, military, economic, and ideological confrontation in a zero-sum game.[73]

After the Cold War, the alliance system among Western powers led by the United States has continued. The disappearance of common enemies, the trend toward global multipolarization, sharpening economic competition among members of the alliance, and dissatisfaction with the U.S. predominance among other Western countries have threatened the cohesiveness of the alliance system. To arrest the deterioration of the alliance system, the United States has attempted to generate new missions for the alliance system through the eastward expansion of NATO and the revision of the U.S.-Japan security cooperation guidelines and to strengthen the partnerships among alliance members. The idea of partnership among Western powers had been raised in the Cold War era, though it has received greater emphasis recently. But the common form of relations among Western states is still the alliance relationship, even the concept of partnerships proposed still contains a strong legacy of Cold War confrontation. This alliance model differs markedly from the partnerships among major powers proposed by Chinese leaders in the post–Cold War era.

China has already established partnerships with most major countries of the world (see Table 10.1) and so have the United States and Russia, France and Russia, and others. According to the idealistic point of view of the Chinese leadership, the basic characteristics of the partnerships it advocates are: partners should be equal, should trust each other, should not form alliances nor engage in confrontation, and should not direct themselves against any third country or damage any third country's interest. Partnerships of this kind symbolize a sense of global responsibility and removal of the Cold War legacy and mentality. The development of constructive relations of mutual trust and cooperation will promote positive reforms of the international political and economic order and help to move away from the traditional zero-sum-game situations among major powers. Partnerships along this line will facilitate cooperation and stable relations among major powers, as well as promote stability, economic development and multipolarization globally. Chinese leaders consider the Sino-Russian strategic, cooperative partnership an experiment and a pioneering effort regarding relations among major powers in the post–Cold War era and in a way a model for the next century.[74]

This idealistic view naturally has to be adjusted to the demands of *realpolitik*, and Chinese leaders have rarely been slow in meeting its challenges. However, the idea of constructive, cooperative partnerships apparently has a broad appeal to the major powers, which can easily accept the idealistic aspects of the concept and find its practical aspects useful. Hence the concept has become common diplomatic language in the fluid transition toward multipolarity.

Conclusion

After the end of the Cold War and especially since the mid-1990s, China has been adjusting its international strategy and redefining its international role in view of the changing global balance of power and China's improvement of its comprehensive national power. Chinese leaders believe that their promotion of multipolarity assumes an important role in this adjustment process. As Jiang Zemin observed, "China enjoys an advantageous position in the adjustment of relations among major powers, China's international status has been obviously improving, and its international role has been strengthening day by day."[75]

In the post–Cold War era, China's international strategic behavior has shown different patterns and characteristics. Compared with the nonalignment model in the 1980s, the partnership model allows China to take more initiative. This is in line with Deng Xiaoping's earlier appeal that China has to show its commitments (yousuo zuowei) to international strategy. This initiative is mainly based on the Chinese leadership's considerations regarding relations among major powers in the next century as well as the implications of China's emergence as a major power.

Although China's network of partnerships is more or less in shape and the partnership model has become a significant trend in the post–Cold War era, China still encounters many restrictions. In the first place, China's capabilities are limited; it is not strong enough to direct other major powers to act in accordance with its strategic principles. More important still, the United States remains the sole superpower, and a genuine multipolar world is at best a distant goal. The United States' attempt to maintain its superior status in a unipolar world constitutes the major obstacle to China's promotion of its new model of relations among major powers.

The breakup of the Soviet Union means that the predominance of the United States faces no formidable challenges,[76] and Chinese leaders believe that it will attempt to shape the international balance in its favor and maintain its predominant position. After the Gulf War victory, the Bush administration proposed a "new world order." In April 1994, the Clinton administration presented a "new strategic framework" for NATO. Its intervention in Iraq and Kosovo, bypassing the UN Security Council, were further indicators of its attempt to build a new global power center—NATO—as its tool to establish a unipolar world. In some ways, the Asian financial crises in 1997–1998 helped to strengthen the American position, and NATO's intervention in Kosovo in the spring of 1999 demonstrated significant support for the United States in the Western world.

The United States' international strategy is perceived by the Chinese leadership as a countercurrent against the trend toward multipolarity. The prospects of partnerships among major powers promoted by China and other countries will depend on the expansion of their aggregate comprehensive national power and the competition between the two forces attempting to build their respective world orders. The process of building a multipolar world will be a long and complicated one. Chinese leaders believe that domestic politics in the United States and its China policy will have a significant impact on China's international strategic goals of peace and development.

Postscript

The crucial factor for China's success in establishing a network of partnerships with the key countries in its foreign policy framework was the change of the China policy on the part of the Clinton administration in its second term. In October 1997, China and the United States declared that they would build toward a constructive strategic partnership in the twenty-first century. Although the United States was not the first country to establish a partnership with China, its status as the sole superpower determined its significant role in China's network of partnerships. In the eyes of the Chinese leadership, if Al Gore had won the presidency in the 2000 presidential election and maintained the China policy of the Clinton administration, China's network of partnership would be developing smoothly. However, George W. Bush's electoral victory and his hawkish policy toward China led to a rapid cooling of the bilateral relationship, with the potential danger of leading to a downward spiral.

In his presidential election campaign, George Bush already regarded China as a "strategic competitor." After his inauguration, Secretary of State Colin Powell emphasized that China was "not a strategic partner, but a trade partner," and the challenge for the United States was to make efforts to maintain a constructive relationship with China in accordance with American interests. Subsequent to the spy plane incident on April 1, 2001, the Bush administration further adopted a series of unfriendly measures: massive arms sales to Taiwan, including strong statements on assisting Taiwan in its self-defense; appointment of a senior State Department official as the Coordinator for Tibetan Affairs; and termination of bilateral military exchanges. These measures almost pushed Sino-American relations to a dangerous situation.

It appeared to Beijing that after Colin Powell's visit to China in July 2001, the Bush administration gradually abandoned the definition of the bilateral relationship as one between competitors. Colin Powell indicated that the Bush administration expected to develop Sino-American relations and treat them seriously, so as to bring the bilateral relationship from a state of high tension to that of relaxation. In view of the terrorist attacks on September 11, 2001, the Bush administration further adjusted its China policy in its promotion of an international coalition to combat terrorism. Indeed the incident provided an important opportunity to improve relations with China. President Bush attended the APEC informal summit in Shanghai in the following September and pledged to develop a constructive relationship of cooperation with China. There were three meetings between the two heads of state in 2002. In view of the redefinition of the bilateral relationship from one between competitors to one of constructive, cooperative partners, the Bush administration reaffirmed the "one-China principle" on the Taiwan issue and resumed bilateral military contacts and exchanges. It further agreed to include the East Turkestan Islamic Movement in its list of terrorist organizations and proceeded to freeze the group's assets in the United States. Sino-American relations have returned to the proper track of contact, dialogue, and cooperation. David Shambaugh even regarded that the two countries had already established a de facto relationship of strategic partnership, and that the bilateral relationship was at its best since the Tiananmen incident.[77]

The practice of unilateralism and the strengthening of its unipolar global strategy on the part of the Bush administration based on the strength of the United States as a superpower, however, have been perceived by Beijing as a serious obstacle to China's construction and development of a framework of strategic partnerships among the major powers. The Chinese leadership appreciates China's limitations and would like to promote partnerships among the major powers on the basis of equality, mutual trust and understanding, mutual respect, and mutual accommodation of their respective interests. With such a foundation, they can then participate in the management of international affairs in an equal and cooperative manner, moving toward the ideal of an order of multipolarity. Thus Chinese leaders anticipate serious obstacles in their promotion of a partnerships network in the twenty-first century. In view of the above, they have been adopting a low profile in recent years in such efforts; they appear to be more pragmatic too in consideration of the balance of the strategic interests among the major powers concerned. After all, the stage of declaration of the establishment of partnerships has already been completed.

Notes

1. Due to differences in cultural backgrounds and perspectives, Chinese scholars and their Western counterparts have offered different definitions of international strategy. Regarding the works of Chinese scholars, see Chen Zhongjing, *Guoji Zhanlue Wenti* (International Strategic Questions) (Beijing: The Current Affairs Press, 1987), p. 2; Jin Yu, ed., *Deng Xiaoping Guoji Zhanlue Sixiang Yanjiu* (A Study of Deng Xiaoping's International Strategic Thinking) (Shenyang: Liaoning People's Press, 1992), p. 3; Liang Shoude and Hong Yinxian, *Guoji Zhengzhi Gailun* (A General Discussion of International Politics) (Beijing: Chinese Central Editing and Translation Press, 1994), pp. 59–60; Shi Zhifu ed., *Zhonghua Renmin Gongheguo Duiwai Guanxishi 1949.10–1989.10* (A History of Foreign Relations of the People's Republic of China, 1949.10–1989.10) (Beijing: Peking University Press, 1994), p. 1; and Peng Guangqian and Yao Youzhi, eds., *Deng Xiaoping Zhanlue Sixiang Lun* (On the Strategic Thinking of Deng Xiaoping) (Beijing: Military Science Press, 1994), pp. 56–57. Regarding the works of Western scholars, see Liddel Hart, *Strategy* (New York: Praeger, 1967); Allan W. Lenter, *The Politics of Decisionmaking: Strategy, Cooperation and Conflict* (Berkeley Hills, CA: Sage, 1976); Colin S. Gray, *Strategic Studies: A Critical Assessment* (Westport, CT: Greenwood Press, 1982); Helmut Schmidt, *A Grand Strategy for the West: The Anachronism of National Strategies in an Interdependent World* (New Haven, CT: Yale University Press, 1985); Edward Luttwark, *Strategy: The Logic of War and Peace* (Cambridge, MA: Belknap Press of Harvard University Press, 1987); Elmer Plischke, *Foreign Relations: Analysis of Its Anatomy* (Westport, CT: Greenwood Press, 1988); and K. J. Holsti, *International Politics: A Framework for Analysis* (Englewood Cliffs, NJ: Prentice-Hall, 1997).
2. Stephen A. Kocs, "Explaining the Strategic Behavior of States: International Law as System Structure," *International Studies Quarterly* 38, no. 4 (December 1994): 535–556.
3. This is the typical periodization of Chinese foreign policy in the Cold War era presented by Chinese officials and scholars. Former Chinese foreign minister Qiao Guanhua summarized Chinese foreign policy as follows: "Uniting with the Soviet Union to deter the United States in the 1950s, combating the two superpowers in the 1960s, and uniting with the United States to deter the Soviet Union in the 1970s." See *Guoji Zhanlue Yanjiu Jijinhui*, ed., *Huanqiu Tongci Liangre* (The Entire Planet Sharing the Same Temperature) (Beijing: Central Documentary Press, 1993), p. 2; Zhang Xiaoming, "Lengzhan Shiqi Xinzhongguo de Sici Duiwai Zhanlue Juece" (China's Four International Strategic Decisions in the Cold War Era), in Liu Shan and Xue Junde, eds., *Zhongguo Waijiao Xinlun* (A New Treatise on Chinese Foreign Policy) (Beijing: World Affairs Press, 1997), pp. 1–20.

4. Zhang Wankun, *China's Foreign Relations Strategies under Mao and Deng: A Systematic and Comparative Analysis*, Working Paper Series 1998, No.2 (Department of Public and Social Administration, City University of Hong Kong), pp. 15–18.

5. For the English literature on China's emergence as a major power, see, for example, William H. Overholt, *The Rise of China: How Economic Reform Is Creating a New Superpower* (New York: W. W. Norton, 1993); David S.G. Goodman and Gerald Segal, eds., *China Rising: Nationalism and Interdependence* (London and New York: Routledge, 1997); and Stuart Harris and Gary Klintworth, eds., *China as a Great Power: Myths, Realities and Challenges in the Asia-Pacific Region* (New York: St. Martin's Press, 1995).

6. This point was raised by Chinese foreign minister, Tang Jiaxuan, when he answered reporters' questions at a press conference held during the second plenary meeting of the Ninth National People's Congress on March 7, 1999. See *Renmin Ribao*, March 8, 1999.

7. Zhongguo Shehui Kexueyuan Yuyan Yanjiusuo Cidian Bianjishi (Dictionary Editorial Office, Language Research Institute of Chinese Academy of Social Sciences), *Xiandai Hanyu Cidian (xiudingben)* (Modern Chinese Dictionary, revised ed.) (Hong Kong: Commercial Press, 1992), p. 514.

8. *The American Heritage Dictionary of the English Language*, 3rd ed. (Boston: Houghton Mifflin Company, 1992), pp. 1320–1321.

9. The "no alliance" and "not directed against or at the expense of a third party" principles were first raised in the Sino-Soviet Joint Declaration in 1989; see *Renmin Ribao*, May 19, 1989. The "no confrontation" principle was first proposed in connection with Sino-American relations in the post–Cold War era; see *Renmin Ribao*, December 1, 1992.

10. Chinese Premier Zhu Rongji mentioned this in his speech at the Massachusetts Institute of Technology during his trip to the United States in April 1999; see *Renmin Ribao*, April 16, 1999.

11. Ibid., December 18, 1997.

12. Ibid., April 26, 1997.

13. "Goujian Xinshiji de Xinxing Guojia Guanxi" (Constructing a New Model of State-to-State Relationship for the New Century), *Renmin Ribao* editorial, December 8, 1997.

14. The Five Principles of Peaceful Coexistence are: respect for territorial integrity and sovereignty, noninterference in domestic affairs, equality and mutual benefit, nonaggression, and peaceful coexistence.

15. Joseph Y.S. Cheng, "Zhongguo de 'Xiandaihau' Waijao Zhengce" (China's Modernizations' Foreign Policy), in Joseph Y.S. Cheng, ed., *Zhongguo yu Yazhou* (China and Asia) (Hong Kong: Commercial Press, 1990), pp. 3–36.

16. Liang Shoude, "Maixiang Ershiyi Shiji de Shije he Zhongguo de Waijao Zhanlue" (The World Approaching the Twenty-first Century and China's Foreign Policy Strategy), in Liang Shoude, ed., *Guoji Zhengzhi Xinlun* (A New Treatise on International Relations) (Beijing: Chinese Social Science Press, 1996), pp. 28–41.

17. *Zhongguo Gongchandang Di Shierci Quanguo Daibiao Dahui Wenjian* (Documents of the Twelfth National Congress of the Communist Party of China) (Hong Kong: Joint Publishing [H.K.] Company Ltd., 1982), p. 23. *Shisida Yilai Zhongyao Wenxian Xuanbian (Shang)* (A Selection of Important Documents since the Fourteenth National Congress of the Communist Party of China, vol. 1) (Beijing: People's Press, 1996), p. 34.

18. *Renmin Ribao*, September 13, 1997.

19. Ibid., May 10, 1999.

20. Deng Xiaoping, "Opening Speech at the Twelfth National Congress of the Communist Party of China (September 1, 1982)," in *Selected Works of Deng Xiaoping (1982–1992)* (Beijing: Foreign Languages Press, 1994), p. 14.

21. Deng Xiaoping, "We Must Safeguard World Peace and Ensure Domestic Development (May 29, 1984)," *Selected Works*, pp. 66–67.

22. Han Nianlong, ed., *Dangdai Zhongguo Waijao* (The Foreign Policy of Contemporary China) (Beijing: Chinese Social Science Press, 1990), pp. 472–475.

23. Ibid., p. 340; Zhang Guang, *Zhongguo de Waijiao Zhengce* (China's Foreign Policy) (Beijing: World Affairs Press, 1995), pp. 46–52.

24. In 1998 and 1999, the authors visited the Chinese Academy of Social Sciences, Beijing University, the Central Party School, the China Institute for International Strategic Studies, and the Shanghai Institute of International Studies. They held extensive discussions with over fifty academics and research workers on Chinese foreign policy. To facilitate exchange of ideas, they will not be quoted directly. Instead, their views will be summarized and presented as those of the Chinese research community on China's foreign policy.

25. For details, see *Yazhou Zhoukan* (Asia Weekly) (a Chinese weekly in Hong Kong), 13, no. 14, April 11, 1999, p. 18.

26. Deng Xiaoping, "The Principles of Peaceful Coexistence Have a Potentially Wide Application (October 31, 1984)," *Selected Works*, pp. 102–104.

27. Deng Xiaoping, "A New International Order Should Be Established with the Five Principles of Peaceful Coexistence as Norms (December 21, 1988)," *Selected Works*, pp. 274–276.

28. Deng Xiaoping, "Seize the Opportunity to Develop the Economy (December 24, 1990)," *Selected Works*, p. 350.

29. Authors' interviews in Beijing and Shanghai in 1998 and 1999.

30. Zheng Suchun, "*Mianxiang Xinshiji de Zhongou Guanxi* (Sino-European Relations Facing the New Century)," *Renmin Ribao* online, October 22, 2000.

31. Robert Ash, "Quarterly Chronicle and Documentation," *The China Quarterly* 154 (June 1998): 475.

32. Joseph Y.S. Cheng, "Jiushi Niandai Zhongguo de Dongmeng Zhengce: Tuidong Diqu Duojihua" (China's ASEAN Policy in the 1990s: Pushing for Multipolarity in the Regional Context)," in Joseph Y.S. Cheng, ed., *Houlengzhan Shiqi de Zhongguo Waijiao* (Chinese Foreign Policy in the Post–Cold War Era) (Hong Kong: Cosmos Book, 1999), pp. 170–210.

33. See, for example, Zhang Dalin, "An Analysis of Japan's Strategy of Returning to Asia," *International Studies* 4–5 (published by the China Institute of International Studies in Beijing) (1994): 1–10.

34. Shang Huipeng, with Lin Liangguan and Han Hua, "Indo-Pakistani Security Dialogue and China's Policy Towards South Asia," in China Society for Strategy and Management Research ed., *Study Reports on International Situation 1997–1998* (Beijing: China Society for Strategy and Management Research, 1998), p. 186.

35. *South China Morning Post* (an English newspaper in Hong Kong), April 28, 1998 and May 5, 1998.

36. *China Daily* (an English newspaper in Beijing), May 16, 1998, and *Sunday Morning Post*, May 17, 1998.

37. *South China Morning Post*, May 13, 1998.

38. *Ming Pao*, September 21, 1998.

39. *Renmin Ribao*, May 30, 2000.

40. Deng Xiaoping, "Let Us Put the Past Behind Us and Open Up a New Era (May 16, 1989)," *Selected Works*, pp. 284–287.

41. Qian Qichen, "Jieshu Guoqu Kaipi Weilai—Huiyi Deng Xiaoping Tongzhi Guanyu Shixian Zhongsu Guanxi Zhengchanghua de Zhanlue Juece" (Terminate the Past, Open Up the Future—Remember Comrade Deng Xiaoping's Strategic Decision on Realizing the Normalization of Sino-Soviet Relations), *Renmin Ribao*, February 20, 1998.

42. Ibid., May 20, 1989.

43. Xia Yishan, "Sino-Russian Partnership Marching Into the 21st Century," *Beijing Review* 40, no. 18, May 5–11, 1997, pp. 9–12.

44. Tai Ming Cheung, "China's Buying Spree: Russia Gears Up to Upgrade Peking's Weaponry," *Far Eastern Economic Review* 156, no. 31, August 7, 1993, p. 24.

45. Ronald Montaperto, "China as a Military Power," *Strategic Forum* (Internet edition) 56, December 1995.

46. *South China Morning Post*, November 28, 1996.

47. Tai Ming Cheung, "China's Buying Spree."

48. *The Washington Post*, March 31, 1993.

49. *South China Morning Post*, November 22, 1997.

50. The Yakhont missile reportedly flies at a low altitude of five to fifteen meters during the last 120 kilometers en route to a target. See *South China Morning Post*, November 22 and 26, 1997.

51. Jennifer Anderson, *The Limits of Sino-Russian Strategic Partnership* (Oxford, UK: Oxford University Press, 1997), pp. 13–25, 53–61; and Yan Zheng, "Haolinju Haohuoban Haopengyou—Zhonge Zhanlue Xiezuo Huoban Guanxi Jiqi Yiyi" (Good Neighbors, Good Partners and Good Friends—the Sino-Russian Strategic, Cooperative Partnership and Its Meaning), *Renmin Luntan* (People's Forum) (Beijing), no. 5, 1998, pp. 16–20.

52. For details of the remarks of Jiang Zemin and Li Peng, see *Renmin Ribao*, April 26, 1997 and December 18, 1996, respectively.

53. Wu Guoguang, "Zai Zhanlue Chongtu Xia Duihua: Bianhua Zhong de Zhongmei Guanxi" (Dialogue in Strategic Conflict: Sino-American Relations in Transition), in Law Kam-yee, ed., *Zhongguo Pinglun 1997* (China Review 1997) (Hong Kong: The Chinese University Press, 1998), pp. 51–78; Yang Xuetong, *Zhongguo yu Yatai Anquan: Lengzhan hou Yatai Guojia de Anquan Zhanlue Zouxiang* (China and Security in the Asia-Pacific Region: The Trends of the Asia-Pacific Countries Security Strategies in the Post–Cold War Era) (Beijing: The Current Affairs Press, 1998), pp. 251–278.

54. See Deborah Lutterbeck, Bruce Gilley, and Andrew Sherry, "Riders on the Storm," *Far Eastern Economic Review* 161, no. 25, June 25, 1998, pp. 10–12; American ambassador to China, James Sasser, also mentioned this point when he attended a meeting in New York on May 1, 1999, see *Ta Kung Pao* [Chinese newspaper in Hong Kong], May 3, 1999.

55. In February 1998, many officials and scholars from the United States and China attended the symposium on "The Sino-American Relations Facing the Twenty-first Century: The Constructive Strategic Partnership" held in Shanghai. Chinese participants emphasized that, if the United States could not handle the Taiwan issue well, the constructive strategic partnership between China and the United States would become meaningless. See *"Mianxiang Ershiyi Shiji de Zhongmei Guanxi: Jianshexing de Zhanlue Huoban Guanxi" Yantaohui Zuizhong Baogao* (Final Report of the Symposium on "The Sino-American Relations Facing the Twenty-first Century: The Constructive Strategic Partnership") (Shanghai, February 16–18, 1998), p. 12. A number of Chinese scholars mentioned this when the authors interviewed them in Beijing and Shanghai in 1998 and 1999.

56. See *Ming Pao* editorial, May 15, 1999.

57. See *Renmin Ribao* editorials of May 21, May 25, May 28, June 2, and June 3, 1999, respectively.

58. Zheng Yu, "Zhong-E Zhanlue Xiezuo Huoban Guanxi de Qianjing" (The Prospects for the Sino-Russian Strategic, Cooperative Partnership), *Guoji Jingji Pinglun* (International Economic Review) (Beijing, September 11–October 1997), pp. 50–53; Anderson, *The Limits of Sino-Russian Strategic Partnership*, pp. 20–24; and John W. Garver, "Sino-Russian Relations," in Samuel S. Kim, ed., *China and the World: Chinese Foreign Policy Faces the New Millennium* (Boulder, CO: Westview Press, 1998), pp. 129–131.

59. U.S. President Bill Clinton confirmed this when he was interviewed by Chinese reporters of Chinese mass media stationed in the United States on the eve of his China visit on June 19, 1998; see, Zheng Yuan, ed., *Kelindun Fanghua Yanxinglu—Meiguo Zongtong de Zhongguo "Jainyan"* (Remarks and Deeds of President Clinton During His China Visit—American President's "Advice" for China) (Beijing: Chinese Social Science Press, 1998), p. 25; and *Wen Wei Po* [Chinese newspaper in Hong Kong], October 31, 1997.

60. Deng Xiaoping, "Excerpts from Talks Given in Wuchang, Shenzhen, Zhuhai and Shanghai (January 18–February 21, 1992)," pp. 358–359.

61. Deng Xiaoping, "With Stable Policies of Reform and Opening to the Outside World,

China Can Have Great Hopes for the Future (September 4, 1989)," *Selected Works*, p. 311.

62. *Renmin Ribao*, August 29, 1998.

63. Deng Xiaoping, "The International Situation and Economic Problems (March 3, 1990)," *Selected Works*, p. 341.

64. Yan Xuetong, "Study Report on China's Foreign Relations in 1997–98: International Security Environment for China's Rise," in *China Society for Strategy and Management Research*, ed., pp. 81–91; Joseph Y.S. Cheng, *"Jiushi Niandai Zhongguo de Dongmeng Zhengce: Tuidong Diqu Duojihua"* (China's ASEAN Policy in the 1990s: Pausing for Multipolarity in the Regional Context); and Zhang Wankun, *"Houlengzhan Shiqi Zhongguo Junshi Zhanlue Fenxi"* (An Analysis of China's Balance of Power Strategy in the Post–Cold War Era), in Joseph Y.S. Cheng, ed., *Houlengzhan Shiqi de Zhongguo Waijiao* (Chinese Foreign Policy in the Post–Cold War Era), pp. 170–210, 100–130, respectively.

65. Joseph Nye, *Bound to Lead: The Changing Nature of American Power* (New York: Basic Books, 1990).

66. Paul Kennedy, *The Rise and Fall of the Great Powers: Economic Change and Military Conflict from 1500 to 2000* (New York: Random House, 1987), p. 540.

67. Huang Shuofeng, *Guojia Shengshuai Lun* (On the Rise and Fall of States) (Changsha: Hunan Press, 1996), pp. 1–5; Bo Guili, *Guojia Zhanlue Lun* (On National Strategy) (Beijing: China Economic Press, 1994), pp. 8–9.

68. The first two rebuilding of the international orders in this century refer to the ones after World War I and World War II. After World War I, China, although a victorious country, still could not resist the decisions of Western powers detrimental to its interests. By the end of World War II, although China was treated nominally as one of the Five Great Powers, China's national interests were still comprised by the secret treaty concluded by the United States and the Soviet Union.

69. Li Baojun, "Zhanlue Huoban Guanxi Yu Zhongguo Waijao de Lishi Jueze" (Strategic Partnerships and the Historical Choice for Chinese Foreign Policy), in Joseph Y.S. Cheng, ed., *Houlengzhan Shiqi de Zhongguo Waijao* (Chinese Foreign Policy in the Post–Cold War Era), pp. 75–99.

70. Joshua S. Goldstein, *International Relations*, 3rd ed. (New York: Longman, 1999), pp. 27–29, 77–88; K. J. Holsti, *International Politics: A Framework for Analysis* (Englewood Cliffs, NJ: Prentice-Hall, 1995), pp. 254–258, 330–331; and Liang Shoude and Hong Yinxian, *Guoji Zhengzhi Gailun*.

71. Robert Gilpin, *War and Change in World Politics* (New York: Cambridge University Press, 1981); Derek McKay and H. M. Scott, *The Rise of Great Powers (1648–1815)* (London and New York: Longman, 1983); Robert Keohane, *After Hegemony: Cooperation and Discord in the World Political Economy* (Princeton, NJ: Princeton University Press, 1984); Paul Kennedy, George Modelski, *Long Cycles in World Politics* (London: Macmillan Press, 1987); Liu Jinghua, *Baoquan de Xingshuai* (The Rise and Fall of Hegemony) (Beijing: China Economic Press,1997); and Zhu Ning, *Xiage Shiji Shuizuiqiang—Jinhou Ershinian Zhongmei Jingzheng de Diyuan Zhanlue Gangyao* (Who Will Be the Strongest in the Next Century? The Geopolitics and Strategic Outlines of Sino-American Competition in the Next Twenty Years) (Shenyang: Liaoning People's Press, 1997).

72. Stephen M. Walt, *The Origins of Alliances* (Ithaca, NY: Cornell University Press, 1987).

73. Joseph Frankel, *International Relations in a Changing World*, 4th ed. (Oxford, UK: Oxford University Press, 1988).

74. See, "Yici Yiyi Zhongda Yingxiang Shenyuan de Fangwen—Zhuhe Jiang Zhuxi Fange Yuanman Chenggong" (A Meaningful and Influential Visit—Congratulations for President Jiang Zemin's Successful Visit to Russia), *Renmin Ribao* editorial, April 27, 1997; and Yan Rong, "The Success of Big Power Relations—A Survey of Sino-Russian Strategic Cooperation Partnership," *Beijing Review* 41, no. 47, November 23–29, 1998, pp. 7–9.

75. Jiang Zemin, General Secretary of the Communist Party of China, mentioned this point

in his speech at the National Organizational Work Conference on December 22, 1997; see *Renmin Ribao*, April 2, 1998.

76. Noam Chomsky, *What Uncle Sam Really Wants* (Tucson, AZ: Odonian Press, 1992), pp. 74, 81–82.

77. Views expressed by Professor David Shambaugh in his address to the Hong Kong Foreign Correspondents' Club on December 11, 2002. See http://www.people.com.cn, December 13, 2002.

11

Ghost of the Strategic Triangle

The Sino-Russian Partnership

Lowell Dittmer

The relationship between the two vast empires astride the Eurasian heartland has been troubled for centuries, despite certain superficial similarities in size and political-economic structure. The Mongol Golden Horde successfully invaded Russia in the thirteenth century, burning Moscow and taking Kiev, and they continued to rule southern Russia and extort tribute for the next two hundred years, leaving a historical legacy of dread. Russia would lag China developmentally for the next several centuries, with a population that did not reach 13 million until 1725 (compared to China's brilliant civilization and ca. 150 million people), and the first visitors to Beijing in the modern era (beginning in the mid-seventeenth century) were obliged to prostrate themselves (*koutou*) before the Qing emperor. Yet the decline of the Manchu Dynasty coincided with Russian industrialization following the defeat of Napoleon, and Russian appetites for trade and territorial expansion led to increasing infringement on imperial China. The Russian imperialist strategy was that of a "free rider": Russian forces typically pressed their claims only when China was preoccupied by more urgent threats. Thus in 1854–1859, while China was engulfed by the Taiping Rebellion (1851–1864), General N. N. Murawjew and twenty thousand troops occupied the delta and north shore of the Amur/Heilong River and the maritime provinces without firing a shot. During the second Opium War, Russian forces made further opportunistic inroads, formalized in the 1860 Sino-Russian Treaty of Beijing. During the Yakub Beg Rebellion in Xinjiang, Russian troops occupied part of the Yili region, formalized in the Treaty of Livadia (later modified slightly in China's favor in the Treaty of St. Petersburg). Completion of the Trans-Siberian railway and the decline of the Qing offered further opportunity for cheap acquisitions, and in 1898 Russia made Port Arthur and Dalian imperial treaty ports, occupied Manchuria in the wake of the Boxer Rebellion, and extended its sphere of influence over China's northeast in 1905. After encouraging the Mongols to rebel in 1910, Russia established a protectorate over Outer Mongolia in the midst of the 1911 Xinhai revolution.

After the Bolshevik Revolution the new Soviet regime renounced its share of the Boxer reparations as well as most imperialist privileges in China, and quickly established diplomatic relations with the short-lived Peking Republic (1924), while also helping to organize and advise the Chinese Communist Party (CCP), and to assist in the reorganization of the Nationalist Party (Kuomintang or KMT), thus insinuating its interests in China through all available avenues. After some three decades of turmoil, during which two tenuous

Communist-Nationalist united fronts fell apart in the process of trying to reunite the country and ward off foreign invasion and the CCP flirted with annihilation, the CCP ultimately drove the KMT from the mainland and turned to the Communist Party of the Soviet Union (CPSU) for help in consolidating its revolution. After prolonged and wary negotiations, Mao and Stalin signed a thirty-year Sino-Soviet Treaty of Friendship, Alliance and Mutual Assistance (February 14, 1950), capitalizing on shared ideological values and a history of revolutionary collaboration to establish a Eurasian partnership in pursuit of worldwide revolution.[1] Moscow agreed to provide a loan of US$300 million over five years, plus construction aid in building fifty (eventually thrice that) massive heavy industrial projects, and ceded most of the concessions it had recently gained in negotiations with the Nationalist regime. But not until Beijing sent "volunteers" into the Korean War, and soon after contributed generous aid and advisors to the first Indochina War (particularly at Dien Bien Phu), was Stalin fully satisfied with the Chinese contribution. Although this massive exercise in transplanting modern (socialist) industrial culture from one country to another was to end badly, for amply documented reasons, the period of cooperation made an enduring contribution to Chinese political economic development, meanwhile establishing "old school ties" of lasting value with the next generation of Chinese and Russian leaders.[2]

The alliance fell apart in the 1960s in the wake of a snarl of perceived betrayals including Khrushchev's renunciation of the cult of personality, China's ideologically presumptuous Great Leap Forward, Soviet fecklessness in defense of Chinese interests in Taiwan and subsequent rescission of its promised help with Chinese nuclear development, diverging priorities over whether to win the Third World through People's War or parliamentary coalition building, and the future direction of proletarian dictatorship in established socialist regimes. The period of friendship segued in the late 1950s into three decades of fratricidal polemics, diplomatic encirclement, and counterencirclement maneuvers, an arms race, and border violence that obsessed both sides at the time and has puzzled them ever since. As we now know, the most sensitive phase of this rivalry was touched off by the Soviet invasion of Czechoslovakia in August 1968, exciting Chinese apprehensions of analogous application of the incipient Brezhnev Doctrine to the PRC and leading to the series of border clashes initiated by Beijing in March 1969. After Mao's death in August 1976 the ideological animus against "socialist hegemonism" began to dissipate, while the rise of Reagan's anti-Soviet crusade led to a new Cold War and a bipolar arms race. Beijing came to interpret this exclusively in "superpower" terms, a status to which China did not then aspire, hence relieving Beijing of some of its security concerns: PRC arms spending was reduced by some 7 percent per annum as a proportion of GDP from 1979–1989, while these funds were invested in economic growth. Whereas the Chinese quietly allowed the thirty-year Sino-Soviet treaty to lapse upon its expiry in 1981, they agreed to discuss mutual problems, and beginning in 1982, after Sino-American normalization and the Third Communiqué, a new series of Sino-Soviet "normalization" talks were held, alternating semiannually between the two capitols in the spring and fall of each year, usually involving the same team of officials on either side. Progress was initially glacial due to Soviet intransigence over what Beijing called the "three fundamental obstacles": heavy fortification of the Sino-Soviet border and in Outer Mongolia, Soviet troops in Afghanistan, and support of the Vietnamese threat to China's southeastern flank (and to China's ally Cambodia). Talks nevertheless continued on schedule, accompanied by gradually increasing trade and cultural exchanges, helping to contain the dispute during the post-Mao and post-Brezhnev succession crises.

When Gorbachev decided to rationalize Soviet foreign policy in the late 1980s, he decided, while retrenching high-risk ventures in the Third World, to try to revive the Sino-Soviet friendship, in hopes of creating a Eurasian socialist redoubt. In speeches at Vladivostok (July 1986) and Krasnoyarsk (September 1988), he proposed a freeze on deployment of nuclear weapons in the Asia-Pacific region, Soviet withdrawal from the Cam Ranh Bay naval facility in Vietnam, and unilateral reduction of the Soviet military by 500,000 troops within two years, nearly half (200,000) of which would come from the region east of the Urals. This Soviet "new thinking" (*novo myshlenie*), according to which Brezhnev's vaunted achievement of "strategic parity" had redounded in few substantial gains at immense cost, eventually satisfied all three Chinese "obstacles." Meanwhile, inasmuch as both countries' economies were running aground on the limits of "extensive development" under command planning—the Soviet Union after years of stagnation under Brezhnev, China after radical Maoism had exhausted itself in the Cultural Revolution—fresh leadership teams in both capitols turned to "socialist reform," an attempt at revitalization referred to as perestroika and *gaige kaifang*, respectively. There was again a sense that both countries, with symmetrically structured and ideologically oriented economies, might learn from one another. Because China had been first to experiment with reform, much of the initial learning was by the Soviet Union, but China also paid close attention to Soviet experiments. In fact, the liberalization that culminated in the 1986 protest movement that in turn led to the fall of Hu Yaobang had been inspired by Gorbachev's prior call for Soviet political reform (as well as Deng Xiaoping's Delphic encouragement). Whereas such "learning" was to be sure selective and ultimately led in divergent directions, the fact that both countries were engaged in analogous socioeconomic experiments and interested in each other's experience helped to orchestrate their détente.[3] Based then on both foreign policy and domestic policy convergence, it had become possible by the end of the 1980s to convene a summit meeting sealing the "normalization" of party-to-party relations.

This summit, held in May 1989 amid student demonstrations at Tiananmen Square that necessitated moving all ceremonies indoors, quite unexpectedly marked both climax and terminus to this process of convergence around socialist reform. The sanguinary Chinese solution to spontaneous student protests, implemented within a fortnight of Gorbachev's departure, led to international sanctions and to a Soviet decision to avoid any analogous crackdown, either domestically or among fellow Warsaw Pact Organization signatories.[4] But without resort to outside force the European socialist regimes could not stand, and by the end of 1990 all but China, North Korea, Laos, Vietnam, Cuba, and the Soviet Union had succumbed to a wave of anti-Communist protest movements. Throughout this period the Chinese leadership, still defending both Marxism-Leninism and their own brutal "solution" to mass protest, deplored this turn of events, criticizing the Gorbachev leadership for "deviating from the path of socialism" and of contributing to the collapse of the bloc. In early 1990, Deng Liqun and the more ideologically self-righteous "leftist" wing of the CCP even advocated a public critique of Soviet errors, which Deng Xiaoping vetoed. No sooner had Beijing become reconciled to cooperation with Gorbachev—after the Gulf War (January–February 1991), some socialist rejoinder to a triumphal American "new world order" was deemed advisable—than was Gorbachev's own survival threatened by the August 1991 coup attempt. Though Beijing came perilously close to supporting the coup publicly before its consolidation, it recovered in time to reaffirm its commitment to noninterference, only to witness with mounting dismay the ensuing dissolution of the Soviet Union into fifteen republics, twelve of which agreed to join the Commonwealth of Independent States (CIS).

Part of the reason for the PRC's quick recognition of this new political reality was that had it not, many alternatives seemed open to the former USSR: There seemed every likelihood of reconciling the old Russo-Japanese territorial dispute (involving three small islands and a tiny archipelago north of Hokkaido) and signing a peace treaty with Japan; South Korea had just granted Moscow a $3 billion concessionary loan in appreciation for its recognition of Seoul; and Taiwan briefly established consular relations with Latvia and very nearly exchanged ambassadors with the Ukraine and Outer Mongolia before being deterred by PRC warnings. The new line in the Kremlin under Yeltsin and Kozyrev, erstwhile *bête noires* of Chinese Kremlin watchers, who plausibly suspected the CCP of supporting the August 1991 coup conspirators, was anti-Communist and pro-American. Indeed, the first bilateral "strategic partnership" to be proclaimed (in 1992) was between the United States and the fledgling Russian federation. Chinese strategic analysts expressed concern lest successful Russian economic reform lure foreign direct investment (FDI) from China, thereby undermining increasingly performance-based CCP legitimacy.

Yet Moscow's new international prospects under bourgeois democracy proved greatly exaggerated. The decisive domestic factor is that the Russian "double bang" of marketization and privatization failed utterly to revive the economy, which went into free fall: Real GDP declined 13 percent in 1991, 19 percent in 1992, 12 percent in 1993, and 15 percent in 1994. Privatization ushered in a two-tiered market, consisting of "oligarchs" at the top, former managers of state-owned enterprises who simply grabbed the country's major industrial assets, and a geographically balkanized patchwork at lower levels, as the country's distribution network collapsed. Under the circumstances, leading Western industrial powers, still overburdened with debt in the wake of the arms race and a worldwide recession following the second oil price hike, were far less munificent with financial support than had been expected; only Germany, now reunified thanks to Gorbachev's refusal to invoke the Brezhnev Doctrine, made substantial subventions to Russian economic development (over US$20 billion in 1993 alone). In the West, after Russian arms were discredited (and a former ally defeated) in the Gulf War (in which Moscow, after voting in support of the enabling Security Council resolution, played no visible role), Russia was demoted from bipolar nemesis to diplomatic nonentity, excluded from any role in resolving the Yugoslav imbroglio, and finally invited to the "Group of Seven" but initially only as an observer. Yeltsin's emergent political rivals, both on the left (Zyuganov and the revived Communist party, the CPRF) and the right (e.g., Lebed) challenged his nationalist bona fides and urged a shift from West to East, arguing on geostrategic grounds in favor of a more "balanced" international posture between East and West. Even in the East, hopes of new breakthroughs were quickly dispelled: negotiations with Japan premised on a territorial compromise realizing Khrushchev's (never implemented) 1958 agreement (provisionally splitting the four, then phasing in a more comprehensive settlement including complete retrocession) aroused unexpectedly passionate military and local opposition, prompting Yeltsin to postpone his visit twice and not even to table the proposal when he finally arrived in Tokyo in October

1993. With regard to Korea, the initial euphoria elicited by Gorbachev's 1988 Krasnoyarsk speech and by the September 1990 establishment of diplomatic relations (to Pyongyang's indignation) did not survive shock at the collapse of the Soviet regime, and though bilateral trade has revived it has not lured much South Korean investment. Thus the 1993 admission of six former satellites to the Council of Europe, and the 1994 proposal to enlarge the North Atlantic Treaty Organization (NATO) to include three former Eastern European satellites, finalized in 1997 in apparent appreciation of American election-year constituency concerns more than any realistically perceived security threat, was but the last in a series of diplomatic setbacks.[5] Having dissolved the Warsaw Pact upon the collapse of the Iron Curtain, Moscow could see no further point to NATO but to push Russia completely out of Europe.

Thus ironically, two nations who had never been able to agree on the same ideology now found it possible in the absence of shared ideological assumptions but under straightened international circumstances (post-Tiananmen Chinese ostracism, the Russian economic meltdown) to converge on a "strategic cooperative partnership" (*zhanlue xiezuo huoban guanxi*), a formulation the Chinese attribute to the Soviets.[6] First proposed in the form of a "constructive partnership" by Yeltsin in September 1994 (at the inaugural presidential summit in Moscow), then elevated to a "strategic partnership for the twenty-first century" during Yeltsin's April 1996 summit in Beijing (a month after China's confrontation with the United States over Taiwan and in the context of Clinton's reaffirmation of a strengthened Japanese-American Security Alliance), the partnership has since become a fungible term of endearment in the diplomatic lexicon of both powers, as China formed partnerships with Pakistan, France, Germany, the European Union, Japan, Korea, and the United States, while Russia claimed partnerships with Japan, Iran, India, as well as the United States. The implication is to vaguely privilege a relationship without making (or demanding) any specific commitments of one's "partner," and one can obviously have an indefinite number of partners at once. Yet for both, the first partnership has remained pivotal, an entry ticket back to what Jiang Zemin calls "great power strategy" (*da guo zhanlue*), precisely because this was the only relationships with sufficient leverage to pose a credible alternative to the lone superpower. The partnership disavows any threat to a third party (i.e., the United States), from whom each stands to gain more in economic terms than from its relationship with the other, in strictly bilateral terms. The United States had long been China's most important trade partner, for example, with some ten times as much trade as with Russia (though this is not true for Russia). But without mutually agreed strategic objectives or opponents, just how meaningful is this "partnership"?

The argument here is that it is far more meaningful than generally credited, held together by dovetailing strategic and material interests and institutionalized convergence. There is perhaps a degree of analytical momentum among observers of the relationship, leading first to a tendency to discount the possibility of scission after the thirty-year alliance was formed, than to a tendency to discount the prospect of reconciliation in the wake of the thirty-year dispute. Yet since its formalization in 1998, the strategic partnership has only deepened, leading the director of the East European, Russian, and Central Asian Studies Institute at the Chinese Academy of Social Sciences to call the decade "the best in the history of bilateral relations."[7] Certainly it is not based on overwhelming mass support. While a December 2000 poll showed the majority of Russian politicians, journalists, and business leaders consider China to be Russia's most important strategic partner (by far outweighing the United States, which was fifth on the list), in the parts of Russia most directly affected by the opening to China, there is a strong negative bias (in public opinion polls held in the south of Primorskiy kray, residents prefer cooperation with the United States, Japan, South Korea, even Germany—but not China). Nor is there much popular enthusiasm (despite the legacy of Soviet assistance in the early years) for Sino-Russian cooperation in China, where Russian bids for contracts or commercial sales are routinely rejected. It is at the elite level that an institutionalized momentum has been established, with seven presidential summits between 1992–1999 and six exchanges of premiers, plus countless other official delegations. This momentum has not been derailed by successions, as Putin made China one of his first destinations (his first in Asia) upon becoming president in early 2000.

The constituency of the relationship has shifted over time, from the committed socialist reformers of the 1980s to a "red-brown" coalition of Communists and Nationalists in the aftermath of Tiananmen to Putin's power pragmatists of the early 2000s. The collapse of the Communist bloc threw both opponents and proponents of the relationship into temporary disarray; whereas, before that time, the relationship had been endorsed by reformers on both sides of the Ussuri and opposed by the old guard, since then there was an ironic reversal of roles. In China, despite the Yeltsin regime's repudiation of the ideological legacy they once shared, two factors now sweeten the relationship: First, the fact that Russia's embrace of capitalism has been so disappointing at correcting the Soviet malaise has made it an effective object lesson for the CCP to vindicate its own hard line. Second, despite its supposed ideological transformation (and subsequent military drawdown), Russia remains the world's most plausible strategic counterweight to U.S. "hegemonism." Global unipolarity and American unilateralism have been feared and resented by both powers since the end of the Cold War. China's reform bloc became more wary of the partnership because, by raising the old specter of Sino-Soviet alliance within a "strategic triangle," it threatened to alienate China from the West—but then most reformers did not survive Tiananmen. In the former Soviet Union, while Tiananmen momentarily disabused Gorbachev and his supporters of their illusions about Chinese reform, their lease on political life also proved brief. Meanwhile, in Russia, the fact that the CCP was able to crush liberal opposition and prevail while communism was self-destructing elsewhere inspired the forces of orthodoxy that had once been among China's most vociferous critics.[8] The pro-China stance of the Communist Party of the Russian Federation (CPRF), since the 1995 elections the most powerful party in the Duma, reflects this group's ideological assumptions. At the same time the former pro-China liberals, including scholars such as Lev Delyusin and former diplomats such as Yevgeniy Bazhanov, though on guard against any nostalgia for fraternal solidarity, remain basically sympathetic with the PRC. The now-marginalized anti-China bloc consists of two quite disparate currents: the radical pro-Western bloc, intellectually led by the Moscow Institute of Foreign Relations (affiliated to the Russian Foreign Ministry) and linked politically to such figures as Yegor Gaydar and the Yabloko movement; and radical nationalists such as Vladimir Zhirinovsky (whose Liberal Democratic Party had an unexpected electoral success in 1993), who regard China as an alien security threat. The local political leaders (now elective) of contiguous regions of the Russian Far East, particularly Primorskiy and Khabarovskiy krays, share some of this radical nationalism in their obsession with the border threat and inflated estimates of the problems of smuggling and illegal migration; but at the same time, the economic prosperity of their domains has become so closely linked to that of the PRC that they cannot but support trade. At the top, a pragmatic majority under first Yeltsin and then Putin has since the mid-1990s favored a "balanced" pro-China tilt.

Certainly one important dimension of the partnership is internal or bilateral, turning their 2,700–mile land border, still the world's longest, from a concentration of troops, tanks, and land mines into an economic thoroughfare and generally improving relations between two of the largest countries in the world.[9] Since 1992, there have been dozens of high-level diplomatic exchanges, and summit meetings have been regularized on an annual basis. These have resulted in hundreds of agreements, among the most important of which were the 1991 agreement to delimit the eastern borders and initiate border demarcation, the 1993 five-year Military Cooperation Pact, the September 1994 agreement for mutual nonaggression, mutual detargetting of strategic weapons, and non-first use of nuclear force, and the 1997 agreements on trade, oil, and gas development and cultural cooperation. Substantial progress on mutual force reduction had already been achieved under Gorbachev, and Yeltsin at the 1992 summit followed suit with a proposal for mutual but gradual demilitarization (thus avoiding the dislocation occasioned by rapid Soviet withdrawal from Eastern Europe) to the minimal number of troops required for peaceful border patrolling (now numbering some 200,000). This has permitted both countries to shift priorities, as China strips troops from the border to face Taiwan and the South China Sea, and Russia redeploys to the security threat created by Chechnya and the expansion of NATO. The most significant development since 1992 has been the set of five-power agreements between China and Russia and the three bordering central Asian republics (Kazakhstan, Kyrgyzstan, and Tajikistan) signed in Beijing in April 1996 and Moscow in April 1997. In the former, both sides agreed on mutual force reduction and military confidence-building measures on their respective borders; while the latter established a "zone of stability" restricting military activity to a depth of 100 kilometers along the frontier and making border security exercises more transparent. Although Moscow guards its strategic interest in these loyal members of the CIS jealously, it has seen fit to chaperone this somewhat unusual negotiating teamwork, thereby facilitating Chinese border agreements with all three central Asian republics (though subsequent border demarcation has lagged in the case of war-ravaged Tajikistan). China has since become Kazakhstan's largest trade partner, agreeing in 1997 to invest US$9.7 billion (the equivalent of half the host country's GNP [Gross National Product], China's largest FDI project on record) to build oil and gas pipelines from the Caspian oilfields to the Xinjiang region and ultimately on to Shanghai. Kazakhstan, in turn, has promised to control Uighur acolytes of an independent "Eastern Turkestan" (whose borders abut ethnically riven Xinjiang province) on its territory. This negotiating exercise between China and the four former Soviet republics was regularized in annual meetings and in the summer of 2001 expanded to include Uzbekistan and formalized as the Shanghai Cooperative Organization (SCO), thereby forming the largest security-related international organization in the world in which the United States was not directly involved. In view of the salience of militant Islamic nationalism to these still-secular dictatorships, the SCO's mandate was expanded to include drug trafficking and terrorism, and an antiterrorist center was subsequently established at Bishkek. Although at least temporarily eclipsed by the incursion of the U.S. forces in the central Asian republics during the attack on the Taliban regime in 2001, Pakistan, the SCO Turkmenistan, and Mongolia have also expressed interest in joining. Yet there is a certain amount of subsurface tension over resources, as the RF continues to regard the region in its own sphere of interest and to view the SCO as a vehicle through which China will attempt to cultivate economic ties with central Asia and build oil and gas pipelines eastward.

Jiang Zemin and Gorbachev, during their April–May 1991 Moscow summit, had already agreed in principle on how to "delimit" the borders (e.g., Moscow accepted the Thalweg or deepest part of the main channel as the "line" dividing the Ussuri/Wussuli and Amur/Heilungjiang rivers). Demarcation was then conducted during the next seven years, over the vociferous objections of local Russian politicians. This resulted, among other things, in giving China access to the Sea of Japan, via the Tumen River, whose development was foreseen in a cooperative development project supported by the UN Development Program (UNDP), also involving North Korea. China also regained sovereignty of one-square-mile Damansky/Chenbao Island, where the 1969 border clash started. At the November 1997 Beijing summit, the two sides signed a demarcation treaty for the eastern sector, including an agreement suspending the sovereignty issue for joint development of three still disputed small islands on the Amur/Heilong River (including Heixia/Black Bear Island). At the November 1998 "hospital summit" in Moscow, both sides expressed satisfaction that both eastern and western sections of the border had finally been accurately demarcated, with the exception of three small islands in two areas in sensitive proximity to the cities of Khabarovsk and Blagovaschensk. Implementation of the 1997 agreement on joint land use remains incomplete, however, while negotiations have turned to the most sensitive issue of all, the ultimate disposition of Heixiazi, an island on the outskirts of Khabarovsk, where many city notables have their dachas.

Bilateral trade made a great leap forward in the early 1990s, to fill the vacuums left on the one hand by the Tiananmen sanctions (the value of all Western investment in China dropped 22 percent during the first half of 1990) and by the collapse of the centralized Russian distribution system. While total Soviet foreign trade dropped 6.4 percent for 1990, Sino-Soviet trade volume increased to $5.3 billion, a quarter of which was border trade. Several Sino-Russian Special Economic Regions were established in emulation of the thriving Special Economic Zones (SEZ) in the southeast, more than two hundred cooperative projects were initialed between localities of the two countries, and China dispatched some fifteen thousand citizens to the Soviet Far East for temporary labor service. But these steep early rates of commercial growth could not be sustained. The 1991–1992 economic crisis in the RFE (Russian Far East) left Russians unable to repay Chinese exporters, and the Russians complained of shabby product quality and disruption of their (hitherto monopolized) retail networks. Visa-regime negotiation in 1993 and Moscow's subsequent imposition of steep border duties, cuts on transport subsidies and restrictions on organizations entitled to engage in foreign trade thus caused Sino-Russian trade to plunge by nearly 40 percent in the first half of 1994. In 1995 it began to recover, reaching $5.1 billion that year and $6.85 billion in 1996, but in 1997 it sank to $6.12 billion, and dropped further to $5.48 million in 1998, particularly after Yeltsin's mid-August devaluation of the ruble and debt restructuring. While this obviously foreclosed Yeltsin's goal to raise bilateral trade to US$20 billion by 2000, as Russian GDP revived over the turn of the millennium, Sino-

Russian trade picked up as well, reaching a record-high trade volume of US$10.67 billion in 2001 and $5.45 billion in the first half of 2002 (an 18.7 percent increase over the same period of 2001). China remains Russia's third-largest trade partner outside the CIS, and long-term demographic trends suggest a potential for continuing growth of commerce and even investment.

The primary beneficiary of expanded trade is the RFE, a resource-rich but climatically forbidding region hosting only about 7 percent of the federation's population but a much higher proportion of the country's subsurface resources. In the Soviet period, the region was subsidized by artificially low transport rates, and by Moscow's financial support for the military-industrial complex (and prison colonies) constructed there. When these subsidies were curtailed upon the union's collapse, the RFE suffered an economic decline even worse than that of European Russia. The RFE experienced its first population contraction of 250,000 in 1992 and has continued to shrink through outmigration, falling by the end of the millennium to some 7.4 million people (versus some 120 million along the Chinese side of the Heilongjiang). Against this background, the sudden influx of Chinese workers or traders (allegedly including large numbers of criminals) excited alarm. According to Chinese statistics, border crossings amounted to 1.38 million in 1992 and 1.76 million at their peak in 1993. For the Russians, however, the central issue was not how many were crossing but how many stayed. Unofficial Russian estimates of Chinese illegal residents ran as high as 1 million in the Far East and 2 million nationally in 1994 (versus Chinese estimates of 1,000–2,000). According to some Russian demographic projections, the Chinese could be the second largest minority population in the Russian Federation by 2050. In the light of these trends, the future seems to contain a contradiction between a growing Russian need for supplemental labor to realize the RFE's economic potential in the wake of continuing population decline and Russian fears of Chinese demographic inundation. For the moment, fear seems to be the controlling factor, with extravagant Russian predictions of future trade growth mocked by border restrictions that result in a steep trade imbalance in Russia's favor. For their part, the Chinese, seeing little progress on the Tumen or other joint border projects since the mid-August ruble devaluation, have been stinting in their approval of Russian investments (such as the failed Russian Three Gorges Dam construction bid, or the Lianyangang nuclear reactor). Yet in 2001, a treaty was signed to build a pipeline to deliver oil from Siberia to northeastern China, projected to deliver some 147 million barrels of oil annually by 2005.

One facet of the economic exchange that has clearly battened on post-Tiananmen sanctions is that of military technology and equipment sales. Deprived of American arms since 1989, the Chinese returned to Russian arms merchants, from whom much of their original hardware came and which hence offered advantages in terms of compatibility of parts. From the perspective of Chinese elite preferences this in fact may have been one of the relationship's saving graces, appealing to precisely those "leftists" otherwise critical of the Gorbachev regime. General Xu Xin, deputy chief of the People's Liberation Army (PLA) General Staff, accompanied Li Peng on his ice-breaking April 23–26, 1990, visit, and on May 30, this was followed up by a military delegation led by Liu Huaqing, vice-chair of the CCP's all-powerful Central Military Commission (CMC), to discuss the transfer of military technology. During Liu's meeting with Soviet Defense Minister Dimitri Yazov (the highest level military contact since the early 1960s), the Soviets indicated that they would be willing to provide help in the modernization of Chinese defense plants constructed on the basis of Soviet technology in the 1950s, at bargain prices. This visit coincided with the Chinese decision to cancel a US$550 million purchase of avionics to upgrade fifty Chinese F-8 fighters, the first such Sino-U.S. deal to be considered since Tiananmen. It was reciprocated on June 1 by the first Soviet army delegation to visit China in thirty years. By the fall of 1990, China had agreed to buy twenty-four troop-carrying helicopters from the USSR capable of operating in high-altitude climates (the United States had refused to consider selling such weapons systems, which seemed ideally suited for operations in Tibet).

Global sales of Soviet arms dropped "catastrophically" in the wake of the Gulf War, which witnessed Soviet arms being eclipsed by high-tech American weaponry. Inasmuch as military equipment had been the second largest item in the Soviet export inventory (after petroleum products), continued Chinese interest was particularly welcome at this point, and Russian monitoring of arms exports relaxed conveniently.[10] In 1993, the two countries signed a five-year military cooperation pact giving China access to advanced military technologies in nuclear submarine propulsion, underwater missile launchers, muffling technology for diesel submarines, technology for increasing the range and accuracy of ICBMs, solid rocket fuel, and so forth. Negotiations for the purchase of Sukhoi SU-27 fighters, under way since early 1990, culminated in the purchase of twenty-six at a "friendship" price of more than US$1 billion (about 35 percent of which China could pay in hard currency, the rest in barter goods), with an option to buy an additional forty-eight fighters. In March 1992, China also took delivery of the sophisticated S-300 antiaircraft missile system and SA-10 antitactical ballistic missiles. The first contingent of Chinese pilots was sent to Moscow in June 1992 to undergo a one-and-one-half-year training course, and by the turn of the millennium some four thousand Russian experts were based in China by "private" contractual arrangement, helping to modernize Chinese nuclear and missile capabilities.[11] The 1995–1996 confrontation over the Taiwan Strait whetted Chinese appetites for further acquisitions, and in November 1996, the two sides renewed their military cooperation pact, allowing China to purchase thirty to fifty SU-30MKK multipurpose fighters, four diesel-powered (Kilo-class) submarines, four Sovremenniy-class destroyers with accompanying Sunburn antiship missiles designed to counter U.S. carrier fleets (two of which have so far been delivered), and a fifteen-year licensing agreement to produce up to two hundred additional Su-27s (as Chinese F-11s) at a production line in Shenyang (with a restriction against exporting them). By early 1997 China was the leading purchaser of Russian arms, machinery, and equipment (surpassing India). Proceeds from arms sales financed over half of Russian military production, and most of this came from China (e.g., in 2000, 74 percent of Russia's arms exports went to Asia, of which 52.6 percent went to China and 18.2 percent to India).[12] Annoyed by the private contract to license Chinese production of SU-27s, the Russian Foreign Ministry reportedly vetoed sales of Tu-22 Backfire long-range bombers and Su-35 fighters, but the Chinese were able to purchase Russian in-flight refueling technology to extend the range of Chinese bombers to more than one thousand miles, as well as Russian space technology. The (Russian and Israeli—denied by the Chinese) media in early 2000 divulged an arrangement whereby China would purchase sophisticated radar surveillance aircraft (similar to the American AWACS), produced in Russia and electronically equipped in Israel, but when Washington induced the Israelis to renege on the deal, the Russians agreed to sell China their own Il-76 electronic surveillance aircraft. The Russians have also endeavored with limited success to interest the Chinese in civilian technology—some 25 percent of the Chinese commercial aircraft pool is now Russian.[13] In interviews, the Russians dismiss Western concerns that their weapon sales might upset the military balance, pointing out that if they do not sell arms to the PRC some other country will, with the worst conceivable consequences for Russian security (perhaps forgetting that the 1969 Sino-Soviet border clash was fought with Russian weapons on both sides).

Building upon such mutually useful interactions, the following set of symbiotic constituencies has emerged to provide institutionalized support for the partnership: (1) Russia's military-industrial complex finds it simpler to continue production runs rather than undergo defense conversion, at least in the short run, and China is their largest customer. Heavy industry more generally (e.g., the machine tool industry, oil and gas companies, the nuclear and hydropower industries) is oriented to market-opening initiatives for similar reasons, though they may be overly optimistic; whereas, the Chinese have been ready purchasers of Russian weaponry, they have by and large not sought to upgrade aging Soviet plant technology, preferring to leapfrog to the most advanced levels, even if that means starting from scratch in many enterprises.[14] Still the Russians may be competitive (certainly price competitive) in hydroelectric power and nuclear energy development projects, provided they are able to swing attractive financing deals. (2) The state trading companies who, since the 1994 Russian tariff and immigration legislation, have regained control over bilateral trade, are now staunch supporters of expanded economic relations. At the same time both Chinese and Russian shuttle traders continue to evade bureaucratic control, and border trade has revived, comprising some 30 percent of the total (with illegal barter trade adding an unknown additional percentage) by the late 1990s, serving a broad market on both sides of the border. (3) Regional governors, though vociferously opposed to territorial concessions and open borders, have willy-nilly come to appreciate their regions' growing dependence on the Chinese economy as a locomotive for their own.

From the Russian strategic perspective, Asia has generally gained importance since the Cold War, following secession of the Eastern European satellites, the Baltic states, Ukraine, and Belorussia. Though the national identity remains preferentially "Western," Russia has since the mid-1990s defined itself geopolitically as a land bridge between Europe and Asia. The Russian Federation survived its experiment with communism in rather ravaged condition—while inheriting four-fifths of the former USSR's territory, it was left with half its GDP (equivalent to half of China's current GDP) and less than half its population (about 148 million, which has since proceeded to shrink). The Soviet military of 6 million men has shrunk to a demoralized force of 2.3 million unable to subdue rebellion in tiny Chechnya. Thus the "big brother" relationship has to a certain extent been reversed. Within Asia, India and China emerge as the twin pillars of Russian foreign policy—one in the south, one in the east. The partnership already provides access to Hong Kong (where Russia now has a consulate), to membership in ASEAN's Regional Forum and (since 1998) to APEC, as Russia still awaits entry into the Asian Developmental Bank. Whereas Kozyrev once lectured his Chinese hosts on human rights, Moscow has since repeatedly used its vote to block China's condemnation by the UN Human Rights Commission in Geneva. The two frequently make common cause against interventionist initiatives on behalf of human rights supported by the United States and Britain, leading to deadlocks on the UN Security Council redolent of the Cold War era. General Leonid Ivashov, head of the Russian Defense Ministry's international cooperation department, recently claimed China as Russia's "ideological ally" with a common interest in rejecting "military *diktat* in international relations," and the two are now coordinating their defense doctrines and staging joint military exercises. Noting that, in the wake of Bosnia, "NATO is being turned into a global organization," they see no alternative but to draw together to oppose U.S. "hegemonist" interventions, and both protested American bombing campaigns in Iraq and Yugoslavia. Joining the Russians in opposition to NATO expansion, the Chinese also oppose any alteration of the Anti-Ballistic Missile (ABM) Treaty or to American plans to install National Missile Defense or Theater Missile Defense on the Asian rimland. All of these mutual strategic interests were drawn together in a twenty-year, twenty-five-point Treaty of Friendship and Cooperation signed in Shanghai on July 16, 2001, reportedly at Chinese instigation. Both sides at the time denied any intended threat to third parties, and a Bush spokesman agreed to its strategic irrelevance, but this is mere diplomatic discretion.[15] While it is true that the treaty is less binding than its predecessor, it does represent an escalation of the relationship since proclamation of the partnership, as it is the first (and only) time China has signed a broad political treaty with any country since renouncing such commitments upon embarking on its "independent foreign policy" in 1982.[16] And although the treaty is not a binding alliance, it contains two articles that approach the commitments typically associated with alliances.[17]

Though primarily concerned with the interests of the two signatories, the partnership also has multilateral implications. Their consistent joint refusal to support international sanctions against the Democratic People's Republic of Korea (DPRK) throughout the decade-long effort to prevent Pyongyang from developing nuclear weapons bespeaks a common interest in retaining a protective glacis of buffer states against the crescent of economically dynamic nuclear threshold states (viz., Japan, South Korea, Taiwan) on their shared eastern rim. Whereas China's relations with Vietnam and India have improved of late, the Sino-Russian reconciliation also deprives such former regional rivals of alternative patronage, taking some of the wind out of the rivalry. For China, neutralization of the Russian threat permits a readjustment of military modernization priorities from the army to the navy and air force, and a shift of forces from the north to the southeast. In the context of a growing PLA budget amply supported by a booming economy, this has strategic relevance for Japan, Korea, Taiwan, and Southeast Asia, all of whom depend on sea lanes of communication through the South China Sea, to which China has made rather sweeping territorial claims. But of most direct and immediate relevance is the security of Taiwan, against which most Chinese arms purchases seem now to be directed. On the Taiwan issue, Russia has played an interesting double game. On the one hand, the Russians have unstintingly endorsed Chinese claims to the island. Not only did Yeltsin parrot Clinton's July 1998 "three nos," denying Taiwanese claims to sovereignty, but Foreign Minister Igor Ivanov recently cautioned Washington not to interfere militarily to protect Taiwan if China were forced to invade.[18] The Russians also have not hesitated to promote the efficacy of the weapons they sell in deterring American aircraft carriers.[19] At the Moscow summit in November 1998, Yeltsin added a fourth "no" to Clinton's three, promising not to sell weapons to Taiwan. On the other hand, like many other countries, Russia has inaugurated informal trade relations with Taiwan (Taiwan opened its trade office in Moscow in 1994, Moscow reciprocated in Taipei in 1996) while formally recognizing only the PRC, and trade relations with Taiwan expanded by 1995 to US$1.2 billion (vs. $5 billion with China); by 1997, Taiwan had become Russia's fourth largest trading partner in Asia, with arrangements under way for direct air links.

Acutely aware of the precariousness of an Asia policy relying predominantly on one "partnership" with a partner whose relative power in East Asia is greater than Russia's own (and whose interests may diverge), Moscow has sought to expand ties with other Asian powers. Russia's noninvolvement with the Asian Pacific Region (APR) before 1992 has been replaced by much greater openness to multilateral trade and investment, and the RF has sought membership in all existing regional forums, including APEC, the ASEAN Regional Forum (ARF), the Asian Development Bank (ADB), and the World Trade Organization (WTO). Moscow's relations with Tokyo improved considerably after the November 1997 Krasnoyarsk "tieless" summit, reviving hopes (thus far unrealized) of Russo-Japanese territorial settlement. Though Japan has become the third biggest aid donor to Russia (after the United States and Germany) and Japanese trade with the RFE increased 40 percent from 1992 to 1995, making Japan the RFE's biggest Asian export market, investment has not followed trade, and the territorial dispute has continued to hamper bilateral relations.[20] For imports of temporary workers as well as commodities, the RFE prefers to rely on South Korea, which has no history of irridentist territorial claims, geopolitical rivalry, or demographic pressure. Russia has also announced its intention to form a strategic partnership with India, an old treaty partner that under nationalist (BJP) leadership has been expanding its traditional rivalry with China into Southeast Asia and the South China Sea. India is Russia's second-largest weapons client, having purchased tanks, aircraft, missiles, and naval vessels, and Moscow has gone so far as to promise that it would take its relationship to India into account in selling arms to the PRC. The RF has also attempted to reactivate its Soviet-era relationships with North Korea, Vietnam, and Iraq—all now in Beijing's sphere of interest. Just as China's grand design for the APR tends to leave Russia out and focus on the Sino-Japanese-American triangle, Russia's grand design diverges from the Chinese preference for unilateral power balancing in its historically rooted interest in multilateral grand designs. Gorbachev first pushed the idea of an Asian collective security treaty in the late 1980s (to little effect), and Foreign Minister Andrei Kozyrev revived the idea in January 1994, advocating step-by-step movement toward a "security community" open to every country in the region. For these purposes, Russia proposed to establish an Asian-Pacific Center for Conflict Prevention and an Asian-Pacific Institute on Security Problems. In the late 1990s, Russia promoted the CBM and demilitarization agreements signed in 1996–1997 by the "Shanghai five" as a security model applicable to the APR, and has held meetings promoting this idea with various countries (e.g., India's December 1996 agreement with China was modeled on the five-nation CBM). Whereas China, despite its own increasing interest in multilateralism as witnessed by its involvement in ARF, the SCO, the four-power talks on Korea and ASEAN plus three, has shown little interest in seeing Russia's initiatives progress further, Japan has shown considerable interest in this type of multilateral forum.[21]

One of the defining hallmarks of the end of the Cold War was, of course, the disintegration of the Soviet bloc, triggered in part by the May 1989 Tiananmen protests. In view of the erstwhile "polar" status of the Soviet Union in a bipolar system, this has thrown the structure of international politics into considerable disarray. Since the 1960s, strict bipolarity was compromised and made somewhat more flexible by the rise of the Nonaligned Bloc on the one hand and by the ideologically based secession of China from the bloc on the other. The latter, more relevant to the contemporary APR, gave rise in the 1970s to the "strategic triangle," which permitted a manipulation of the balance of power to facilitate Soviet-American strategic arms control and reduction as well as China's policy of "reform and opening to the outside world." Triangularity meant essentially that the three players arranged their bilateral relationships to take into account their respective relationships to the third player. It began with Washington playing the pivot to mediate Sino-Soviet hostility (and advance American interest in détente), permuting during the Reagan era to permit Beijing to take advantage of the revived Soviet-American arms race to pursue an "independent foreign policy" (and cut its own defense budget). The consensus impression of strategic analysts has been that the triangle collapsed at the end of the Cold War, for at least two reasons. First, the USSR bowed out of the game, both because its national heir, the RF, repudiated communism and disclaimed any further interest in pursuing worldwide revolution, and because the RF lost the economic, demographic, and national wherewithal to sustain a bipolar confrontation in any case. Second, the ideological deradicalization that proceeded simultaneously in the two "revolutionary" powers since the 1960s, particularly in the wake of China's Cultural Revolution, deprived them of a plausible pretext for the use of nuclear weapons (except for deterrence), depriving the triangle of much of its strategic raison d'être. Yet despite these severe blows to the relevance of this conceptual device, the spirit or "ghost" of the triangle lingers on, breathing life into such diplomatic documents as the 1996 strategic partnership and the 2001 friendship treaty. That spirit is the notion that these three powers, despite the transformation of the world that has followed the collapse of the bloc, can still "tilt" the international balance of power. To repeat, this is only the "ghost" of the triangle and not the triangle itself, because triangular relationships are no longer mutually exclusive and there is extensive socioeconomic interpenetration, and the members in any event no longer represent distinctive approaches to political reality. Yet all three are permanent members of the UN Security Council and major powers on the world stage, and if they cannot tilt the balance, who can? Which is to say that the ghost of the triangle is a functional artifact of an alternative spirit: the specter of global unipolarity and U.S. unilateralism. At least until some more plausible equilibrating counterforce emerges (the European Union?), the triangular ghost may continue to be invoked by this alternative.

The future of the Sino-Russian strategic partnership entertains three contrasting scenarios. First, the partnership is a snare and a delusion, an elaborate diplomatic pretense undermined by continuing suspicions and divergent national interests, whose future is very bleak indeed. Second, the partnership is real, a successful diplomatic effort to craft international cooperation, but will remain limited in relevance to bilateral relations without significant strategic implications for the West, Japan, or the APR. Third, the partnership is real, and may be expected to affect not only bilateral relations but also strategic relations among the major powers of the world and certainly within the APR, perhaps in a direction adverse to Western interests. The most realistic possibility, in my judgment, is the second, though the ultimate choice is not an analytical but a political one, to be decided by the actors involved depending on a host of incalculable future international developments. In essence, the partnership represents the attempt of two large and precarious multiethnic continental empires to form a mutual help relationship useful to them in the face of a potentially hostile international environment. Both countries have traditionally been "garrison states" or "developmental dictatorships" ruled by a hierarchically disciplined national security apparatus (while this tradition was temporarily broken in the Russian case, it seems to be returning to the mold). In a post–Cold War world lacking strategic structure or balance, both feel threatened by de facto American hegemony. Without international help, they fear being ostracized, sanctioned by international regimes, torn asunder by ethnic cleavages, even possibly assaulted by national or international expeditionary forces engaged in humanitarian intervention.

Although the strategic partnership does represent a challenge to specific American ideological objectives (such as human rights) and foreign policies (such as National Missile Defense [NMD] and [TMD]), it is not necessarily a threat to world peace or to vital American national interests. The partnership is not a Comintern pact organized around a programmatic vision for a new world hierarchy, but is essentially defensive, designed to enhance the national interests of the two participants. Each partner has demonstrated a willingness to use violence to achieve those interests, but the threat to peace that this entails is for the most part (e.g., Chechnya, Xinjiang, Tibet) localized.[22] And beyond the endorsement in principle to their right to use violence on behalf of sovereign interests, neither partner necessarily feels obliged to come to the aid of the other in that instance. For example, the limits to Chinese efforts to prevent NATO expansion, to persuade Japan to forfeit the four northern islands, or to include Russia in any resolution of the Korean embroilment are fairly clear. And Russian interest in blocking the post-1997 expanded definition of the Japanese-American Security Alliance, or in forcing Taiwan to negotiate reunification on Chinese terms, has been mostly rhetorical. Neither partner has, nor do they share, either an ideology or a coherent international vision beyond their endorsement of multipolarity. It remains to be seen whether the relationship should evolve in a future direction conducive to the realization of more ambitious objectives, but for now it is limited not only by implicit conflicts between specific national priorities but by the interest each partner retains in closer relations with the center of international economic and political gravity in the West.

Notes

1. See the massive study of this period by Dieter Heinzig, *Die Sowjetunion und das kommunistische China 1945–1950: Der beschwerliche Weg zum Buendnis* (Baden-Baden: Nomos Verlagsgellschaft, 1998).

2. Jiang Zemin himself was trained at Moscow's Stalin Autoworks, a showcase of Soviet industry. Li Peng was a graduate of the Moscow Power Institute, and Admiral Liu Huaqing, a prominent advocate of increased purchases of Russian weaponry, was trained at the Voroshilov Naval Academy in Leningrad. Jennifer Anderson, *The Limits of Sino-Russian Strategic Partnership* (London: International Institute for Strategic Studies, Adelphi Paper no. 315, 197).

3. Lowell Dittmer, *Sino-Soviet Normalization and Its International Implications* (Seattle: University of Washington Press, 1992).

4. It is fairly clear that Gorbachev's visit played some role in exacerbating the Tiananmen protest, but China may have also played a role in the collapse of Soviet and Eastern European communism, first by contributing a "demonstration effect" to European protesters, and second

by discountenancing mass repression as a politically feasible option. See Nancy Bernkopf Tucker, "China as a Factor in the Collapse of the Soviet Empire," *Political Science Quarterly* 110, no. 4 (Winter 1995): 501–519. The negative significance of the Chinese "solution" should not be overstated, however. Gorbachev already began backing away from his Eastern European commitments after Reikjavik and the Intermediate Nuclear Force (INF) agreement revived détente in December 1987. The significance of the INF was to remove American power from Western Europe (from whence U.S. Pershing IIs could pulverize Moscow in less than ten minutes), enabling Gorbachev to dismantle Soviet security forces there and put the relationship on a cash basis. In March 1989, in a meeting with Hungarian Premier Grosz, Gorbachev stated his opposition to Soviet intervention in WPO members' affairs, in effect rescinding the Brezhnev doctrine. Richard C. Thornton, "Russo-Chinese Détente and the Emerging New World Order," in Hafeez Malik, ed., *The Roles of the United States, Russia, and China in the New World Order* (New York: St. Martin's Press, 1997), pp. 221–238.

5. Admitted to the Council of Europe, a loose confederation of future candidates for the European Union, were Hungary, the Czech Republic, Slovenia, Romania, and the Baltic Republics of Lithuania and Estonia.

6. Personal interview, Chinese Academy of Social Sciences in Beijing.

7. Li Jingjie, "Pillars of the Sino-Russian Partnership," *Orbis* (Fall 2000): 527–539.

8. The Institute of the Far East (IDV, in its Russian initials) in the Russian Academy of Sciences, previously led by Oleg Rakhmanin, now by his former deputy Mikhail Titorenko, still the largest Moscow research center for Chinese studies, has shifted from its critical stance toward Maoist ideology to an ardent embrace, largely the CCP has avoided privatization and political reform while successfully regenerating socialist economic performance. Alexander Lukin, "Russia's Image of China and Russian-Chinese Relations," *East Asia: An International Quarterly* 17, no. 1 (Spring, 1999): 5ff.; see also Evgeniy Bazhanov, "Russian Perspectives on China's Foreign Policy and Military Development," in Jonathan Pollack and Michael Yang, eds., *In China's Shadow: Regional Perspectives on Chinese Foreign Policy and Military Development* (Santa Monica, CA: Rand, 1998), pp. 70–91.

9. The Sino-Soviet border was some 7,000 kilometers long. Since the disintegration of the USSR, it has contracted to 3,484 kilometers, while the Sino-Kazakh border stretches for about 2,000 kilometers, the Sino-Kyrgyz border for 1,000 kilometers, and the border with Tajikistan is about 500 kilometers long.

10. Russia's export of tanks in 1992 dropped 79-fold, sales of combat aircraft fell 1.5 times in comparison to 199, leaving warehouses of the military-industrial complex overstocked with unsold weapons. China was the principal buyer of Russian weapons in 1992, making purchases worth US$1.8 billion. Pavel Felgengauer, "Arms Exports Continue to Fall," *Sogodnya* (Moscow), July 13, 1993, p. 3.

11. Sharif M. Shuja, "Moscow's Asia Policy," *Contemporary Review* 272, no. 1587 (April 1998): 169–178; Jamie Dettmer, "Russian-Chinese Alliance Emerges," *Insight on the News* 16, no. 13 (April 13, 2000): 20; Jyotsna Bakshi, "Russia-China Military-Technical Cooperation: Implications for India," http://www.idsa-india.org/an-jul-100.html.

12. A. Kotelkin, former head of the main Russian arms exporter, Rosvoornezhenie, as cited in Alexander A. Sergounin and Sergey Subotkin, "Sino-Russian Military Cooperation: A Russian Perspective," *Regional Studies* 15, no. 4 (1997): 24; Associated Press, Moscow, February 15, 2001.

13. See Peggy Falkenheim Meyer, "Russia's Post–Cold War Security Policy in Northeast Asia," *Pacific Affairs* 67, no. 4 (Winter 1994): 495–513.

14. Sherman W. Garnett, ed., *Limited Partnership: Russia-China Relations in a Changing Asia* (Washington, DC: Carnegie Endowment for International Peace, 1998), pp. 22–23.

15. "Just because Russia and China have entered into an agreement does not necessarily mean it's something that would be adverse to the interests of the US," Ari Fleischer said. AFP, July 17, 2001.

16. Interfax (Moscow), December 26 and 28, 2000.

17. Article 8 specifies that each country shall refrain from any foreign or defense policy that would jeopardize the interests of the other, or join an alliance or take an action that would undermine the sovereignty, security, or territorial integrity of the other. And Article 9 pledges that if one of the parties should face the threat of aggression, the other should immediately consult with it "with the aim of removing the threat."

18. Dettmer, "Russian-Chinese Alliance Emerges," p. 21.

19. China has reportedly also been negotiating for Russian satellite intelligence information on strategic facilities in Taiwan. Igor Korotchenko, "Moscow and Beijing Are Building Up Their Strategic Ties," *Nezavisimoye voyennoye obozreniye* (Moscow) 1, 11/10–16/00, #42 (215).

20. During Prime Minister Obuchi's November 1998 visit, the two signed a joint "Moscow Declaration on Building a Creative Partnership between Japan and the Russian Federation," which provided for the establishment of two subcommittees, one to discuss border demarcation, the other to study joint economic activities on the four disputed islands without prejudice to the two countries' legal claims. They also agreed to establish a joint investment company, strengthen economic cooperation, and promote intellectual and technical cooperation and exchanges. Japan has also facilitated Russian entrance into APEC. See Peggy F. Meyer, "The Russian Far East's Economic Integration with Northeast Asia: Problems and Prospects," *Pacific Affairs* 72, no. 2 (Summer 1999): 209ff.

21. Anderson, *Limits of Sino-Russian Strategic Partnership*, p. 68.

22. The Taiwan issue, in the Chinese case, is a conspicuous exception, but far too complex to be reviewed in this compass.

12

The Clash of Ideas

Ideology and Sino-U.S. Relations

Zi Zhongyun

John Hay's note on open-door policy can be regarded as the first U.S. quasi-policy toward China after which the two countries entered a stage of direct dealings with each other. Apart from *realpolitik* considerations, Sino-U.S. relations have been characterized by the impact of both sides' ideological traditions. The hundred-year history of Sino-U.S. relations have been permeated with the interflow and clash of ideas; however, it should be pointed out that from the outset the interflow of ideas was by no means even, rather, it has been a one-way track with American ideas exerting influence on China and seldom vice versa.

Ideological Heritage in the Foreign Relations of Both Countries

On the American Side

According to American historian Michael H. Hunt in his incisive book, *Ideology and U.S. Foreign Policy*, by the early twentieth century, three core ideas began to wield a strong influence over U.S. foreign policy, namely, quest for national greatness, racial hierarchy in attitudes toward other peoples and "limits of acceptable political and social change overseas" which, in more direct terms, implies reservations and misgivings toward revolutions in other countries.[1] As Hunt's conclusions coincide by and large with the observations of this author and as it is also valuable to compare the corresponding ideas on the Chinese side, this chapter will start with borrowing these three points as an outline to discuss the problems concerned, but the interpretations and explorations are the author's.

Quest for National Greatness

There is a sense of quest for national greatness inherent with the birth of the United States of America. It is an idea rooted in the thinking of its founding fathers that existed before the founding of the nation. Different from other big powers, this sense of national greatness did not necessarily stem from power and wealth; as, in fact, in its early days the United States was neither powerful nor particularly rich. Due to the unique historical conditions and ways in which its nation building has been accomplished, the U.S. sense of national greatness bears the following characteristics:

First, it was connected with ideological expansion from the very beginning. The group of thinkers represented by people like Thomas Paine who provided the theoretical basis for American Independence all came from Europe and inherited both the European Enlightenment tradition and Calvinism. They came across the ocean to this new land to put into practice their ideals, which met with many obstacles on the declining Old Continent. To the masses of ordinary immigrants, here was a land of opportunity to build up home and wealth through hard labor, while for the thinking elites, this was a place to build up an ideal republic according to the philosophical ideas, religious beliefs, and moral values they had inherited from their homeland. The gifted natural conditions of this land plus auspicious historical opportunities enabled both groups of people to successfully realize their dreams. This overwhelming success and the attribution of material success to their religious faith developed a sense among the people that they had received a "mandate of heaven" to accomplish on the earth by communicating this formula for "success."

Second, this dream of national greatness has come true progressively in keeping with the growth of national strength. In less than a century from 1776 to the end of the Civil War in 1865, the United States of America accomplished its expansion into forty-eight states on the continent (to be formally joined by Alaska and Hawaii in 1959) and then embarked on expansion overseas. Except for the case of the Philippines, this expansion did not take the form of the occupation of the colonies but had been one of economic interests and ideological influence taking place either in turn or synchronously. This was a natural process in which demand and capability were in timely coordination and which, therefore, advanced with little difficulty.

Third, by the end of the nineteenth century when the United States started its expansion overseas, it represented the most progressive and vigorous social system and rapidly developing productive force in the world. This is a fact recognized by almost all the European progressive thinkers of the time, including Karl Marx. So, the youthful Americans, full of self-confidence, looked around and found no other people but themselves who were fit for fulfilling the task of building a new and better world. The sense of being the "elites" of God born with the mandate to change the world in their own image found further ground in the real world.

Racial Hierarchy

In general, this attitude of racial hierarchy meant white supremacy toward colored peoples, but the Americans carried it further to Anglo-Saxon supremacy, which put other Caucasian peoples on a hierarchical scale. Despite their protestation for basic human rights, the early freedom fighters of America did not believe in equal rights for all races or else they would not have treated the indigenous Indians the way they did nor supported or condoned the slavery of black people. Even people belonging to the white race were differentiated in hierarchical orders: the French, Germans, Spanish, and Italians were considered inferior to Anglo-Saxons and further down the hierarchy were the Jews. Next came Asians of the yellow race. The Exclusion Movement against the Chinese and Japanese in the 1880s bore a distinct mark of racial discrimination. The attitude of Americans toward the different races at home would naturally be carried overseas. This was nothing new. Again the ideas came from Europe, particularly England as John Stuart Mill mentioned explicitly in the introduction of his classical work, *On Liberty*, that his doctrine was meant to apply only to "human beings in the maturity of their faculties." Children and young people below the age of manhood fixed by law would naturally be excluded and "those backward states of society in which race itself may be considered as in its nonage" should also be left out of consideration. He went further to say: "Despotism is a legitimate mode of government in dealing with barbarians, provided the end be their improvement and the means justified by actually effecting the end."[2] It is known to all that the old European colonialists justified their conquests with the theory of distinction between superior and inferior races. The American way differed from the European one in that it put more emphasis on cultural influences. To educate and change the inferior races besides conquering them by force constituted a component part of their "mandate of heaven." China and Japan ranked a little higher than other colored peoples in the hierarchy because they were considered to be more likely to change through education. By the end of the nineteenth century, Japan did best in learning from the West and was regarded as the "good student" while China remained a promising student, once given the benefit of an American education.

It goes without saying that this concept of racial hierarchy has greatly changed with time, but to a certain degree it still exists, albeit, not so explicitly. As late as in the latter part of World War II, President Franklin D. Roosevelt and Prime Minister Winston Churchill were talking about the establishment after the war of a world peace institution led by "English-speaking democracies." The very existence of the term "WASP"—White Anglo-Saxon Protestant—to indicate the mainstream of elites in American society is self-explanatory, whatever qualifications may be put on it. Conceptually, nowadays the Jewish people and people of European origin have merged into the mainstream culture of America, while the situation of other ethnic groups is more complicated.

Reservation toward Revolutions of Other Peoples

The reason that Americans have reservations about revolutions in other countries may be hard to understand for those who would ask: Since the Americans won their own independence through a revolutionary war and take upon themselves the promotion of democracy and freedom outside the country, why should they take such a conservative attitude toward revolutions waged by other people who also share the ultimate goal of achieving democratic rights? The first revolution overseas confronted by the United States after its founding was the French Revolution. Chinese historians usually put the French Revolution and the American War of Independence on par as examples of "bourgeois revolution." While it is true that both were inspired by the same ideas of enlightenment and both built up a republic as a result, the process and the nature of the two revolutions are very different. The American one could be rightly called a revolution by the bourgeoisie in that it was led by well-educated property owners, and the core issue was free trade and freedom from unjust taxation by the British. The immigrants on the new continent wanted to be free from the British yoke to make their own fortune without restraint. Class divisions were not yet distinctive among the population. The French Revolution, however, had already the participation of the *tier état*, including the *sans culottes* and, along with the overthrow of the monarchy, the demands of the revolutionaries also included social reform and economic equality. The storms, slaughters, and the shakeup of the social order from below by mob actions were beyond the acceptance of the Americans. Therefore, after initially showing sympathy for its aim of opposing a despotic monarch, even Thomas Jefferson distanced himself from the French Revolution. (Here I am only talking about the ideological aspect of the American attitude. The diplomatic issue concerning relations with Great Britain and France and the controversies between Madison and Jefferson are not involved.) Alexis de Tocqueville in his famous work *On American Democracy* made ample comparisons between the conditions of the American and French revolutions and said: "The French democracy overthrew everything that confronted it on its way, shaking what it could not destroy. . . . It has been marching continuously among the disorders and agitations of a fighting," while the American Revolution "seemed to have almost attained its natural limits; it was operated in a simple and smooth way, or one could say that this country has seen the results of a revolution that occurred in ours but without having a revolution by itself." In this way, "the republic exists in America, without fighting, without opposition, without proof by tacit agreement, a kind of 'universal consensus.'"[3]

By the mid-nineteenth century, waves of revolution occurred in Europe and various doctrines of socialism emerged with Marxism at its peak. The situation in America was very different. For the upper social strata, the need was to consolidate the existing social order. As for the people underneath, although the disparity between rich and poor was increasingly visible, they still had "new frontiers" to explore with great potential for both geographical and social mobility. If in Europe, the petty bourgeois was feeling the constant threat of falling into the rank of the proletariat, in America, on the contrary, the have-nots still harbored great hopes to gain property and enter the ranks of the bourgeoisie. Therefore, socialist ideas did not find suitable soil in America. After the Civil War, with the rapid growth of industrialization and the exhaustion of new frontiers, class contradictions began to grow and became especially acute after the first great economic crisis in the 1890s. At the same time, the United States began to have interests overseas. Under such circumstances, it was quite natural that the American mainstream elites were unfavorably inclined toward radical revolutions in other countries both for the sake of maintaining overseas interests and for fear of their impact on their domestic social order.

On the Chinese Side

Great Nation of "Heavenly Dynasty"

China has always been a great nation; moreover, for thousands of years, the Chinese had embraced only the concept of a united world "under heaven" with China (meaning the Han people) in the center and barbarians in the periphery. Beyond that, everything was in the mist. The Chinese sense of cultural superiority had been absolute and there had only been a history of "barbarians" being assimilated into the Chinese culture and not a single case of the reverse. This sense coincided with that of the Americans but took the opposite trajectories when the two encountered each other. While the United States saw its dream of national greatness fulfilled with the continuous rise of its national power, China, on the contrary, witnessed the shattering of its dream as a great nation in a sharp downfall of its national strength, and the process was unbearably shocking and painfully humiliating. At the turn of the century, after several rounds of struggle, most thinking Chinese had abandoned the dream of recovering the glory of past dynasties and began to seek a way out through reform in order to catch up with the outside world. Their mentality was a complex mixture of nostalgia of the old with hopes for the new, a sense of superiority mixed with feelings of inferiority, admiration of but resistance toward Western culture, and a deep reluctance to admit to the backward and weak state of China. Generally speaking, by that time the posture of China in dealing with foreign countries had changed from one of a superior dynasty to one of an underprivileged nation striving for the position of equality. In fact, since then, the Chinese have always been highly sensitive as to whether they are treated as equals in dealing with more powerful and more advanced countries.

Complex of Racial Superiority and Inferiority

This issue of racial superiority and inferiority is connected with that of national greatness. Historically, the sense of racial superiority depended on the longevity and richness of the culture. Throughout their history, the Han people had often been conquered by force by "barbarians" who in several cases established their rule in the heartland of China. But before long they all assimilated into the Han culture, with the people subdued by force becoming teachers of civilization and the dynasties established by various national minorities merging into the long stream of Chinese history. The Manchus of the Qing Dynasty were no exception. Thus, when they, as heir of the great Chinese civilization, first encountered the aggressive Westerners, the British in particular, they regarded them as but another "barbarian" race with "red hair" that could be dealt with according to the age-long tradition. The result was a real shock in that culturally the situation was upside down. As Guo Songtao, the first Chinese minister to Europe discovered and wrote in painful frankness, "They [the Europeans] look at China now just as China in the heyday of the Three Dynasties [Xia, Shang, Zhou] used to look at the barbarians."[4] For a short period after that there was an anti-Manchu feeling among certain Han people. The Taiping Rebellion built its legitimacy on an anti-Manchu position and Dr. Sun Yat-sen also raised the slogan of "expel the Tartar bandits" at the initial stage of the rebellion, as if the decline of China had been solely due to Manchu domination. But he soon changed his position and the slogan became "a republic for all five nationalities in a common struggle against foreign powers." In general, the struggle of Chinese people against imperialism has been connected with one against white supremacy and racial discrimination. Racial consciousness as a colored people was brought out along with national consciousness under the impact of invasion from white imperialists. This racial consciousness took different forms under different circumstances among different groups of people, revealing sometimes an inferiority complex and at other times a high sensitivity and strong resentment toward the slightest hint of discrimination. There was another complex consciousness in relation to other colored peoples. It was a mixture of sympathy, a sense of superiority, and a fear of falling into the same plight of complete slavery. The political consciousness of uniting with all the oppressed peoples in the world to fight imperialism came only after the introduction of Marxism in China.

The Demand for Reform and Revolution

In contrast to the United States, China has experienced great upheavals, drastic changes, and wave upon wave of revolutionary movements ever since the mid-nineteenth century. Two features stood out with these movements: first, inspiration from Western influence and second, violence in reform. Even peaceful reform movements ended up in head-cutting and bloodshed. In this sense, the Chinese revolutionary movements were closer to the French Revolution. This was due to the special conditions in China at that time. Externally, the pressure from foreign powers was so heavy and the gap of national strength was so wide that China had no time to take it easy and to do things step by step. Domestically, social contradictions were too acute and the ruling dynasty was too corrupt, ignorant, and despotic to allow for peaceful and incremental reform. All progressive intellectuals and reformers of various trends of the time took their inspiration from Western ideas of democracy and freedom, while striving against oppression and exploitation by Western powers. This, in fact, represented the basic aspirations of the Chinese nation, which were summed up as "antifeudalism, anti-imperialism." Thus, learning from the West in terms of social and ideological revolution and opposing the Western countries in defending national dignity and interests constituted a complex mentality of Chinese intellectuals and other reformers. The attitudes of rulers of successive governments were to the contrary. More often than not, they were too weak and too corrupt to defend national interests against foreign usurpers and yielded to foreign pressure, but they went all out to resist the influence of Western ideas of democracy and freedom as a dangerous threat to their rule. Accordingly, the advocates of Western ideas were targets of suppression. As for the United States, much as it was eager to spread the American ideas to change China in its own image, its actions in China often went against its moral principles by standing against the revolutionary people.

Those were the ideas that wielded influence on the relationship between China and the United States since their first encounter at the turn of the twentieth century, ideas that contributed to the weaving of an extremely complicated and intriguing picture in history.

Mutual Hopes and Disappointments in Practice

It goes without saying that the Chinese first encountered Western ideas through Europeans. However, the Americans followed suit quite soon and in terms of ideological influence took the lead before long. Compared with the relationship between other countries, Sino-U.S. relations over the last hundred years have carried more clashes of ideas, more sentimental factors, and more hopes and disappointments as a result of each other.

On the part of the Chinese, the hopes placed on the United States by various governments were different from those by intellectuals and reformers. The latter turned toward the Americans for support in their struggle for a more democratic and modernized China, while the former almost invariably regarded the United States as a possible countervailing force against the immediate threat to China. This chapter deals only with the ideological aspect and will leave aside the diplomatic activities.

The Beginning of the Twentieth Century to the Victory of the Northern Expedition in 1926

In this period of time, three categories of Americans played the most important roles on the Chinese scene, namely, the missionaries, diplomats, and businessmen. In addition, there were nonmissionary educators, representatives of philanthropic institutions, journalists, and other cultural figures who wielded various levels of influence. All these people did not hold the same views, but whether out of idealistic belief or pragmatic interest considerations, they all showed an enthusiasm in influencing and changing China and remolding the thinking of the Chinese. Among them, the churches were by far the most active elements. Starting from the arrival of Elliaj Bridgemen and David Abeel in 1830, and facilitated by the "Wangxia Treaty," American missionaries had already made efforts over more than half a century by the time the "open-door" policy was implemented and had turned the emphasis of their efforts from direct proselytizing, which was not very effective, to running schools. The far-reaching role played in Chinese modern education by the American Christian colleges, the returned students from the United States through the fund from the "Boxer's indemnity," and other American resources like the Rockefeller Foundation are known to all and need not be explored in detail.[5]

By that time, the American economy had been developed enough to demand overseas expansion, but the United States was not powerful enough to contest the vested interests of other powers that had come to China before it. The "open-door" policy opposed in principle but recognized in practice the spheres of influence of foreign powers in China. What the United States demanded was only equal opportunity without discrimination. But this principle was eroded by Japan, which became increasingly aggressive in China after the Russo-Japan War of 1905. Under such circumstances, American superiority in the cultural aspect stood out more visibly, and that was entirely in conformity with the ideology related to its foreign policy as mentioned at the beginning of this chapter. And China in its transmutation provided an excellent opportunity for American influence. It was a period of dazzling changes with trends of ideas and political forces contesting each other. Almost each of them had in one way or another tried to win American support and more often than not the U.S. policy and attitude disappointed the democratic forces. When Chinese reformers and revolutionaries, inspired by Western ideas, rose to struggle for a democratic China, the U.S. government's policy supported the conservative governments or political forces. Following are a few salient examples.

The 1911 Revolution

This was the first revolution ever in Chinese history, which put an end to monarchial dynasties. Dr. Sun Yat-sen and his comrades were undeniably inspired by American ideas. The Declaration of Independence was first translated and published in Chinese in 1901 in *Guomin Bao,* a journal run by Chinese students in Japan. In 1904, Sun Yat-sen wrote an article in English entitled "The True Solution of Chinese Question: An Appeal to the People of the United States," which was published in the form of a pamphlet and distributed in New York and other cities of the United States. This was a direct appeal to the American people and ended with the following words:

We . . . must appeal to the people of the civilized world in general and the people of the United States in particular, for your sympathy and support either moral or material, because you are the pioneers of Western civilization in Japan; because you are a Christian nation; because we intend to model our new government after yours; and above all because you are the champion of liberty and democracy. We hope we may find many Lafayettes among you.[6]

However, the United States took a negative attitude toward the revolution. One of the important reasons was that the revolution exploded on the demand by Chinese businessmen to retrieve the right for building the Huguang Railroad, which directly clashed with American interests. The U.S. government hoped that the Qing court could suppress this movement. Seeing that the Qing dynasty was actually falling apart, it formally assumed a neutral position and advocated peaceful negotiations between the North and the South, a position which was in fact favorable to the Qing dynasty. After the fall of the dynasty, the United States preferred Yuan Shikai to Sun Yat-sen, because it needed a strongman to hold the situation together and protect American interests and also because few Americans in China reported favorably on Dr. Sun. An interesting episode in 1915 revealed American attitudes toward the idea of revolution in China. In June of that year, Frank J. Goodnow, American constitutional advisor to President Yuan Shikai, wrote a memorandum at the request of Yuan on a comparative study of political systems in different countries in which he said something to the effect that a constitutional monarchy might be more suitable to China, but added quickly that to put this into practice there must be certain necessary conditions that he did not see existing in China. This memorandum was used by the "Chou An Hui," an organization to promote the restoration of the monarchy in China and to enthrone Yuan Shikai as the emperor, which in its declaration published in August quoted Goodnow out of context saying that even a great American political scientist was of the opinion that a monarchial system suited China better than a republic. Goodnow was chagrined by this propaganda and issued a statement denying this allegation together with the publication of the full text of his memorandum. The whole movement of "restoration" proved a farce in Chinese history. It would be unfair to conclude merely from Goodnow's memorandum that Americans had supported the movement; however, the official reports of the American legation in China showed a gross overestimation of the forces supporting the restoration movement. According to Charge d'Affaires MacMurray, the majority of officialdom, even including foreign-educated young officials who hitherto connected with the republican movement, were in favor of the idea, the opposition was "wholly academic, giving no evidence of any likelihood of serious resistance," "the Government has the situation well enough in and to assure that no revolutionary movement is likely to gain headway" and "the great bulk of the people . . . indifferent to the question of the form of government." As for the attitude of foreign powers, provided the event would not lead to disorders detrimental to trade or foreign interests, many felt that "if the Chinese find it possible to reinstitute their traditional form of government . . . so much the better."[7]

During the period of confrontation between the Beiyang government of the war lords and the Guangzhou revolutionary government led by Sun Yat-sen, the United States had all along recognized and supported the North. So much so that in 1923, when Dr. Sun decided to reclaim the share of the remaining tax payment duly belonging to the Guangzhou government that had been denied him by the Beijing government, the United States took the lead to send a fleet to the Bay of Guangzhou in a saber-rattling gesture. This action deeply embittered Dr. Sun toward the United States and he had this to say in another "Appeal to the American Citizens":

> Ever since we first started our revolution to overthrow the despotic and corrupted gov-ernment and install a republic in China, we had taken inspiration from and followed the example of the United States. We had hoped for a Lafayette to fight on our side for this just cause. However, in the twelfth year of our struggle for freedom, it is not a Lafayette, but an American Commander of Fleet who led more warships than any other country to enter our territorial sea, in an attempt to jointly crush us and eliminate the Chinese Republic. Is it true that the fatherland of George Washington and Abraham Lincoln has abandoned outright its lofty belief in Liberty and degenerated from a liberator to an oppressor of people fighting for their freedom? We cannot believe it. We hope the offic-ers and soldiers of your fleet will seriously think over this question before shelling on us although their guns have already targeted Canton, an open city.[8]

The disappointment in the United States may explain partly, though not wholly, the decision of Sun Yat-sen to ally with the Soviet Union.

The Versailles Peace Conference

When Woodrow Wilson was elected president, his idealism and liberalism had a strong appeal to Chinese intellectuals, especially because he appeared as a defender of the prin-ciples of self-determination and equality among nations. The canon of the League of Na-tions drafted by him was much more popular in China than in the United States. Chinese left-wing intellectuals including people like Chen Duxiu, founder of the Communist party in China, also had held him in high esteem. Chen Duxiu wrote in December 1918 in his opening words for the first issue of *Meizhou Pinglun* (Weekly Review) shortly before the Versailles conference:

> All the speeches made by President Woodrow Wilson of the United States are open and above board. He can be regarded as the best man in the world. He has said many things, among which two points are most important: First, countries are not allowed to infringe upon the freedom and equal rights of others by the exercise of power politics; Second, governments are not allowed to infringe upon the rights to equality and freedom of their own people. Do not these two doctrines constitute the stand for justice and against power politics? Therefore, I say he is the best man in the world.[9]

A week after, he published again in the same magazine an article entitled, "The Awak-ening and Demands of Eastern Nations after the European War," in which he further advo-cated that representatives of the Eastern nations at the coming Paris conference should unite and strive for the adoption of a resolution on "Equality among All Mankind against Discrimination," based on the principles of President Woodrow Wilson's Fourteen Points for the League of Nations. The essence of the resolution was to oppose unequal treatment of Asian peoples by European and American nations in international relations and the domination by militarists and war lords in domestic politics. Here again, the U.S. system was cited as an example.[10] This line of thinking was quite common among broad circles of Chinese at the time, not only intellectuals. At that historical juncture, the most ardent aspiration of the Chinese nation as a whole was to redress the national humiliation imposed on her by the 1915 treaty with Japan (known as the "21 points") and to recover its sovereignty over territories in Shandong from the hands of the defeated Germany. They had expected with all trust to find support from the United States because of the declared principles of the latter. Yet, to their utter disappointment, it was President Wilson himself who made the decision and instructed the U.S. delegation at the Versailles conference to accommodate Japan's demands over Shandong at the expense of China's legitimate rights. This was a heavy blow to China, which ignited the famous May Fourth Movement, a movement of far-reaching consequences in Chinese modern history, after which the Chinese revolution entered a new stage. Once again widespread disillusion of the United States took over the Chinese people, progressive intellectuals in particular. It is not an exaggeration to say that this fact constituted an important factor in pushing the Chinese intellectuals to the left; they ultimately abandoned Western democracy for Marxism.

The May Thirtieth Movement of 1925

The May Thirtieth Movement was a large-scale, anti-imperialist mass demonstration with students at its core and joined by the working class. It started from the international concession in Shanghai and was met with armed suppression by the police, thus leading to bloodshed. The United States took the same stand as the British and other foreign powers and sent marines with warships down the Yangtze River. The U.S. counsel to Shanghai justified the killing of demonstrators in his report to Washington, which was published by the Department of State and naturally caused a wave of indignant protest from Chinese students in the United States.[11] The action of the United States in this event not only infringed upon the sovereign rights of China but also belied the democratic rights stipulated in the first amendment of the U.S. Constitution. This could be explained in two ways: (1) Between moral principles and selfish interests, the latter always prevailed over the former in U.S. foreign policy; (2) Out of the concept of racial hierarchy, the Americans did not deem the Chinese people were qualified to enjoy the same rights as the American people nor did most of the administrations of American missionary schools show more tolerance and democratic spirit. The expulsion by Hawks Pott, president of the St. John's University in Shanghai, of a large number of teachers and students, which led to their creating another university, was a well-known case.

The United States' behavior in the May Thirtieth Movement intensified the then-rising nationalism among the Chinese, strengthened their demand for taking over the control of education, and rendered more convincing the denouncement by Marxism, which was newly introduced to China, of the hypocritical nature of Western democracy.

The Period of National Government

The Northern Expedition Campaign of 1926 led by KMT and CCP in cooperation took the form of a radical revolution. The United States at first continued to side with the North against the revolution as before. Seeing the trend was irreversible, it tried to protect American interests as much as possible under the circumstances and started to seek relatively moderate groups among the revolutionaries with whom it could cooperate. Before long, Chiang Kai-shek managed to take over power and decided to make alignments with the United States the core of his foreign policy. The United States, for its part, saw in Chiang Kai-shek the "strongman" that it had looked for who could unite China and at the same time provide conditions for the Americans to protect their interests and promote ideological influence. So, at this point, after wavering among different factions of power in China in uncertainty for over twenty years after the turn of the century, the United States settled on Chiang Kai-shek, and this policy was to last for the next twenty or so years. In the ideological aspect, this policy corresponded with the lines of thinking mentioned above. In the surge of revolutionary movement against feudalism and imperialism in China, Chiang represented the "moderate" faction that could alleviate the U.S. misgivings. The declared nation-building theory of the Republic of China being based on the principles of Western democracy plus the fact that both Chiang himself and the Soong family were Christians made China appear more susceptible than ever to American influence. The reality of one party dictatorship practiced by the KMT was assuaged by the existence of the doctrine of "tutorship period." For the Americans, that was an opportunity to "educate" and remold China. In fact Ambassador John Leighton Stuart talked about replacing one party "tutorship" of the KMT by American "tutorship" and sought to "coach" the KMT.[12] Finally, the United States and Chiang found a common cause in anticommunism following the campaign of purging the Chinese Communists in April 1927. Therefore, the United States was willing to accommodate to a certain extent the nationalism of the Chiang government. For example, during the campaign for the "Sinonization" of education in China at the end of the 1920s, all American schools, willingly or reluctantly, made changes following the educational policies of the national government and by and large acted in conformity with the trends of the times. This move had tremendous impact on Chinese education, and the smooth transition played a positive role. The American expectations of China reached an all-time high during the "New Life Movement" advocated by Chiang Kai-shek in 1936. Many Americans and other Westerners thought that this was the most hopeful period for China to embark on the road to Western-style development, but at this time they were interrupted by Japanese aggression.

During China's resistance war against Japan, the U.S. government persisted in its policy of neutrality for a significant period of time, but the U.S. public sympathy with China was constantly on the rise, and this to a certain extent facilitated the change of stance by the Roosevelt government. After Pearl Harbor, in an effort to encourage the Chinese government to actively fight the Japanese, the United States made a series of steps toward treating China as an equal among major powers, including the signing of a treaty to relinquish extraterritoriality, the repeal of the Chinese Exclusion Acts, the recognition of the status of China as one of the great powers in the process of preparing for the building of the United Nations, and the support of China's rights to recover all territories occupied by Japan after the war. This was a period where moral principles coincided by and large with real interests in the U.S.-China policy.

After the end of World War II, when China was split by the civil war, with the deepening of dependence of the KMT regime on the United States, the relationship between the two became more and more unequal. The United States showed increasing dissatisfaction with the KMT for its incompetence, corruption, and repressive ways. While continuing its aid to Chiang in his fight against the Communists, the United States constantly pressured him to carry out some sort of democratic reform, and the argument by Chiang for failure to do so had always been that China had special conditions to which a Western style of democracy might not be applicable. This relationship continued until the KMT regime was forced out of the mainland, or in a broader sense, until Richard Nixon's visit to China. The Chiang regime in Taiwan became a de facto protectorate of the United States, and the relationship was more unequal than ever. But by and large, Chiang kept his Nationalist stand in persisting in his "one China" position. On the other hand, in the name of the need against Communist subversive activities, he did not carry out the democratic reform asked for by the Americans. Therefore, during Chiang Kai-shek's time, the conservatives, mostly Republicans in the United States, had more sympathy with Chiang while the liberals were more critical of him and had more sympathy with the "self-determination" school. This situation underwent a change after the Nixon visit to China and further changed after normalization of relations between the PRC and the United States. In sum, historically, the United States had good relations with the KMT government longer and more consistent than with any other Chinese government. This relationship also united to the highest possible degree the ideological principles and practical interests in U.S. foreign policy, despite frictions and tensions between the two. Even today, most Americans regard the present regime in Taiwan as their good students whose economy presents a good example of "graduation" from American aid and who, at last, embarked on the road to "democracy" pushed for so long by the United States.

Relations with the Communist Party of China

Ideologically, the mainstream of Americans are against communism, and between the CCP and the KMT, their natural sympathy would be with the latter. But in practice such a statement risks being oversimplified. In the long course of history, there were twists and turns and the attitude of ordinary Americans was not always in conformity with the government's policies. The position of the American Communist Party and left-wing elements were not involved in the discussion, because their influence in forming the American opinion was insignificant. Until World War II, the bulk of the American people had very little knowledge of China as a country, much less the Chinese Communists. In the 1920s and 1930s, a few idealistic intellectuals and journalists had visited some revolutionary bases and most of them left with good impressions of the CCP. But whatever they had written was passed almost unnoticed by the American public. The most well-known book among these writings was of course Edgar Snow's *Red Star over China*. But even this book had become widely read only after the 1960s. When it was first published in the 1930s, the Chinese translation of the book had much more influence on Chinese readers under the KMT domination than in the United States. Actually, China came within the scope of vision of Americans only after the beginning of the war against Japanese aggression.[13]

During the War

By the latter period of the war, the Americans became increasingly dissatisfied with Chiang's KMT, because of its ineffectiveness in fighting the Japanese and incompetence and corruption in running the government. On both points, the CCP showed a contrast and Americans who visited Yenan all reported favorably of the situation there. For a certain time, a quasi-consensus was formed by Americans concerned in China that the U.S. aid for war efforts should also go to the Communist troops. At that time the ideological factor did not play an important role, and the Chinese Communists were called "agrarian reformers" by some Americans.[14] President Roosevelt himself was relatively liberal toward left-wing people at home partly due to his own open-mindedness and partly due to the fact that the Soviet Union was still an ally.

The CCP leaders also had favorable opinions of the United States under President Roosevelt and for a certain period of time placed their hopes on the "democratic elements" in the United States to provide them with a share of U.S. aid in the resistance war and, after the war, to wield its influence toward dissuading the KMT from launching armed suppression of the CCP forces and instead encouraging the realization of a coalition government with the participation of the CCP. That is why at the initial stage of the Marshall Mission, the CCP welcomed him more enthusiastically than the KMT; however, the positions were reversed at a later stage. From an ideological perspective, the following words of Zhou Enlai in relaying Chairman Mao's message to Marshall upon his return to Chongqing from Yenan at the end of January 1946 are very interesting:

> We believe that the democracy to be initiated in China should follow the American pattern. Since in present day China, the conditions necessary to the introduction of Socialism do not exist, we Chinese Communists, who theoretically advocate Socialism as our ultimate goal, do not mean, nor deem possible, to carry it into effect in the immediate future. In saying that we should pursue the American path, we mean to acquire U.S. styled democracy and science, and specifically to introduce to this country agricultural reform, industrialization, free enterprise and development of individuality, so that we may build up an independent, free and prosperous China.

Zhou Enlai then continued to relate an anecdote:

> It has been rumored recently, that Chairman Mao is going to pay a visit to Moscow. On learning this, Chairman Mao laughed and remarked half jokingly that if ever he would take a furlough abroad, which would certainly do much good to his present health condition, he would rather go to the United States, because he thinks that there he can learn lots of things useful to China.[15]

A certain degree of tactical consideration was not excluded from this conversation, but there is no reason to doubt Mao's sincerity about the road he deemed appropriate for China. The idea that China could not practice socialism immediately was also expressed in his works prior to that, for example in "On the Coalition Government" and "On New Democracy." As for how much he understood the implication of "U.S. styled democracy," there is no hard evidence. At least he was not entirely opposed to it at that time as he was to show in later years. All these indications plus the obvious independent road to revolution taken by the CCP from the Soviet Union gave the United States hope, and for a certain period of time, there was constant discussion among U.S. policy-making circles as to whether the Chinese Communists were more "Chinese" or more "Communist." But this exploration remained within the inner circle discussions. What the CCP actually experienced from U.S. policy was one of increasingly open support of Chiang against the Communists following the latter stage of the Marshall Mission. It may be concluded that complete disappointment on the part of the CCP toward the United States started at the time of failure of the Marshall Mission. From then on, the CCP propaganda always put the United States in a negative light, not only its foreign policy, but also its domestic situation, including its system and ideology as a whole. It is hard to say, if the United States had adopted a different policy toward the Chinese civil war, whether or not the CCP and Chairman Mao would have viewed the American style of democracy differently.

Ideological Confrontation in the Cold War

After Mao's "lean-to-one-side" policy statement, the United States gave up trying to influence China's policy but remained hopeful of their potential to have an ideological influence. In his letter of transmittal to the president at the head of the *China White Paper*, Dean Acheson expressed hope on the reassertion of "the democratic individualism of China."[16] The phrase brought strong reactions from mainland China. Chairman Mao personally wrote a series of articles (published at the time under different names) severely criticizing the idea. In these articles, "democratic individualism" was personified in the Chinese version by adding to it words like "defenders of" or "elements of." This implied that Dean Acheson had in his mind a group of people in China on whom he placed hope and who were called by Mao Zedong the "social basis" for American imperialist influence. This made Western educated intellectuals very nervous, and they found it necessary to make clear their position by publicly denouncing Acheson's words and professing their support for the "lean-to-one-side" policy.[17] Immediately after that, or using this as a motive force, a large-scale campaign of "ideological reeducation" and "elimination of imperialist ideological influence" was carried out, mainly in institutions of higher learning, the first of a series of political and ideological campaigns to be carried out in the coming decades.

In the 1950s, with the intensification of the Cold War, ideological confrontation between the two countries had no way to improve. On the U.S. side, ideological struggle was clearly regarded as an important instrument for the "containment" of communism. In the now well-known NSC68 series of policy papers, alongside political, economic, and military programs, there was a program of "Foreign Information and Education" that laid out in detail both ends and means in terms such as: "to wage overt psychological warfare calculated to encourage mass defections from Soviet allegiance," "operation by covert means in the fields of economic warfare and political and psychological warfare with a view to fomenting and supporting unrest and revolt in selected strategic satellite countries,"[18] and "the peoples under the domination of the Soviet Union are potential allies whose hope for ultimate liberation should be nourished. . . . This is particularly true of intellectuals in governments and out."[19] This policy of the United States, which could be literally qualified as one of "ideological subversion," might have had a certain influence in Eastern European countries, such as was seen in the role played by Radio Free Europe in the Potznan and Hungarian events of 1956. As for China, the consequences bore strong Chinese characteristics. There was no evidence that the leaders of the Chinese Communist Party were aware of the existence of NCS68, which was top secret at that time, but they did know of the existence of a U.S. policy of "psychological warfare" and had always kept high vigilance against American intentions in the ideological field. This situation only pushed Chinese domestic policies further to the left and aggravated the mistrust by Mao Zedong of Chinese intellectuals, even to include those among veteran revolutionaries as was demonstrated by the series of political campaigns from the antirightist movement in 1957 to the peak of the Cultural Revolution.

The Sino-Soviet split became open after Krushchev took power. The CCP was on the left of the Communist Party of the Soviet Union (CPSU) and criticized the Soviets for deviating from Marxism-Leninism, weakening the struggle against imperialism, and failing to support revolutionary movements of oppressed nations and peoples. Finally, the Soviet Communist Party was labeled as having taken on a road of "revisionism." In 1957, John Foster Dulles, in answering questions at a press conference, said something to the effect that there would be an "evolutionary" change in the Soviet Union rather than "revolutionary" and that there was already a trend in the Soviet Union toward somewhat greater personal freedom. He also argued that, "If Krushchev goes on having children and they have children's children, his posterity will be free."[20] The CCP took full advantage of the statement made by Dulles, suggesting that the United States place its hopes on the third and fourth generation of socialist countries for "peaceful evolution" into capitalism, and it has been repeatedly quoted in all kinds of propaganda and educational materials so that this notion is familiar to almost all those who had a certain knowledge of international affairs even to this day. In the 1960s, the CCP published nine "Open Comments on the Letter by the Central Committee of the CPSU," the last one of which was on the danger of "peaceful evolution" and constituted actually the ideological base for the Cultural Revolution. After that, the CCP produced a Twenty-five Point Program for Building a United Front against U.S. Imperialism, which was the height of anti-Americanism. During that period of time, class struggle in the ideological field was emphasized and pushed to a new height until the emergence of the slogan "to assume total dictatorship in ideological area" arose out of the Cultural Revolution. The disastrous results were known to all.

An interesting phenomenon appeared in the United States in terms of reaction to what happened in China during this period. While policy wise, the Kennedy and Johnson administrations regarded China as more rigid and dangerous than the Soviet Union, for the first time, revolutionary China became appealing to broad circles of intellectuals, young students in particular. It was the time when the anti–Vietnam War movements led to the rise of counterculture trends, and there arose a wave of doubts about the basic American system. News about what was happening to China seemed to provide a brand-new alternative, and the Cultural Revolution was also idealized among quite a number of young Americans. Consequently, the interest in learning more about China greatly increased, and it was during that time that the study of contemporary China from new perspectives began to develop and thrive. Although disillusion about the Cultural Revolution came before long with the revealing truth by the Chinese themselves, the interest in understanding China and the whole field of Chinese studies lasted and developed into an important area in the academic world, which was to play a role in the promotion of normalization and beyond.

The 1970s and 1980s

It was known to all that the breakthrough in Sino-U.S. relations made by Mao and Nixon was based on strategic and geopolitical considerations as they chose to set aside ideological differences. However, soon afterward China started an era of reform and opening-up that meant a drastic change. Policies that had lasted for decades were criticized and rejected, and the trend toward the emancipation of minds was quite spectacular. In contrast, the Soviet Union appeared more rigid, and it was the Soviets who, in turn, criticized the Chinese Communists for embarking on the capitalist road. Naturally, Americans both in and out of the government warmly welcomed this change in China and once more the hope to influence China's course of reform to American's liking rose rapidly. In negotiations between the two, the Chinese government was more interested in technological transfer and trade, while the American side was more enthusiastic in promoting cultural and educational exchanges. For masses of ordinary Chinese, upon the opening of the door to the outside world after years of being closed, once more the gap with the developed countries was painfully felt, and there was a strong desire to catch up in the shortest period of time. To learn from the West was no more a taboo and the United States naturally became the most favorable mirror for modernization. In human relations, the age-old and widespread ties of families, relatives, teacher-students, and schoolmates across the Pacific were soon restored and American cultural influences reentered China more rapidly than had been expected. The attitudes toward this phenomenon differ among different circles of Chinese. Some intellectuals and young students looked at the United States with admiration and, somewhat like their predecessors in the early years of the century, turned toward the U.S. example when they were dissatisfied with the situation at home. American democracy and freedom had a strong appeal to them. For ordinary people, the propaganda against the United States in the old days had a backlash effect and without much knowledge, they had a vague idea of a wonderful land where everybody could become wealthy. Some veteran revolutionaries and some of the more conventional intellectuals began to feel uncomfortable with these trends, and those in power at different levels felt their authority threatened by the desire for Western-style democratic reforms. Thus, the need for resisting the erosion of Marxism by Western liberalism again appeared on the official ideological agenda in themes like opposition to "spiritual pollution" and "bourgeois liberalism." This backlash in conjunction with continued pressure from the United States for ideological reform and increased misgivings on the part of Chinese officials at the impact of American ideological influence developed a growing tension that climaxed in the events of 1989.

The Situation in the 1990s

After 1989, the friction between the two countries was mainly ideological, with the United States focusing on the issue of "human rights," and China vehemently resisting interference into its internal affairs. It is true that there were a series of other problems ranging from intellectual property rights to nuclear proliferation, but in the pre-1989 atmosphere, conditions could have been less harsh on China, particularly when there were no emotional factors involved in terms of domestic pressure in the United States. For the Americans, the 1989 events were a heavy blow to American expectations of China, which they had thought were becoming "more like us." After his visit to China in 1984, President Ronald Reagan referred to China as the "so-called Communist," to justify his change to a more friendly attitude toward the country. After 1989, especially with the disintegration of the Soviet Union and the fall of the Berlin Wall, China again assumed the role previously held by Russia as the "bad guy" in the eyes of the American public. In recent years, with the rapid development of the Chinese economy, American opinion shifted again from expectation of a collapse of the present regime to concern about the "potential threat" of a China too strong to be handled. As a result, the question frequently discussed among elite opinion circles has become how to "contain" China.

On the Chinese side, the government has made a point to play down ideological factors in dealing with Western countries ever since the beginning of reform and the implementation of the open-door policy. The principle that different political systems should not affect relations between countries was reiterated by Chinese leaders right after the 1989 events. Yet, the chronic raising of the "human rights" issue with China as the main target, both in U.S.-China relations and in international stances, reminded the Chinese leaders of the U.S. policy of "containment" toward China during the Cold War. The "peaceful evolution" attitude of Chinese intellectuals and young students toward the United States has also undergone a great change between the pre- and post-1989 period. It would be an exaggeration to say that anti-Americanism is rising among Chinese youth, but one can say quite objectively that the admiration and favorable views of the United States have faded considerably. People no longer turn to American examples and support when they are dissatisfied with the domestic situation. To a certain extent, this is a healthy development in that the Chinese people are more mature and free of the frustration that develops out of feeling that as Chinese they were victims of unfair treatment by Americans. Here the gap in perceptions is very wide. While the Chinese feel that they have barely stretched their back after a hundred years of bending, the Americans are already talking about constraining China to prevent it from growing too tall. While recognizing the trend toward globalization, the Chinese, with a history of having their sovereign rights violated for a century by foreign powers, attach much importance to the concept of sovereignty in international relations. Yet, the theory that the traditional concept of sovereignty has become obsolete is now in vogue in the West. While it is true that generally speaking, the present younger generation in China is more pragmatically than idealistically inclined, it is also true that the pursuit of a strong and prosperous China remains the common and deep-rooted national aspiration prevailing among Chinese of all ages and social strata. In this line of thinking, Chinese identify themselves more with the establishment as opposed to the United States as compared with pre-1989 years. To Americans it may certainly seem a paradox, but the post-1989 environment has at least partly been the result of American pressure on China during the last few years on various issues in the name of promoting democracy in China.

In conclusion, a strong ideological character is inherent in U.S. foreign policy. As a rule, any foreign policy statement by any U.S. administration invariably contains the principle of promotion of liberty and democracy, though perhaps in different wording. In practice, however, the decisive factor has been *realpolitik* considerations, based primarily on strategic and economic interests and therefore double or multiple standards have characterized American moral principles toward different countries in different cases. This is exemplified in the United States willingness to ally with undemocratic and even despotic regimes. Of all the ideologies alien to Western style democracy, communism is considered the most irreconcilable rival. That means it is unlikely that U.S. foreign policy can be entirely free from any element of anticommunism, and it is also unlikely that the United States would cease entirely to interfere with other countries' internal affairs. In dealing with China over the last hundred years, Americans have rarely been free from the urge to change China or to influence China to their liking; hence, the constant shift between expectation and disappointment. The Chinese attitude toward the United States also has a certain consistency despite all the upheavals and changes of regimes throughout the years. For most intellectuals and reformers, American democracy and America's relatively reasonable and efficient social structure, not to mention advanced science and technology, have always been appealing. While American arrogance and the impact of American power politics have hurt their national pride time and again. Sometimes, practical interests have also caused resentment. For various Chinese governments, good relations with the United States have often been important for geopolitics in a complicated international situation and/or for the need of economic development, but more often than not, American political and cultural influences are regarded as a threat to stability and the consolidation of their rule. Especially, the habitual way the United States interferes in matters considered by China to be internal affairs has the potential to heighten Chinese vigilance to the point of affecting normal cultural exchanges. Therefore, in the relationship between China and the United States, apart from difficulties that are normal between any countries from a clash of interests, there are additional problems in the ideological aspect of Sino-U.S. relations, bearing an emotional character that is rare with relationships between other countries. The interaction of ideas is unbalanced, with the Americans on the offensive and the Chinese on the defensive. This situation is not likely to go away in the foreseeable future and much wisdom is needed to handle it properly in order to develop a healthy relationship.

Notes

1. Michael H. Hunt, *Ideology and U.S. Foreign Policy* (New Haven: Yale University Press, 1987), pp. 17–18.

2. John Stuart Mill, *On Liberty* (originally published in 1859), edited by Elizabeth Rapaport (Indianapolis: Hackett, 1978), pp. 9–10.

3. Alexis de Tocqueville, *De la Democracie en Amerique* (Paris: Robert Laffont, 1986), pp. 47, 49, 365, translated from French by the author.

4. Guo Tongtao, *Guo Songtao Riji* (Guo Songtao Diaries), vol. 3 (Changsha: Hunan People's Publishing House, 1982), p. 439.

5. References among others: Jessie Lutz, *China and the Christian Colleges, 1850–1950* (Ithaca, NY: Cornell University Press, 1971); Arthur Wardron and Zhang Kaiyuan, eds., *Chinese Western Culture and Christian Colleges* (Wuhan: Hubei Education Press, 1991; Gu Xuejia, Arthur Wardron, Wu Zonghua, eds., *Zhonguo Jiaohuei Daxueshi Luncong* (Papers on the History of Christian Colleges in China) (Chengdu: Chengdu Technology University Press, 1994); Zi Zhongyun, "The Rockefeller Foundation and China," *American Studies in China*, 1995, pp. 84–121.

6. Photocopy of the original printed in *Zongli Quanji* (Complete Works of the Prime Minister), in Hu Hanmin, ed, *Minzhi Shuju*, vol. 4 (Shanghai 1930), p. 368.

7. *Foreign Relations of the United States, 1915* (Washington, DC: USGPO), pp. 49–61.

8. Chen Duxiu, "Zhi Meiguo Guomin Shu" (Address to the American Citizens), December 17, 1923, *Sun Zhongsan Quanji* , vol. 8, pp. 389–391.

9. *Duxiu Wencun* (Works of Chen Duxiu) (Heifei: Anhui People's Publishing House, 1987), p. 388.

10. Ibid., pp. 389–391.

11. Tao Wenzhao, *Zhongmei Guanxishi* (A History of Sino-U.S. Relations), 1910–1911 (Congqing: Congqing Publishing House, 1993), pp. 105–106.

12. *Foreign Relations of the United States, 1946*, vol. X, pp. 593–594.

13. For details, see Kenneth E. Shewmaker, *Americans and Chinese Communists, 1927–1945: A Persuading Encounter* (Ithaca, NY: Cornell University Press, 1971); Stephen Mackinnon and Oris Frieshen, eds., *China Reporting: An Oral History of American Journalism in the 1930s and 1940s* (Berkeley: University of California Press, 1987); Peter Rand, *China Hand: The Adventures and Ordeals of American Journalists Who Joined Forces with the Great Chinese Revolution* (New York: Simon and Schuster, 1995).

14. Two famous books in point are: J. W. Esherick, ed., *Lost Chance in China* (New York: Random House, 1974); and Theodore White and Annalee Jacobe, *Thunder Out of China* (New York: William Sloane Associates, 1946).

15. *Foreign Relations of the United States, 1946*, vol. 9, pp. 151–152; the Chinese version exists in *Zhou Enlai Yijiusiliunian Tanpan Wenxuan* (Selected Documentation of Zhou Enlai's Negotiations in 1946) (Beijing: Zhongyang Wenxian Publishing House, 1996), pp. 92–93.

16. Letters of Transmittal, *The China White Paper*, reedited with introduction by Lyman P. Van Slyke (Stanford, CA: Stanford University Press, 1967), p. xvi.

17. The five articles at the end of the fourth volume of *Selected Works of Mao Zedong*, starting with "Cast Away Illusion and Prepare for Struggle," were all targeted at the *China White Paper* and Acheson's Letter of Transmittal in particular. Except for one, namely "Friendship or Aggression?" all the other four articles criticized the notion of "democratic individualism" in strong terms. The Chinese newspapers from August 12 to the end of September carried statements almost every day by famous intellectuals and members of democratic parties or group meetings denouncing the *China White Paper*, and Dean Acheson's words in particular.

18. *Foreign Relations of the United States, 1950*, vol. 1, p. 285.

19. Ibid., p. 454.

20. Secretary Dulles' News Conference of July 2, *Department of State Bulletin*, July 22, 1957.

13

Making the Right Choices in Twenty-first Century Sino-American Relations

Liu Ji

We are at a key turning point in history. The turbulent events of the twentieth century have ended with the close of the Cold War. It is likely that the twenty-first century will be one of more intense competition and power shifts. At the same time, the dawn of peaceful development for all mankind is also in sight. Both the United States and China, respectively the richest and most developed country and the most populated developing country, bear a large burden of responsibility for the future of humanity. The whole world is focusing on the prospects for Sino-American relations in the twenty-first century, and opinions vary widely. However, human beings are intelligent creatures and possess the power of choice; decision-making is perhaps their most important activity. To make these choices and carry them out is to make history. *The Coming Conflict with China* by Richard Bernstein and Ross Munro is one of several works making analyses and predictions about the future of Sino-American relations. Are the analyses correct? Are the predictions accurate?

My answer is yes if people make irrational wrong choices. My answer is no if people make rational correct choices. As a scholar, I will talk now about a choice that is possibly both rational and correct.

The Countries That Are the Most Prone to Mistakes

The first of such countries is China for the following three reasons:

1. Since the end of the Cold War, China has become the only big country to maintain a Marxist, socialist system. The doctrinaire Marxists take it for granted that China should take responsibility for world revolution. Many people of bullied and humiliated nations also expect China to lead them in confronting hegemonic politics. Thus, China is liable to be in an opposing position to the United States for ideological and moral reasons.

2. The time-honored and unique cultural tradition of China is a force for (national) integration; it includes an ardent spirit of nationalism. The nationalism of the cultural tradition of China has a unique characteristic. In peaceful times, there are often "internal struggles" within the nation. As soon as foreign invasions occur, however, the nation unites. China never initiates aggression. This has been proven over thousands of years of Chinese history and especially in recent centuries. Yet China is inclined to powerful passive resistance to aggression. It is for this reason that China has survived the invasions of various international powers and never yielded. Therefore, this nation is sensitive to international currents of anticommunism and hegemonic politics and, when confronted with such senti-

ments, is prone to respond with a narrow kind of nationalism. A book published in China, *China Can Say No*, reflects these emotions.

3. Under the leadership of Deng Xiaoping, China has undergone a great socialist reform movement and has achieved well-acknowledged success. As the economy experiences continuous two-digit growth, the socialist production forces of China, the nation's overall strength, and the standard of living of the Chinese people have entered a new phase. This is encouraging but at the same time may bring about arrogant attitudes that lead to various mistakes.

Having carefully studied the post–Cold War situation and taken lessons from history, the Chinese government has set forth principles for dealing with international relations, especially relations with the United States. These principles are defined as "avoiding confrontation, reducing problems, strengthening understanding, and furthering cooperation." Thus, the risk of making the above-mentioned mistakes is reduced. If the United States were to adopt the same principles, the Sino-American friendship would be strengthened.

As a learned American said, "If one treats China like an enemy, China could become an enemy." I would like to add two sentences to his. The first is: If one is hostile to China, one will find that China is not only an unconquerable enemy, but also a most steadfast enemy. The second is: Truly treat China as a friend, and you will have a real friend. I hope the United States makes no mistakes in dealing with China, otherwise, it will certainly be disastrous not only to China and to world peace and development but also to the United States.

The second of these countries is the United States for the following four reasons:

1. The United States became the sole superpower in the world after the Cold War. Some in the United States believe that it has obtained a thorough triumph, but they fail to recognize the wounds it suffered during the Cold War. Dizzy with success, the United States is prone to arrogance. As a result, "an arrogant army will necessarily lose," and "sole self-importance" (in international affairs) is the attitude of hegemony, which leads to isolation.

2. The United States became a world power this century during the two "hot wars" and the Cold War. Several generations grew up confronting definite enemies. Therefore, there is a sense of loss after the Cold War—a loss of direction in policies and in many aspects of work. This should be an opportunity for developing new situations and for putting aside Cold War interests, yet some continue to long for the past and continue to think in accordance with Cold War logic. For that reason, they seek to identify a new enemy, and in many circles, China appears to be the natural choice.

3. Some in the United States make ideology a priority in international affairs. This attitude was widespread in the Cold War. The former Soviet Union is a typical example. It tried to spread its Communist model in many ways. This led to the disunity in the socialist camp that was celebrated for a while and became a major cause of her loss in the Cold War. Now the United States, as the victor of the Cold War, is attempting to spread the American model of democracy. By using ideology as leverage in diplomacy, it is possible to repeat the mistakes of the former Soviet Union.

4. The United States is a very insular country. In this way it moves against the tide of history and is prone to mistakes in judgment. The entire world has knowledge of the United States, but the United States has little knowledge of the world. Perhaps it does have some knowledge of Europe, but it fails to have any deep understanding of developing countries, of China, and of Asian cultures. Some in the United States have the attitude that it is not

necessary to understand them. Even if they try at times to do so, they always make analyses and judgments of others in accordance with their own framework of values. How could it then avoid committing mistakes?

The Coming Conflict with China emanates from these four factors.

Lessons from the History of Sino-American Relations

There are lessons to be learned from two wrong choices in Sino-American relations. Reviewing the past helps one to understand the present. By drawing lessons from the past, one can avoid repeating mistakes.

The first wrong choice was made after the end of World War II. Since China and the United States were allies, Sino-American relations should have been entering a new era of friendly cooperation. At that time, there were two major political forces in China: the National party led by Chiang Kai-shek and the Communist party led by Mao Zedong. During the 1940s, Mao Zedong got along well with the coordination team of the U.S. Army at Yan'an.

As is well known, Mao Zedong objected to Stalin's intervention in the Chinese revolution; hence, he was not trusted by the Communism International (Comintern) and the Communist party of the Soviet Union. In some articles Mao Zedong wrote in the 1940s, he says that the future goal of the Chinese revolution is to build up a democratic republic similar to that of the United States. When Huang Yanpei, a famous proponent of democracy, visited Yan'an, he asked Mao Zedong how to assure that the Communist party would not repeat the cycle of rise, decline, and collapse of the previous dynasties in Chinese history. Mao Zedong's answer was very firm: democracy.

In 1945 and 1946, Mao expressed through certain channels his intention to visit the United States. Due to ideological considerations, however, the American government chose Chiang Kai-shek's government even though it represented the privileged class and was already corrupt. The United States fully supported Chiang Kai-shek's initiating the civil war, which proved to be catastrophic for the Chinese people.

Finally, the Chinese people chose the Communist party and supported it. With millet as their main provision and rifles as their main weapons, the Communists defeated in only three years 8 million of Chiang Kai-shek's troops who were armed with American equipment. For the history of this period we can look to the memoir of Averell Harriman, then the American ambassador to the Soviet Union, and to the White Book by Dean Acheson. Due to the wrong choice made by the American government, the Communist party and the Chinese people led by Mao Zedong were pushed into an anti-American position.

The second wrong choice was made when the Chiang Kai-shek dynasty collapsed. At that time, American Ambassador John Leighton Stuart stayed in Nanjing, an action with deep implications. In addition, through private contacts he made during his teaching career at Yenching University, he had already been in touch with the Chinese Communist Party (CCP). When the People's Republic of China was established in October 1949, Mao Zedong had planned that the United States would be the first country with which to establish foreign relations. The state council of the United States also held a conference on Far East issues in which such "most intelligent thinkers" as John Fairbank, and A. Doak Barnett participated. The participants almost unanimously agreed to recommend that the United

States recognize the new government of China. This would have been a good opportunity to better Sino-American relations. President Harry Truman, however, did not heed the opinions of these experts, and the good opportunity disappeared immediately.

Not long afterward, civil war broke out in Korea, and the United States became involved in it. At that time, the newborn People's Republic of China (PRC) had suffered severe war wounds and various abandoned industries needed restoration. Pondering how to deal with the United States, the number one superpower in the world, the CCP, and the Chinese people faced a most difficult choice.

The archives of the former Soviet Union have recently come to light. From these materials, one can find that, in the beginning, Mao Zedong and other leaders of the CCP did not propose to "resist America and assist Korea," or at least they did not want direct military involvement. This is quite understandable for it is responsible in light of the interests of the country. But the American hawks, represented by General Douglas MacArthur, were dizzy with military success. They declared openly that the Yalu River was neither a boundary between Korea and China, nor an end to the U.S. army's advancement. The American air force had already bombed China's northeast region in an attempt to kill the newborn People's Republic in the cradle. As a result, the Chinese government and the Chinese people came to a realization that "when the lips lose, the teeth will feel cold." In order to defend their own homes and country, they fought directly against the United States for three years.

Without a doubt, the Korean War was a huge drain on China's national strength, leading to a great delay in the process of Chinese modernization. By making the wrong choice, the United States brought disaster on the Chinese people and stimulated their nationalism; the whole country shared bitter hatred of an enemy. China forced the U.S. Army to sign a peace agreement. As a weak nation declaring victory over the United States, China became an exception to the rule. American Commander Mark Clark, a cosigner of the peace agreement, sadly acknowledged that he was the only general in American history to sign a nonvictorious agreement. Because of the wrong choice made by the United States, the opportunity for better Sino-American relations was lost. After the war, China and the United States became hostile for an entire generation.

My purpose in revisiting these unpleasant events is by no means to stimulate Sino-American confrontation but to spell out how important it is to make the right choices at turning points in history. I want to show the harm in oversimplified ideologization, in neglecting the opinions of experts, and in allowing oneself to become dizzy with success.

The influence of a generation of hostility is deep-rooted, and one cannot expect that everything can be put right in a single day. Since formal diplomatic relations were established, Sino-American relations have not developed evenly. They are sometimes good and sometimes bad. This is quite logical. At this significant historical moment at the dawn of the twenty-first century, the question is whether or not Sino-American relations can begin a new era of stable and friendly cooperation. This is a vital choice for both China and the United States.

The Wide Basis for Friendly Sino-American Cooperation

Friendship and cooperation cannot be built on hope alone. An adequate material basis is essential. And Sino-American relations have such a basis.

1. China and the United States do not have any geographic elements for confrontation. China and the United States are remote neighbors and have no common borders besides the Pacific Ocean; therefore, there cannot be any territorial disputes. As is well known, geographic conflict is the most common cause of national disputes; most wars in history have been territorial in nature. It is fortunate that there is no such element for conflict between China and the United States.

2. Over the long-term, Sino-American relations have generally been good. In Chinese minds, memories of British, Russian, and Japanese invasions in recent times have been slow to fade. By contrast, the United States did not ever directly invade China with major force. Although there have been such unfair agreements as the Wangsha Agreement and the 1900 Indemnity, these have mainly been of interest to historians and have not made wide and deep impressions on most people. On the contrary, China and the United States fought together twice, each time in defense against an aggressor. During World War II especially, they became close allies in antifascism. When Chinese people fought the war against Japan, the United States aided China greatly and left long-lasting good impressions on the minds of the Chinese people. It is worth noticing that after President Richard Nixon's visit to China in 1972, China and the United States actually established strategic relations. China made special contributions to the end of the Cold War, which had lasted for as long as half a century. These two experiences of alliance furthered the interests of both China and the United States and promoted world peace as well. This should be recognized and remembered.

3. China and the United States complement each other economically. This is not only true in terms of natural resources (for example, such strategic mineral resources as tungsten, stibium, and manganese are what the United States lacks) but also in terms of manmade economic resources. For example, China cannot compete with high-tech American products. On the other hand, China possesses the ability to produce basic, everyday products that meet the needs of average American consumers. This situation is completely different from American-Japanese and American-European trade relations. The economic activities of China and the United States go on at different levels in terms of quality. Therefore, the relationship is not one of competitors but of counterparts. This unique situation is conducive to friendly Sino-American cooperation.

Strategic Interests of China and the United States in the Twenty-first Century

The three above-mentioned bases for Sino-American relations are firm and stable, but they form no more than a possible general foundation. To turn possibility into reality and to achieve friendly Sino-American cooperation, historical analysis is a must. A historical opportunity is also needed, some powerful turning point in the historical process. Is the twenty-first century such an opportunity? Where is the turning point? The answer depends on the strategic interests of China and the United States. I do not want to talk here in detail about common strategic interests in maintaining regional and worldwide stability (for example, disarmament and prevention of the spread of nuclear weapons). Instead, I want to emphasize the most fundamental strategic interests of China and the United States and whether or not they can be joined.

1. The most fundamental strategic interest of the United States is to remain the number one superpower in the world. In human history, superpowers that have maintained supremacy for over a century have been rare. It is a general historical principle that each superpower will lead the world for about a century. Can the United States remain the number one superpower and create a historical exception?

After the Cold War, although the United States remained powerful, its wounds were actually severe. Relatively speaking, its international status declined. The world is becoming multipolarized. Japan and Europe have risen because their strength allows them to say "no." American military force remains number one in the world; however, it had to rely on Japanese and European financing in the Gulf War. The most important factor is that the era of the great fleets has passed and will never return. It is economic strength that will decide the number one superpower of the twenty-first century. The twenty-first century will be a century of intense economic competition. Can the United States win this competition?

2. The most fundamental strategic interest of China is to modernize. With five thousand years of unique civilization and a population of 1.2 billion, China, though not a superpower, is indeed a "super society." In terms of the level of economic development, China and the United States belong to different categories, and the United States is much more advanced. In terms of social development, China and the United States also belong to different categories, and the situation for China is much more complicated.

The difficulty and complexity of Chinese modernization is unprecedented in human history. Although great successes have been achieved in the last two decades of reform and opening up to the outside world, there is still a long way to go. The most fundamental strategic interests are to accord with Deng Xiaoping's theory as well as his great plan, to grasp the rare opportunity of history, to complete the great socialist reform, to concentrate on economic construction, to realize the four modernizations, and to build a wealthy, strong, democratic, and civilized socialist country with Chinese characteristics. For this reason, China really requires a peaceful and cooperative international environment.

The significance of friendly cooperation between the United States, the strongest developed country in the world, and China, the most populated developing country in the world, cannot be overestimated. Deng Xiaoping once pointed out, "Sino-American relations must be made good." This is a profound and prescient statement. I believe that the two countries' fundamental strategic interests have the potential to come together.

If the United States wants to maintain its economic power, the first thing to do is to look at how it became the number one economic force in the world. There were many reasons. For example, it has enjoyed exceptional advantages in geographic location and natural resources, along with the special advantages of a country of immigrants. But it is also indisputable that during the two world wars, while there were no battles on American soil, the United States became a magnet for talented, highly skilled people and became a worldwide supplier of weapons and products. This situation no longer exists and this opportunity will never reappear.

There is another factor that is often ignored. After World War II, many other countries lay in ruins. Economic restoration and development required funds and technologies possessed by the United States alone. The United States, however, also needed restoration and development. Thus, it faced a choice. To review history, there was a significant strategic decision made in the development of the Marshall Plan. The United States gave Europe

$50 billion with which European countries could purchase advanced American technology, equipment, and products. In so doing, the money returned to stimulate the restoration and development of the American economy. With European development, its capacity to purchase advanced American products increased. This turned out to have benefits all around during this historical period. A similar case can be found in American-Japanese relations, which stimulated continuous development during the postwar period. Historical experience implies that the United States became the number one superpower in the world because American funds and technologies combined with the urgent needs of a prosperous developing market.

Now history is turning a new page. Where is there a prosperous growing market with urgent needs for funds and technology? China is undoubtedly one. The "super society" with a population of 1.2 billion is naturally a great market. If 10 percent of the Chinese population achieves the Japanese standard of living, it will form a market the same size as Japan. If 20 percent of the Chinese population achieves the American standard of living, it will form a market the same size as the United States. In addition, the Chinese market is not just a potential one, for China's economy has been growing for nearly twenty years at the leading rate in the world. This makes it a substantial market for the twenty-first century.

On the other hand, to realize modernization in the twenty-first century, China will still need funds and technology most of all. Thus, the country that cooperates with China for mutual benefit will obtain the number one position of economic power. Now, unlike in the past, the United States is no longer the only country capable of providing such funds and technology. We hope that the United States can make the right choice and win the competition.

The Taiwan Issue

The Taiwan issue is a sensitive aspect of Sino-American relations and has become a huge obstacle to friendly Sino-American cooperation. It is often grounds for the theory of a "Sino-American conflict." Regarding this issue, I believe that, in order to find a road to better Sino-American relations, several basic concepts must first be made clear.

1. The unification of China is a matter beyond dispute and bargaining. Anyone with a little knowledge of Chinese culture knows that unification has been an essential tradition and the basis for natural establishment throughout Chinese history. Chinese history is a history of fighting disunity and reinforcing unity. Any person or political group that maintains Chinese unification and territorial integrity wins the people's support and the appreciation of historians. Any person or political group that tries to divide China, to surrender the territory of our motherland to others and, thus, to harm the integrity of our motherland, will be cast aside by the people and condemned from generation to generation. The Chinese Communist Party and the government of the People's Republic of China's insistence on reunification is very natural. Do not expect that the Chinese government will make any concessions on this issue.

2. The Chinese government, in dealing with the Taiwan issue as a problem that developed in its recent history, has set forth the principles of "peaceful reunification, one country, two systems, and unconditional dialogue between the two sides of Taiwan Strait." What does it mean if one rejects "one country, two systems"? It means either that Taiwan

unifies the mainland in capitalism or that the mainland unifies Taiwan in socialism, which will necessarily lead to military conflict. Therefore, "one country, two systems" is the only realistic approach for peaceful reunification.

As for how to realize "one country, two systems," and to decide on the type of "one country, two systems," such detailed issues have to have mutual understanding; common ideas must be reached through negotiation. The interests of both sides should be fairly considered in order to achieve the necessary concessions. Considering that Taiwan authorities fear that the negotiation will simply be "bigger annexes smaller," the Chinese government has set forth the principle of "unconditional dialogue." In human history, perhaps there is nothing more fair-minded and reasonable than the concept of the "unconditional." Therefore, any unprejudiced person may find that in terms of "peaceful reunification," the proposal set forth by the Chinese government has reached the maximum possibility. Now the ball is in the court of Taiwan's authorities.

The reason why Taiwan's authorities have never responded honestly is because some people in Taiwan attempt to divide China, to propose "two Chinas," "one China and one Taiwan," or "Taiwan independence." These proposals will never be accepted by the Chinese government.

3. Chinese people expect that reunification will be realized in the near future, and they hold to the adage: the sooner, the better. An earlier unification will assist China and Taiwan's development and progress; however, we are patient enough to wait for reunification. Disunity has lasted for several decades, and of course one cannot expect the two sides to suddenly shake hands and embrace in the space of a single day. There is a gap between the two sides in terms of their economies and living standards, and some of Taiwan's common people may therefore have concerns. We fully understand and make allowance for this.

One should be aware, however, that the time is advantageous for mainland China. Under the leadership of the third generation of CCP leaders centered around Jiang Zemin, mainland China will complete the great socialist reform in accordance with Deng Xiaoping's theory on the socialist construction with Chinese characteristics and will, therefore, continue the trend of development that has already lasted for two decades. Then, in the twenty-first century, China's economy and social progress will develop even further. When the economy of mainland China comes close to, equal to, or surpasses the level of Taiwan, there will be no need for negotiation; the issue will be settled naturally. We believe that people, and people alone, are the masters and creators of history. Thus, today is the best time for Taiwan's authorities to conduct peaceful talks with mainland China, because there are still a few trump cards in her hand. With the passage of time, her position in negotiation will decrease. This is the general trend of history.

4. The day when Taiwan declares independence will be the day war begins. Take a quick look at history since 1949: "the war of resisting the United States and assisting Korea" when the People's Republic had been just recently established; the "boundary conflict with India" which received support from both superpowers; "Treasure Island Defense" against the former Soviet Union, which was armed to the teeth; and the "Boundary Self-defense" against Vietnam, which had once been a brother country of China. As far as the integrity of Chinese territory is concerned, even if the territory involved is just a small, deserted island in the Heilongjian River, the CCP and the Chinese government will not hesitate in waging war in defense of it, let alone Taiwan, the valuable island.

The Chinese Communist Party and Chinese people have been entirely firm patriots, and the proof is a history written in blood. Do not have any illusions on this issue. I think Americans understand this very well. Americans waged the famous Civil War to maintain integrity when the South attempted independence.

Of course, there will be various arts and measures of war when "war is declared." According to the Art of War of Master Sun, "best of all is to make an army yield without a single battle." Considering that Taiwan is but a small island, a military blockade may be an option. Since Taiwan's economy is foreign-trade oriented and lacks inner resources, the declaration of war will cause the hike of import and export insurance. For this reason alone, Taiwan's economy will collapse. What is most fundamental is that we believe that the masses of Taiwan people will not tolerate the ambition of a handful of Taiwan politicians who attempt to "make Taiwan independent." Therefore, it is not necessary to settle the issue with bloodshed. For over a hundred years, Chinese people have suffered too much bloodshed; we want there to be no more bloodshed.

5. The Taiwan issue is entirely a Chinese internal affair, in the same way that what happens in Hawaii or California is an internal affair of the United States. Therefore, the United States should not itself involve in the Taiwan issue. Not to become involved in the Taiwan issue is the most intelligent approach in light of American national interests. In terms of social justice, this accords best with the "democratic spirit" of the United States. For historical reasons and due to the complexity of the international system, however, the United States has unfortunately been involved in this whirlpool. This involvement becomes an uncertain factor in Sino-American relations as well as a stirring element in the domestic-political struggles of the United States. Facing the opportunities of the twenty-first century, the United States may have the following choices:

a. The first option would be to maintain the current situation. On one hand the United States correctly announces that is supports the state policy of "one China"; on the other hand, it often plays the Taiwan card, so that Sino-American relations will from time to time have ups and downs. Perhaps the United States can reap temporary benefits from the conflict between the two sides; however, after ten or twenty years pass, and the Chinese economy develops and Taiwan returns, how will the Chinese people, after peaceful reunification, assess the role of the United States? This is thus a bad choice.

b. The second option would involve either openly or secretly supporting Taiwan's authorities in attempting to establish "one China, one Taiwan" or even "Taiwan independence," as it did in the past with Chiang Kai-shek's group. In so doing, the United States may even intervene militarily, which will necessarily lead to a war between China and the United States. Of course, since the United States is the number one military force in the world, the war will be a disaster for China. However, China is by no means Panama or Iraq, and the United States cannot obtain victory over China. The Korean War and the Vietnam War were previous lessons. In such a case, the United States will suffer economic distress and a loss of power during a protracted war. Thus it is a disastrous choice, the worst of bad policies

c. The third option would require that the United States withdraw completely from involvement in the Taiwan issue, leaving all the problems for the Chinese people themselves to settle. Of course, China will welcome such a policy. Yet, it may cause unnecessary disruption in American politics. From the angle of American interests, it is not the best policy.

    d. A fourth option would allow the United States to use its involvement to serve as an active element for China's peaceful reunification. In the case that the United States accelerates the peaceful reunification of China, it will not only get rid of a "hot potato," but it will also lay a basis for friendly Sino-American cooperation in the twenty-first century. This is the best policy, and the only right choice.

For a long time, people have wanted to settle the Taiwan issue as a way to achieve a friendly Sino-American relationship. In a situation where China and the United States do not have good relations, how could the United States give up playing the Taiwan card? How could China then trust the signals in hope of cooperation issued by the United States? One may think of this problem in a "reverse way." For example, China and the United States should first focus on their most fundamental strategic interests of the twenty-first century and thereby build up a friendly Sino-American partnership. Could the Taiwan issue be settled without difficulty as soon as China and the United States become friendly?

Ideological and Cultural Differences

Ideological and cultural differences between China and the United States are often the theoretical basis used by those who believe that conflict between China and the United States is unavoidable. It is, however, both theoretically and practically groundless.

    1. What is better for ideology and culture? Diversity or uniformity? Human history has shown that diversity is better. A world of many colors is a beautiful world. Competition and assimilation together have furthered civilization. The United States is a country of diverse ideologies and cultural orientations. Why can it not allow Chinese culture and socialism with Chinese characteristics to thrive?

    2. On one hand, there are of course conflicts between different cultures and their ideologies. On the other hand, there are also convergences. Convergences may produce a kind of "hybrid vigor," which is just what we should choose. In five thousand years, Chinese culture has assimilated countless elements from other cultures. "Incorporating things of diverse nature" is a distinct feature of Chinese culture. This is the reason why its vitality has lasted for five thousand years. The cultural history of mankind shows that the extremes of cultural exclusion may prevail in certain periods but are bound to be short lived.

    3. Both culture and ideology belong to the category of spirits or ideas. While ideas do not have national boundaries, revolution cannot be exported. Human history shows that the spread of advanced ideas and truths cannot be stopped by national boundaries or by force. At the same time, the attempt to export one's own culture and ideology by revolutionary, military, coercive, or administrative means will necessarily end in failure, even if the culture and ideology is advanced. People have the right to choose. The peoples of various nations have equal rights to choose. Even if people make a wrong choice, it is up to people themselves to come to the right choice. No one should force people or make choices for them. Is this not the real spirit of democracy? Is this not social proof of the real democratic spirit?

4. Why did Chinese people choose socialism with Chinese characteristics? In 1949, Mao Zedong concluded that Chinese people had learned from the West, and chose the road of capitalism and the capitalist democratic republic system. But the teachers always humiliated and oppressed the student; they did not allow China to learn. Then, the Chinese people chose the Chinese Communist Party and the socialist road that led to the success of revolution. Why did the Chinese Communist Party and socialism succeed in China? A major reason is that, from the very beginning, China differed from international communism and from the Communist party of the former Soviet Union, maintaining China's own approach.

The basic goal of the Chinese revolution is to get rid of weakness, which plagued China for a hundred years, and to make China wealthy, powerful, and modernized. The way to realize this goal is to begin with the reality of China, to seek truth from facts, and to combine the basic principles of Marxism within the context of Chinese reality. To sum it up with a single phrase, it is to have a practical attitude. For things foreign, our principle is to learn whatever is good and put it into Chinese practice and never to imitate indiscriminately. To sum up, we will accept whatever can make China modernized and can make Chinese people well off. This is what we understand as socialism. Chinese people believe that their choice is right, and history has shown that the chosen road is a right one.

As long as one does not have ideological prejudice and the fixed Cold War thinking pattern, one will recognize the nature of socialism with Chinese characteristics, and will understand, sympathize with, and support it.

5. The United States has full confidence in its own culture and ideology, and we fully appreciate and respect that. A nation without self-confidence does not have a future. Of course, "every household has its own special difficulties." The United States has its various social problems. The United States is far from an ideal society. The best choice for the United States is to concentrate on self-improvement. If the United States is truly good, we should have faith in the power of truth and Chinese people will automatically learn from it. China is a nation that is good at learning from others. In modern China, almost all influential trends in thinking have prevailed in turn in China. In today's China, all the important theoretical works have already been, or are being, published in China. Chinese people will pick up the good points to be followed. However, if the United States does not devote itself to self-improvement, while forcing China to accept all it prefers, Chinese people will naturally ask: "Could this be the human rights and democracy of the American style?"

Do not expect China to swallow bitter fruit forced down its throat by others.

Will China Be a Threat to the United States after It Becomes Prosperous and Strong?

1. Must prosperity and strength be a threat? At present, the United States is the number one superpower in the world. Does this imply that she is threatening the rest of the world? Japan and Europe are also wealthy and strong. Do they threaten the United States? China is a peace-loving country, and this has been proven over thousands of years. China is a good-hearted nation. Tens of millions of Chinese have immigrated to other parts of the world, and these overseas Chinese have always been friendly to the people of the countries

they reached. They have been honest workers and have never established any colonies. In modern times, China was humiliated again and again, and hence it especially cherishes world peace and friendship. Hegemony goes against the nature of socialism, and the pseudosocialism that tried to maintain hegemony has eaten the bitter fruit it sowed. This is the reason why China has openly announced that it will never seek hegemony. In conclusion, from any meaning and any angle, China cannot form a threat to any nation.

Perhaps we may illustrate this with some quantitative comparison. If 10 percent of the Chinese population, or 120 million people, are well off enough to tour the United States and other parts of the world, will it not become a catalyst for American and world economic development. On the contrary, if China suffers poverty and political unevenness, and 10 percent of its population thus flees to other parts of the world, the impact would be immeasurable.

2. China is a country with a large population. Its per capita resources are not abundant, and its historical burdens and debts are heavy. Therefore, although at present China has made huge progress, it is still a very poor and backward developing country. According to Chinese standards of poverty, there are still 60 to 70 million people who are not well clothed and fed. China has to concentrate on economic construction and social progress. Even so, not until the middle of the twenty-first century will China be able to reach the level of moderately developed countries. Although the Land Company predicted in 1995 that the Chinese gross domestic product (GDP) will outgrow that of the United States in 2006, the Chinese population is still five times as large as that of the United States. Reaching the level of the United States in terms of per capita income will take at least one hundred years.

China is clearly aware that there are not only quantitative differences, but also qualitative differences that are even more significant. Our output value, though rapidly increasing, is still basically the value created with "industrialization," while what the United States is creating with advanced techniques is the output value of "informationalization." This is a gap between two historical periods. Therefore, Deng Xiaoping said, Chinese people should prepare for continuous efforts through several generations or even several tens of generations to catch up. As for military capacity, what China has is limited defense capacity. From nuclear weapons to laser-oriented weapons and from electronic systems to aircraft carriers, China has nothing that compares to the United States. The absolute defense expenditure of China is as little as one-thirtieth of that of the United States, and per capita defense expense is even less. Anyone of ordinary intelligence will never believe that Chinese military forces threaten the powerful United States.

To say the worst, China will not have the capacity of threatening the United States until the twenty-second century. That would be a problem our grandchildren and great-grandchildren would face. We should have faith in our grandchildren because, one hundred years from now, they will have superior wisdom and more rational thinking. They will find a way to solidify and further develop friendly Sino-American cooperation.

The problem is what legacy we should leave for them. This is really a historical choice. As the generation that spans the beginning of the century, if we leave a firm and stable basis of friendly Sino-American cooperation for our descendants, then we believe they will certainly build up a great house of friendly Sino-American cooperation on the foundation we laid.

Ushering in a New Century of Friendly Sino-American Cooperation

In the twenty-first century, a thoroughly friendly cooperation should be the choice for Sino-American relations. This fulfills the fundamental interests of both countries in the twenty-first century. This is based on a fully rational analysis, and is therefore a practical choice, perhaps even the most practical one. What is needed here is adequate political vision and resolve.

In fact, the trend toward friendly Sino-American cooperation has been developing in a silent way. For example, the trade between the two sides increased from $20 billion in 1990 to $57.3 billion in 1996, an increase of about one and a half times in a five-year period. If the politicians advance in accordance with the trend, the prospect of friendly Sino-American cooperation is foreseeable. If politicians irrationally choose confrontation, it would be detrimental, even disastrous, to both China and the United States.

14

The Making of China's Periphery Policy

Suisheng Zhao

Although China has been seen as a rising power with global significance in the post–Cold War world, its security relies heavily upon maintaining good relations with neighboring countries in the Asia-Pacific region. A Chinese strategist, Wu Xinbo, admits that although China has a great power self-image, it does not have adequate strength to play the role commensurate to this self-image. Therefore, "China is still a country whose real interest lies mainly within its boundaries, and to a lesser extent, in the Asia-Pacific region where developments may have a direct impact on China's national interests. . . . In terms of interests and resources, it is fair to say that China is a regional power with some limited global interests."[1] With limited capacity in traditional economic and military terms, China has to focus its resources on the Asia-Pacific region. Over one decade ago, Steven Levine stated that "outside Asia, China's role is determined more by what China may become than by what it already is."[2] This statement is still true today. Thus Beijing has to formulate its foreign policy particularly in response to issues close to home.

For a long time in the early years of the People's Republic of China (PRC), however, China was "a regional power without a regional policy."[3] The tensions with many of its neighboring countries became an important source of threat to China's national security. Beijing was in constant alert against possible invasions of hostile powers via its neighboring countries and fought several wars with neighbors or with hostile powers in neighboring countries to defuse the threat. China's relations with Asian-Pacific countries began to improve gradually in recent decades after it made a comprehensive periphery policy. This improvement has a significant impact on its security environment.

This chapter tries to analyze the recent changes in Beijing's policy and relations with its Asian neighbors and its impacts on China's security environment. It starts with an examination of Beijing's periphery policy, then goes on to explore the subsequent changes in Beijing's security environment. The last section will analyze the challenges to Beijing's periphery policy in the Asia-Pacific region.

Beijing's "Good Neighboring Policy"

China often calls its Asian neighbors "periphery countries" (*zhoubian guojia*). Although it was always aware of the importance in maintaining stable relations with these periphery countries for its national security, Beijing was never able to make an integrated policy toward neighboring countries. There were many factors responsible for the absence of China's regional policy. One was the frequent domestic turmoil and policy change, which severely limited China's ability to make any coherent foreign policy, including regional policy. The second was China's traditional cultural complacency and the legacy of Sino-centrism, which took China as the center of Asia for granted. The third was what Levine called China's ambiguous position in the region: "more than merely a regional actor, but still less than a global power," which left China in an uncertain relationship with its Asian neighbors. The fourth was China's unique position in the bipolar Cold World setting, which forced Beijing to see its security in global rather than regional terms.

Most of these factors began to change after China launched market-oriented economic reforms and began opening up to the outside world in the early 1980s. Externally, while China was confronted with no serious security threat throughout the 1980s, its strategic position in the triangular relationship between the United States, the USSR, and the PRC declined along with the détente between the two superpowers and the end of the Cold War. Internally, Deng Xiaoping and his reform-mined colleagues were determined not only to halt the domestic political turmoil that characterized the early years of the PRC but also to create a favorable international environment for economic modernization. In response to the new situation, the reform-minded leaders in Beijing made a deliberate effort to devise an integrated regional policy, known as *"zhoubian zhengce"* (periphery policy) or *"mulin zhengce"* (good-neighboring policy), to cope with the changes that challenged China's understanding of its relations with neighboring countries.

A study by You Ji and Jia Qingguo pointed to three new trends in Asia that led reform-minded leaders to pay special attention to its periphery.[4] The first was the prospect of a "pacific century," which Beijing embraced with the hope that fast economic growth in the Asia-Pacific region could offer new energy to China's economic prosperity. Taking the opportunities created by the restructuring of the world economy, China was determined to integrate its economy with the rest of the region. The second was the emergence of "new Asianism," which claimed that the success of Asian modernization was based on its unique values. This concept resonated in the hearts of many Chinese leaders, reformers, and conservatives alike, because it challenged Western ideological and economic centrality. Chinese leaders wanted to help drive this evolving trend of Asianism by working closely with its Asian neighbors. The third was the development of regional or subregional blocs following the collapse of the bipolar system. Beijing decided to take advantage of the collectivism that might provide new mechanisms useful for China to face the West. "Multilaterialism, albeit highly limited, developed as China started cooperation with neighboring states on transnational security problems (e.g., environmental pollution, illegal immigration, drug-trafficking, organized cross-border crime, etc.)."[5]

In light of these new developments, Beijing's leaders began to make a periphery policy that would help China to achieve the goal of creating a regional environment conducive to its economic modernization and national security. According to Liu Huaqiu, director of the Foreign Affairs Office of the State Council, the objectives of the good-neighboring policy was to "actively develop friendly relations with the surrounding countries, preserve regional peace and stability, and promote regional economic cooperation." To carry out this policy, Chinese leaders showed a benign face to negotiate with neighboring countries over a series of disputes. As Liu stated,

> China advocates dialogues and negotiations with other countries as equals in dealing with the historical disputes over boundaries, territorial lands, and territorial seas and

seeks fair and reasonable solutions. Disputes that cannot be settled immediately may be set aside temporarily as the parties seek common ground while reserving differences without letting those differences affect the normal relations between two countries.[6]

Specifically, Beijing's periphery policy was aimed at exploring the common ground with Asian countries in both economic and security arenas by conveying the image of a responsible power willing to contribute to stability and cooperation in the region.

It is important to note that Beijing's good-neighboring policy, just like any other parts of its foreign policy, was closely related to its reform-minded leaders' objective of economic modernization. In order to achieve a high rate of economic growth, these leaders looked for common ground in cooperation with neighboring countries in order to take a share of the rapid economic growth in the region. This economic motivation was very influential in guiding China's periphery policy, evident in their attempt to make diplomacy a means of "serving domestic economic construction" (*waijiao fuwu yu guonei jingji jiangshe*) after the inception of market-oriented economic reforms.[7] Their efforts have been very successful. The overall foreign policy goal since the late 1970s has been to maintain a peaceful international environment for economic modernization. Reform-minded Chinese leaders argued that economic power was both a means and an end of foreign policy. They defined economic modernization in terms of economic security for the nation. A Chinese scholar explained that economic security

> underscores the safety and survivability of those economic parts or sectors vital to the country's growth, the livelihood, and its whole economic interests. Indispensable for achieving such a security are elements ranging from favorable internal and external business environments, to strong international competitiveness, to status and capabilities in world politics. The final goals for economic security are to enhance national economic power, to secure domestic markets while expanding external ones, and to guarantee national interest and advantage in competition and cooperation abroad.[8]

In order to obtain economic power and security, Beijing was interested in the economic development models of Japan and other successful East Asian newly industrial countries (NICs) and tried to lure economic gains from increased trade and investment between China and these Asian countries. As a result of this policy effort, China traded more and more with Asian countries than with other regions after the 1980s. The total trade amount with Asian countries was only $16.6 billion in 1980, but it reached $48.5 billion in 1990 and $175.69 billion in 1995, right before the Asian financial crisis, which counted for over 60 percent of China's total foreign trade.[9]

In the traditional security arena, political leaders in China have made every effort to prevent their neighboring countries from becoming military security threats. A number of periphery countries were perceived as posing such threats due to ongoing border disputes and their relations with outside powers hostile to China. Accordingly, Beijing's periphery policy has two security goals in mind. The first is to settle border disputes "through consultations and negotiations,"[10] and the other is to prevent alliances of its neighbors with outside powers hostile to China. The first policy goal has involved China in a search for secure boundaries and peaceful settlements in land and maritime territorial disputes. The second policy goal has involved China developing strategic relationships and finding common ground with Asian countries in resisting pressures on market penetration and human rights issues from Western powers. Under these circumstances, China's periphery policy has been shaped principally by the dynamics of two major sets of relations. One is the dynamic between China and the global power of the United States. The second is the interaction between China and regional rivals such as Japan and India.

Beijing's special attention to its periphery does not mean that China is willing to turn its back on the world and simply cultivate its own garden. As You and Jia argue, Beijing's periphery policy "is not just about putting the backyard in order. PRC's diplomatic history has clearly showed that troubled relations with the surrounding nations in the past had seriously narrowed China's foreign policy options, especially toward major powers. . . . Closer integration with Asia has become a must in China's regional foreign policy initiatives."[11] In other words, establishing good relationships with neighbors is aimed at providing China with a more secure environment in its periphery as a leverage to increase its influence in world affairs.

Improvements of China's Security Environment in the Asia-Pacific Region

Indeed, China's security environment in its periphery improved greatly since it began to formulate and implement its good-neighboring policy. In his study of China's peripheral security environment (*zhoubian anquan huanjing*), a Chinese security expert, Yan Xuetong, asserted that as China has found more and more common security interests (*gongtong anquan liyi*) with peripheral countries, China has begun enjoying a more peaceful environment together with its neighboring countries in recent years. The result is increased mutual trust between China and its neighbors and enhanced national security for China.

Yan's assertion was based on his observation that China and its neighbors shared two fundamentally common security interests. One was to prevent a world war and a new cold war from taking place, and the second was to avoid regional military conflict. According to Yan, although some countries might be concerned about China's potential to upset the balance of power, most countries agreed with China's terms of strategic balance in the region. Yan divided China's periphery countries into three categories according to the degree of their agreement with China's terms of strategic balance. The first category of countries shared China's interest in developing a regional multipolarization in which China would be one of many important strategic powers playing a balancing role. These countries included Pakistan, North Korea, Burma, Nepal, Cambodia, Malaysia, Singapore, Russia, and central Asian states as they hope to see China becoming powerful enough to reduce the likelihood of the United States intervening in their internal affairs. The second category of countries included Australia, Canada, Indonesia, Thailand, the Philippines, Vietnam, New Zealand, and India, which hoped to maintain the current strategic balance in which the United States had the strategically advantaged position. These countries did not want to see China become a balancing power to the United States, but they did not have major conflicts of interest with China. The third category of countries, mainly the United States and Japan, concerned over the rise of China, wanted to establish a multilateral mechanism in China's periphery to prevent China from becoming a security threat to their interests. Yan believed that because China shared common strategic interests with most of its periphery countries, China had enjoyed and could continue to enjoy a favorable security environment within its periphery for the near future.[12]

It was indeed a prudent policy for China to make the effort to find shared strategic interests with neighboring countries. As a result of this policy, China has improved relations with most of its periphery countries in recent decades. This improvement came roughly in two chronological stages: the late Cold War period of the 1980s and the post–Cold War period of the 1990s.

The first stage started at the time of China's economic reform and opened up to the outside world in the early 1980s. Two policy shifts were significant in this period. The first was to abandon ideology as the policy guide and to develop friendly relations with neighbors regardless of their ideological tendencies and political systems (*buyi yishi xingtai he shehui zhidu lun qingsu*). This policy shift was stated in a Chinese foreign policy history book as one of the five major policy shifts in the 1980s.[13] The second significant policy shift was to change the practice of defining China's relations with its neighbors in terms of their relations with either the Soviet Union or the United States (*yimei huaxian, yisu huaxian*). China would develop normal relations with neighboring countries regardless of their relations with the Soviet Union and the United States. These policy changes resulted in an improvement of China's relations with some periphery countries previously in tension during that period.

One good example was the normalization of its relationship with Mongolia, which had long been perceived as a Soviet satellite in China's northern frontier. A border agreement between the two countries was signed in November 1988. Another example was China's efforts to improve its relationship with India. This effort resulted in the ice-breaking visit of Indian Prime Minister, Rajive Gandhi, to Beijing in December 1988, the first such visit after the Sino-India border war in 1962. China declared that this visit marked the beginning of a normal relationship between these two countries.[14] Maintaining a good relationship with North Korea and, at the same time, improving its relationship with South Korea was a third example. The so-called "traditional friendship" between China and North Korea had always been delicate, as North Korea had swung between Moscow and Beijing for many years. This relationship became particularly difficult in the 1980s when differences in ideology, economic and political systems, and foreign policy were growing between these two Communist neighbors. However, Beijing managed a fairly good relationship with the North while successfully establishing close relations with the South at the end of the 1980s. The improvement of its relationship with Taiwan was still another example. To create a peaceful international environment for its modernization drive, Deng Xiaoping decided to shift Beijing's policy from "liberating Taiwan" by force to a peaceful reunification offensive by suggesting talks between the Chinese Communist Party (CCP) and the Nationalist Party (KMT) in the early 1980s. Yie Jianying, the chairman of the National People's Congress Standing Committee, specifically proposed *santong* (three links; i.e., commercial, postal, and travel) and *siliu* (four exchanges; i.e., academic, cultural, economic, and sports) between the two sides.[15] Although the Taiwan government suspected that Beijing's policy was merely a "united front" tactic at first, it responded to Beijing's peaceful inducement under domestic pressure and began to ease restrictions on trade, investment, and travel to the mainland after 1986.[16] By the end of the 1980s, economic and cultural exchanges across the Taiwan Strait developed rapidly.

The Tiananmen massacre in 1989 and the subsequent end of the Cold War marked the beginning of the second stage of improvement in China's relations with periphery countries. The massacre led to economic sanctions by and deterioration of relations with Western countries. However, it had little negative impact on China's relations with its Asian neighbors as the human rights records in most of these countries were not better than that in China. To a certain extent, these Asian governments were sympathetic to China's authoritarian rule and struggle against pressures from the Western countries. To improve China's international environment after Tiananmen, Beijing's leaders made a series of policy adjustments. A former Chinese diplomat characterized these adjustments in twelve Chinese characters: *wendingzhoubian* (stabilizing periphery), *kaiduowaijiao* (expanding diplomacy), and *liuzhuanjumian* (altering the situation). Specifically, reform-minded Beijing leaders decided to further reduce the role of ideological factors in China's foreign relations and to stop drawing lines according to a country's social-political system or attitudes toward China. The emphasis was on the improvement of relations on China's periphery in order to establish a secure and stable periphery environment.[17] As a result of this effort, it was really ironic that while China's relations with Western countries soured, its relations with its Asia-Pacific neighbors improved after the Tiananmen incident.

Taking note of the improvement in relations between China and the Southeast Asian countries, China normalized diplomatic relations with several influential Southeast Asian countries in the early 1990s: Indonesia (August 8, 1990), Singapore (October 3, 1990), Brunei (September 30, 1991), and Vietnam (November 1991). In spite of the rising concern over the China threat among many Southeast Asian countries, they not only sided with China in opposition to U.S. pressure on the human rights issue but also accepted Beijing's position stating that reunification with Taiwan was solely China's domestic affair. China was invited to attend the ASEAN (Association of Southeast Asian Nations) post–Ministerial Conference in 1991 and became a member of the ASEAN Regional Forum (ARF) in 1994 and ASEAN's comprehensive dialogue partner in 1996. Since then, China has actively participated in what they called *"shanhui jizhi"* (three meeting mechanism) of ASEAN Foreign Ministerial Meeting, the Enlarged ASEAN Foreign Ministerial Meeting, and the ARF.[18]

The Asian financial crisis that started in the summer of 1997 provided a good opportunity for China to further improve its relations with Southeast Asian countries. Although China was not immune from its effect, it withstood the crisis better than many of its neighbors. As the World Bank Global Economic Prospects and the Developing Countries (1998/99) indicated in the aftermath of the crisis, "China's growth is one source of stability for the region."[19] Beijing's leaders took a policy of "stand-by-Asia" and even sent several billion dollars in aid to afflicted Southeast Asian economies. In response to the speculation that China would have to devaluate its currency, Renminbi, under the economic pressure, China's premier, Zhu Rongji, repeatedly promised to maintain a stable currency. A Chinese devaluation would have set off competitive devaluations across the region. This "beggar thy neighbor" competition could have hampered Southeast Asian countries' recovery efforts and would have had devastating economic and political consequences for the whole region. China's positive response to the crisis in comparison to Japan's paralysis helped China gain very significant influence in the region. Because Southeast Asia now depended on China's ability to stabilize its currency, to a hitherto unprecedented degree, China significantly improved its relations with Southeast Asian countries, particularly in economic terms. During the Asian financial crisis, China obtained the power to shape Asia-Pacific development in ways that it never had previously. As a result, at the first ASEAN-plus-1 summit meeting between the leaders of nine ASEAN members and Chinese president Jiang Zemin in Kuala Lumpur in December 1997, a joint declaration was published to establish a good-neighboring and mutual-trust partnership between China and the ASEAN oriented toward the twenty-first century.[20]

In the 1990s, China also significantly improved relations with its neighbors in the north and northwest. A formal diplomatic relationship with South Korea was established on August 24, 1992, which marked the success of China's policy to secure a balanced relationship with both South and North Korea. To continue maintaining the balanced relationship, Beijing was very careful not to anger North Korea by refraining from developing any military contact and security relations with South Korea. Instead, it focused on bilateral trade and investment relations with the South while making every effort to keep strategic relations with the North. In this position, China hoped to better defend its interests in the future reunification process of the two Koreas. As You Ji indicated, "this strategy envisages the likely orientation of China's strategic interest in a Korea that is reunified, peacefully. China hopes to see that there will not be a U.S. military presence after Korea's reunification. It hopes to see that the reunified Korea will take a position that is pro-China rather than pro-Japan."[21]

Following the disintegration of the Soviet Union, China also secured a good start with the newly independent central Asian states of Kazakhstan, Tajikistan, Kyrgyzstan, Uzbekistan, and Turkmenistan in 1992. Three of the five central Asian states share borders of more than 3,000 kilometers with China. Securing its relations with these countries is crucially important to ensure stability along China's borders and to its energy supplies, or as a *Wall Street Journal* report indicated, to "buttress the twin pillars of its future economic growth: political stability and plentiful energy."[22] According to this report, China shared with these countries the concern that radical Islam, inspired by the example of Taliban-controlled Afghanistan, would stir ethnic and popular revolt. China's westernmost region is inhabited by Turkic-speaking Muslim Uighurs, and Beijing fears the possibility of fundamentalist fervor erupting in that region. Russia has been bogged down in guerrilla war with Muslim nationalists in Chechnya, and Tajikistan, Kyrgyzstan, and Uzbekistan have cast militant Islam as their main enemy. This common concern led these countries to work with China to contain ethnic fundamentalism. Eager to prevent Islamic militancy from fueling separatism in Xinjiang, China has dispatched waves of senior politicians and military delegations to Central Asia. It gave parachutes, medicine, and other supplies to airborne forces and border guards in Tajikistan, which has been convulsed by civil war. It also pledged military aid to Uzbekistan, which has been raided annually by an Afghanistan-based opposition group called the Islamic Movement of Uzbekistan. Another important issue on the agenda for China's relations with these central Asian states is energy supplies, as these countries could be an excellent source for China's future energy supplies. According to this report, China's own conservative estimates show oil imports filling half the 6 million barrels a day it will need by 2010, up from just under a third today and zero less than a decade ago. Its need for natural gas, too, is projected to soar fivefold, to 3.9 trillion cubic feet. To meet demand, China has been exploring frantically in Xinjiang, a search

Exxon and other energy giants have joined at various times, and has also invested in two oil fields in Kazakhstan. [23]

To maintain good relations with bordering central Asian states, Chinese president Jiang Zemin took a lead to have the "Treaty of Enhancing Military Mutual Trust in the Border Areas" signed by China, Russia, Kazakhstan, Tajikistan, and Kyrgyzstan in Shanghai in April 1996. This group has since been known as the "Shanghai Five." At first, the Shanghai Five was designed as a forum to encourage dialogue on minor issues of borders and territory among China and its Central Asian neighbors. Yet in a few short years, the group has begun to address political and military questions and to share problems like organized crime. The five countries signed a "Treaty of Mutual Reduction of Military Forces in the Border Areas" in April 1997. At its June 2001 meeting in Beijing, the Shanghai Five accepted a new member, Uzbekistan, and agreed to meet under a new name, the Shanghai Cooperation Organization. These six countries agreed on political, military, and intelligence cooperation for the purpose of "cracking down on terrorism, separatism, extremism" and to maintain "regional security." Reportedly, Iran, Mongolia, India, and Pakistan are interested in joining the organization, which Chinese president Jiang called the "Shanghai Pact." A Western observer believed that the use of the Shanghai Pact perhaps intended "to evoke the former Warsaw Pact." According to this observer, "Together, the Shanghai Pact countries have a population of 1.5 billion; they control thousands of strategic and tactical nuclear weapons, and their combined conventional military forces number 3.6 million."[24] It was from this perspective that an Australian newspaper report stated that, "The newly formed Shanghai Co-operation Organization, bracketing China, Russia and four Central Asian republics, is poised to emerge as a potent force against United States influence and the rising tide of Islamic militancy in the region."[25] A reporter of *The Christian Science Monitor* also believed that "The Shanghai meeting of Russia, China, and four Central Asian nations is an effort to develop an organization that could one day offer a modest geopolitical counterweight to Western alliances. Its timing and substance are considered significant, with Russia and China already drawing closer after signing a number of bilateral agreements."[26]

Indeed, Beijing's relationship with Russia has improved spectacularly in recent decades. There were two important factors that brought these former Communist giants close. First, the Soviet/Russian search for security and economic reform matched China's yearning for stability in the aftermath of Tiananmen. Second, both countries were resentful over the high-handed behavior of the United States. While growing American pressure increasingly irritated China, the Western reluctance to provide large-scale assistance disappointed Russia. Thus, "there was an increasing convergence of views between Moscow and Beijing over the need to create alternative poles in world affairs to the one dominated by the U.S."[27] This convergence of views paved the way for the improvement of Sino-Russian relations.

Following Boris Yelstin's first official visit to China in December 1992, Beijing and Moscow institutionalized a twice-a-year summit meeting system at president and premier levels. A Western observer found that this sustained summit diplomacy was similar to "the Gorbachev-Reagan (the Yeltsin-Bush) summits" in 1985–1992, which helped put an end to the Cold War.[28] Jiang and Yeltsin met five times in six years by the end of 1997. In spite of Yeltsin's critical health situation, the sixth summit was held in a Moscow hospital in November 1998, and Yeltsin visited Beijing for the seventh summit in December 1999.

The 1999 Sino-Russian summit culminated in a landmark joint communiqué criticizing American hegemony and denouncing the use of "human rights interventions" by foreign countries. China officially declared Chechnya a matter for Russia stating that, "The Chinese side supports the government of the Russian Republic's action in fighting terrorism and splitism forces." For its part, the Russians officially supported China's sovereignty over Taiwan. The communiqué criticized the United States both for its attempt to ratify the 1972 Anti-Ballistic Missile (ABM) Treaty as well as for its refusal to ratify the Comprehensive Test Ban Treaty. The communiqué states, "The Russian side supports the Chinese side in opposing the position of any country under any form of bringing the Chinese province of Taiwan into a [antimissile defense] plan."[29]

The Sino-Russian relationship was first defined as a "constructive partnership" in 1994 and "strategic" was added in the Sino-Russian joined communiqué published on April 25, 1996, four days after the publication of the U.S.-Japanese joined security statement on April 17, 1996. This relationship was finalized as a "strategic cooperative partnership oriented towards the 21st Century" in 1997. After the retirement of Yeltsin, Chinese leaders continued the partnership with Yeltsin's successor, Vladimir Putin. At the July 2001 summit in Moscow, the presidents of the two former Communist powers signed the Good Neighborly Treaty of Friendship and Cooperation to defend mutual interests and boost trade. Although the treaty may be a marriage of convenience after both China and Russia faced pressure from the only superpower, the United States, after the end of the Cold War, it was significant in terms of the improvement of bilateral relations between these two former Communist giants. As one observer indicated, "since their first meeting in 2000, Putin and Jiang have met eight times to coordinate what the new treaty describes as their 'work together to preserve the global strategic balance.'"[30] A Reuters' report also indicated that this friendship pact "provided a legal framework for friendship now re-established after decades of mistrust over border and ideological disputes."[31]

The Sino-Russian partnership resulted in the declaration of the demarcation of the eastern section of the Sino-Russian border at the Jiang-Yeltsin Beijing summit in November 1997. A *China Daily* commentary celebrated that "after six years of close cooperation on the basis of consultation, mutual understanding and mutual concessions, the 4,300-odd-kilometer frontier will become a landmark of peace and friendship between the two countries and peoples, laying a solid foundation for stronger good-neighborly relations and regional stability and prosperity."[32]

As a result of successfully implementing the periphery policy, China's security along its borders has been substantially improved. China becomes more confident in its security environment when there is less immediate military threats to China's security. A Chinese official publication proudly declared that "Today, let's look around our neighboring areas from east to west and from north to south. We have basically established a relatively stable periphery environment around our neighboring areas. Our country has established good neighboring relations with all our neighbors. This is the best period since the founding of the PRC."[33]

Challenges to China's Security Environment in the Asia-Pacific Region

Indeed, China's periphery policy has served its interest in establishing a stable regional environment and promoting economic modernization along pragmatic lines. There is no reason to doubt the sincerity of Beijing's leaders in furthering good relations with neighbors when Jiang Zemin, in his report to the Fifteenth CCP National Congress in October 1997, stated that "China's modernization requires a stable international environment" in Asia as elsewhere.[34] China's ability to develop good-neighboring relations has certainly been enhanced by the successful power transfer from Deng Xiaoping to Jiang Zemin at the Fifteenth CCP's National Congress in September 1997. This new leadership has made it clear that China's best interest is to promote cooperation rather than conflict in Asia.

However, China still faces many serious challenges if it wants to further improve the security environment in its periphery. Although Beijing's leaders have declared the victory of the good-neighboring policy, the success has largely been along its land borders in the north and west. The security environment along China's east and south frontiers, particularly the coastal frontiers where China's major economic and political centers are located, have not been improved as much. Some new problems have also been occurring in these areas. In particular, there are at least four sets of challenges with which Beijing's leaders have found it difficult to deal. One is the divisive territorial dispute with several neighbors; the second is the rivalry with other regional powers; the third is the fear of its weak neighbors of the threat of China; and the fourth is the management of its most important relationship with the United States.

China's Border Disputes with Neighbors

The first challenge is to settle border disputes in the following three categories.[35] The first category is over land boundaries. Before the arrival of Western imperialist powers, territorial boundaries along China's frontiers had little significance under the tributary system. After the decline of the Chinese empire in the nineteenth century, Western powers not only took over many of China's tributaries but also pushed the frontiers forward into areas that China would have preferred to control itself. These new frontiers were often institutionalized in what China called "unequal treaties." As a result, after the founding of the PRC, Communist leaders in Beijing found themselves in a series of territorial disputes with its neighbors. The second category is over so-called "lost" territories, which include Hong Kong, Macao, and Taiwan. The third category is over maritime boundaries involving both bilateral and multilateral relations. Bilaterally, China is in dispute with Vietnam over the demarcation of the Tonkin Gulf and with Japan over the Senkaku Islands (Diaoyutai in Chinese), a group of rocky islets lying on the edge of the continental shelf about a hundred miles northeast of Taiwan. Japan considers these islets as part of the Ryukyus; whereas China claims them as part of the Taiwan province and therefore part of China. The major multilateral dispute is over the Spratly/Nansha islands in the South China Sea where there is believed to be a potentially resource-rich region possessing oil deposits. Beijing draws a maritime boundary running from Taiwan southwestward virtually along the coasts of the Philippines, East Malaysia, and Brunei and then northward more or less along the coast of Vietnam. The Philippines, Malaysia, Brunei, and Vietnam have disputed this claim.

Border issues are related to sovereignty, which is Beijing's most important concern in its foreign relations. Beijing has been firm in negotiations over all three categories of territorial disputes. Progress toward the settlement of these disputes has been made, albeit very limited. The most important progress is over land disputes with Russia, which shares a total of an approximately 4,300-kilometer border: about 4,245 kilometers along the Manchurian sections and 55 kilometers between Kazakhstan and Mongolia. China made progress in this settlement primarily because most of the areas in question do not, in general, contain Chinese populations. China signed the first border agreement in May 1991 with the USSR on the Manchurian sections of the common border. In September 1994, the second agreement was reached on the delimitation of the Sino-Russian border in the Altai region between Mongolia and Kazakhstan. Ratified in 1996, it led to the concrete demarcation in 1997. The Sino-Russian joint statement in November 1997 promised that the two sides "will complete demarcation of the Western section of the Sino-Russian border within the agreed period of time."[36] The Chinese-Russian Border Demarcation Commission was disbanded upon its work drawing to an end in April 1999, when more than 2,084 border signs and markers along their borders were set up and 2,444 islands on the border rivers were divided. About 1,163 of these islands went to Russia, and the others went to China. The only unsettled territories are three large islands: two on the Ussuri River and one on the Argun River, all of which are currently controlled by Russia.

The Chinese government has been extremely firm on the second category of territory disputes because it believes that people in these territories are ethnically and historically Chinese. Taking back these territories involves not only the vital security interest of China but also the legitimacy of the regime. Beijing recovered Hong Kong in 1997 and Macao in 1999. Taiwan is now the focus of dispute in this category. Since the early 1980s, Beijing has tried to use the same method to recover Taiwan as it did with Hong Kong and Macao by economic inducement and the proposal of a "one country, two systems" formula. But this peaceful offensive has not achieved its objectives. While total trade across the Strait rose from $5 billion in 1990 to $25 billion in 1997 and to $32.386 billion in 2000, making Taiwan the second (after Japan) largest supplier to the mainland and China the third largest market for Taiwanese goods, the political relationship between the two governments remains officially nonexistent and hostile. In frustration and to show its determination, Beijing launched a series of military exercises in the Taiwan Strait in 1995–1996.[37] However, military coercion has also not stopped Taiwan's political centrifugal tendency. President Lee Teng-hui proposed a "special state-to-state relationship" in July 1999 in spite of Beijing's military threat. Chen Shui-ban, the candidate of the proindependence party, the DPP, was even elected as Taiwan's tenth president in the fiercely contested 2000 election. Although Beijing has made it clear that it is willing to fight a war if necessary to recover Taiwan, it still has to concern itself with the reactions of the United States and other Asia-Pacific countries as well as the resultant rupture of China's economic development. Beijing's leaders have been left very little room to maneuver in this effort to achieve national reunification.[37]

Progress in the third category dispute is extremely limited. In the cases of dispute over the Tonkin Gulf and the Senkaku/Diaoyutai islands, no agreement or compromise has been reached with Vietnam and Japan. In the South China Sea, while Beijing has showed a certain degree of flexibility by suggesting "shelving the disputes and working for joint development" (*gezhi zhengyi, gongtong kaifa*), China's maritime neighbors have been very assertive in contesting Beijing's sovereignty claims. As one study indicated, "Although China has offered join development to other claimants, its concept of joint development seems to involve joint development of the producing oil and gas fields on other claimants' continental shelves—and then only after China's sovereignty has been recognized." In addition, as the same study pointed out, Beijing has continued to "insist on bilateral solutions and its interest and sincerity in participating in a multilateral cooperative solution remains in doubt."[39] China's position has been criticized and even ridiculed by other claimants in the South China Sea. Although Beijing and Hanoi reached an agreement in defining their disputed 1,300-kilometer land border after Chinese Premier Zhu Rongji wrapped up his visit to Vietnam in December 1999, no resolution was found over the two large island groups—the Spratlys and the Paracels (or Xisha and Zhongshao), which China has occupied since 1974 and over which they had a military clash with Vietnam in 1988. Vietnam still occupies most of the Spratlys to which China claims sovereignty. The Philippines has stepped up its claims over the Spratlys in recent years. Protesting against Chinese construction activity on Filipino-claimed Mischief Reef in 1998, the Philippine military set up patrols in the disputed archipelago and announced that the Philippine navy will fire warning shots if Chinese vessels get closer than five miles to challenge its patrols.[40] Beijing has not given up its claims of sovereignty over these islands because they are extremely important for China's security and energy supplies. Sovereignty over these islands keeps all of China's options open regarding resources, should any be discovered. However, Beijing's sovereignty claim may eventually bring China to the fore with all countries in Southeast Asia. Were that to happen, "China's ability to use force" would be "constrained by the possible reactions of the United States, Japan, and ASEAN, which would probably view such action as an attempt by Beijing to dominate the South China Sea."[41] Although exercising sovereignty over the disputed territories is crucially important for Beijing's leaders, it is certainly an extremely difficult decision for them to squander China's military resources and their political capital to seize these barren specks of land.

Rivalry with Other Regional Powers

The second major challenge to China's peripheral security is how to work with other regional powers, namely, Japan and India, to secure peace and stability together in the Asia-Pacific region.

For obvious historical and geopolitical reasons, China's relationship with Japan has always been difficult. Japan was China's most cruel and destructive enemy for a half century between 1895 and 1945. Japan also allied with the United States to contain China before these two countries established diplomatic relations in 1972. Although a Sino-Japan peace treaty was signed in 1978, this formerly friendly relationship has been largely superficial. While China has regarded Japan as a successful model of economic modernization and tried to lure Japanese trade and investment, Japan has been unwilling to build up a potential rival unnecessarily. Japanese loans and investment have come to China on a very lavish scale for years. According to one study, in the period between 1980 and 1996, the total amount of Japanese investment in China was $14.6 billion in comparison to $34.6 billion from the four small East Asian Tigers of Hong Kong, South Korea, Singapore, and Taiwan.[42]

Disappointed Beijing leaders have thus blamed Japan for its arrogant and unfair trading practice. In particular, they have been extremely alert for any signs of Japan's remilitarization. Beijing has played the "guilt card" as a weapon. During Jiang Zemin's visit to Japan in

November 1998, the first visit by a PRC president, Jiang reminded Japanese leaders at almost every public occasion that the past is far from forgiven or forgotten. Beijing has done everything to discourage Japan from aspiring to leadership of the region or taking on a greater global or regional political role. At a seminar on Northeast Asian security held in Shanghai, Chinese strategists asserted that "Northeast Asia is the only region where China has a strategic advantage. One of China's strategic goals should be to delay Japan's advancement toward becoming a major military power."[43] However, China's strategy has not been very effective. When President Jiang visited Japan expecting to dominate the scene, the Japanese refused a formal written apology over the war atrocities even though South Korea received this apology earlier and refused to exclude the Taiwan Strait from its security agreement with the United States. Japan also refused to utter the "three no's" (no support to Taiwan independence, no support to one China, one Taiwan, and no support to Taiwan's bid to join the United Nations) on the Taiwan issue although President Bill Clinton previously made the same pledge in his visit to Beijing. In the meantime, no matter what China thinks, Japan has taken a more and more critical position on China's military modernization efforts. As June Teufel Dryer indicated, "China's economic growth was accompanied by increases in the defense budget that averaged 12–13 percent each year. Given the absence of any external invasion threat and the presence of many domestic problems, this worried the PRC neighbors. Japan began to complain about the lack of transparency in Beijing's defense decision-making."[44] In response, Japan has made China the major target of its national defense strategy. It is hard, in this case, to be optimistic about the future relationship between these two important countries in the Asia-Pacific region. This rival relationship is certainly a major challenge to the success of China's periphery policy.

India is another budding rival of China in Asia. Although these two countries made friends by working together in the promotion of the national independent movement in the Third World during the 1950s, they became enemies and encountered a military clash in 1962. Sino-Indian relations began to improve in the late 1980s. China and India signed two agreements on maintaining peace and tranquility and confidence building measures in the border area in 1993 and 1996, respectively. President Jiang Zemin visited India in November 1996. As Beijing was cultivating a more friendly relationship with New Delhi, it was shocked by India's going nuclear in May 1998 and Indian Defense Minister George Fernandes' characterization of China as "a major threat to Indian security."[45] Beijing was furious and accused India of "running against the international trend of peace."[46]

However, this event was not a bolt from the blue, because Sino-Indian geopolitical rivalry has never stopped in the following three issue areas. The first is the Pakistan issue. While China has sought to improve relations with India, it has maintained a long-term strategic partnership with Pakistan, which India has waged three wars against in the five decades since its independence. New Delhi believes that China has used the Sino-Pakistan alliance to check the growing influence of India in Asia. Tibet is the second issue. Although India has publicly affirmed Beijing's position that Tibet is part of China, it did not welcome China's incorporation of Tibet, which otherwise may have served as a buffer between the two countries. China, on the other hand, has been irritated that India has allowed the Dalai Lama and his exile government to reside in Dharmasala and to campaign for Tibetan independence. The third issue is the two segments of the 2,500 mile Sino-India border that are still disputed: the southeastern Himalayan Mountains, now administered by India, and the Aksai Chin Plateau, through which a major Chinese highway linking Tibet and Xinjiang runs. Solutions to these troubled issues have to be found in order for China to establish a truly good-neighboring relationship with India.

The China Threat to Its Weaker Neighbors

With the rapid economic growth it has been experiencing, China has been regarded as the coming superpower in the twenty-first century. This realization has given rise to the speculation about a "China threat," particularly among China's weak neighbors. They worry that, after China modernizes, Beijing would like to have East Asia as its exclusive sphere of influence, a modern equivalent of the traditional tributary system. Beijing has denied this speculation and offered repeated assurances that "China will never seek hegemony."[47] However, this assurance has not eased the fear of China's weaker neighbors.

To be objective, military expansion cannot be considered a serious Chinese objective at least in the foreseeable future, because China lacks the military power and would face immense internal and external challenges if it were to take an expansionist policy in Asia. The presence and influence of the United States and the strength of dynamic regional powers, such as Japan and India, have defined and will continue to define the boundaries within which China's power may be asserted. Unless unforeseeable vacuums of power develop in Asia, it is difficult to conceive of China acquiring the hegemonic role that the United States long had in the Western Hemisphere or that the USSR played in Eastern Europe.

However, there is a perception gap between China and its weaker neighbors: "China, viewing itself in global terms, does not always realize how strong it is when placed in a regional context. The rest of Asia, viewing China in a regional perspective, does not always realize how weak it is on a global scale."[48] In this case, although China remains relatively weak in many of the measurable indicators of power, China's weaker neighbors see China in terms of its considerable potential and the long historic record of cultural and political domination of the region. China is slow to understand and to properly respond to the suspicions and fears of its weaker neighbors. Talking about China's relations with Southeast Asian countries, for example, a Chinese scholar believed that "as soon as the mutual trust is established, good neighboring relations and partnerships would be able to complete." He asserted that "the foundation for the mutual trust has been laid. The only problems are to overcome some barriers to the mutual trusts."[49] However, the nature of the relationship is much more ambivalent than this understanding. As one Western scholar indicated, "while most of ASEAN's policies towards China are guided by the economic perspectives of a huge Chinese market, explaining ASEAN's constructive engagement strategy towards China, Beijing's ambiguous foreign and security policies are simultaneously a major concern in the region." According to this scholar, "The rapid modernization programs of China's armed forces (including its nuclear arsenal), Beijing's territorial claims in the entire South China Sea, and its gunboat policies towards Taiwan have raised widespread concern over irredentist tendencies in China's foreign security agenda."[50] The fear of the perceived China threat is partially responsible for stimulating a rush to arms in many Asia-Pacific countries after the end of the Cold War.[51] Many of China's weak neighbors have determined to enhance their defense capacities and military preparedness to better deal with regional contingencies.

This ambivalent relationship has been further complicated by China's traditional attitudes toward its weaker neighbors. While pursuing a good neighboring policy, it continues to see periphery countries with a degree of condescension. Taking the example of China's attitudes toward Vietnam, a weaker neighbor that China ruled for centuries in history, the Chinese still view their influence in Vietnam as generous and civilizing. Deng Xiaoping justified China's invasion in 1979 in terms of teaching a lesson to the disrespectful Vietnamese who enjoyed substantial Chinese support during the war against the Americans but sided with the Soviet Union and challenged Chinese sovereignty over the islands in the South China Sea after the war. Vietnam joined ASEAN, a high-profile international organization, to reduce its isolation and vulnerability to the Chinese threat in 1995. Apparently, Chinese leaders have been displeased to see Vietnam's membership in ASEAN. However, they have no option but to live with the fact that its weaker neighbors could forge alliances to cope with China. How to handle this situation is certainly a serious challenge to China's periphery policy.

Relations with the United States

Managing its relationship with the United States, the only superpower in the post–Cold War world, is the most significant challenge for China as it seeks to establish a positive security environment in its periphery. The United States is a forwardly deployed power in the Asia-Pacific region, maintains formal security treaties with Japan and South Korea, and is the primary weapon supplier to Taiwan. Whether or not Beijing can successfully handle the other above-mentioned challenges, to a large extent, depends on the nature and level of American involvement in Asia and on the U.S. policy toward China. The United States has the ability to influence China's periphery policy by affecting the policies of some of China's neighbors and the overall security environment in the Asia-Pacific region. However, China's ability to influence U.S. policy is very limited, because it does not have much leverage over U.S. policy, which is largely a hostage to its domestic politics.

The Sino-U.S. relationship has been in a state of roller coasting since Tiananmen. It deteriorated after 1989 and warmed up briefly during the period between President Jiang's visit to Washington in October–November 1997 and President Clinton's visit to Beijing in June–July 1998. However, Sino-U.S. relations have been under strain again since late 1998. The U.S. Congress was angry over charges of China's nuclear spying and political payoffs, the Clinton administration's careless handling of technology transfer to China, reports of China's new suppression of political dissidents, import obstacles as the Chinese economy slowed, and PLA missile buildup along the Taiwan Strait. As domestic criticism over the Clinton administration's engagement policy increased, the Sino-U.S. relationship was forced onto a difficult path. Chinese Premier Zhu Rongji's visit to Washington in April 1999 failed to halt this downturn. The NATO bombing of the Chinese embassy on May 7, 1999, further moved these two countries toward the edge of confrontation. Although the downward spiral was temporarily arrested, the relationship, as a Chinese scholar indicated, was "stabilized only at a low level" in the last months of the Clinton administration.[52]

When George W. Bush took the White House in January 2001, Chinese leaders became especially concerned about the outlook for U.S.-China relations as Bush had termed China as a "strategic competitor" rather than as a partner, as previously termed by President

Clinton. This concern was confirmed as a series of incidents in the first four months after Bush took office seriously strained Sino-American ties. The most important one was the incidental midair collision between a U.S. Navy EP-13 spy plane and a Chinese fighter jet over the South China Sea on April 1, 2001. Although this incident was finally resolved without lasting impact, Sino-U.S. relations were further strained. As a *Washington Post* article reported, as a result of the crisis, "China's leaders are increasingly concerned that Washington and Beijing are headed for a confrontation as China emerges as an economic and military power in Asia."[53] A series of actions taken by the Bush administration in its first one hundred days of tenure seemed to confirm Beijing's concerns. Bush backed a national missile defense system, which China feared would negate its nuclear deterrent. Over Chinese objections, the U.S. government permitted Taiwanese President Chen Shuibian unprecedented access to the United States, allowing him to stop twice in America and to meet lawmakers in his transit visit to New York City and Houston in June 2001. Bush also shelved the peace process of the Korean Peninsula and hosted the Tibetan spiritual leader, the Dalai Lama, at the White House. He approved a multibillion-dollar weapons package for Taiwan in May 2001, including, for the first time, offensive weapons such as submarines. A pro-China, Hong Kong magazine listed all these actions and believed that "although it is still hard to say that the United States is ready to invade China, all this showed that the United States was indeed openly creating a tense military confrontation atmosphere and raising the political pressure against China for the purpose of driving China into a war panic."[54]

The deterioration of Sino-U.S. relations is certainly not conducive to China's efforts to secure its periphery. China's foreign policy establishment has been worried that foundations are being set now for long-term aggressive competition with the United States. This is not something that Chinese leaders want to see. However, to handle relations with the United States has always been difficult for Chinese leaders since the end of the Cold War. Many Chinese leaders, particularly those in the PLA, have been deeply suspicious about U.S. containment policy clothed as engagement. They have pointed out that while preaching strict Chinese compliance with international arms control agreements in the name of regional stability, the United States has increased its arms sales to Taiwan and strengthened its defense links with Japan and South Korea. China has been particularly alarmed over the U.S. announcement to build the Theater Missile Defense (TMD) system covering the Asia-Pacific following a surprise North Korean rocket test in 1998. TMD may not only be used to protect Taiwan from Chinese missiles but could also give Japan a more active role in regional security matters. The U.S. military intervention in Yugoslavia has seriously fed China's paranoia about Taiwan and Tibet. To add to this concern, China has seen the decision by the Philippine senate on May 28, 1999, to reopen its territory for joint exercises with U.S. forces as another link in an American chain of containment against China. China has been particularly apprehensive about the shift of the U.S. military and strategic focus from the Atlantic to the Pacific, which would have serious consequence on China's periphery policy. An article in Beijing's *Liaowang Zhoukan* (Outlook Weekly) speculated that in the Asia-Pacific region, "following the 'new U.S.-Japanese defense guidelines,' the United States will step up its deployment of Theater Missile Defense and strengthen its military alliance, and the possibility of the establishment of an 'eastern NATO' cannot be ruled out." It further indicated that "in line with the shift of the focus of U.S.

military strategy from Europe to Asia, the U.S. military will redeploy its forces in the west Pacific, step up activities by military planes and ships, expand the scope and frequency of its aggressive reconnaissance against China, and step up activities for advance military probing."[55]

China cannot change the United States' forward deployment and its web of alliances in Asia at least in the foreseeable future. Therefore, working with the United States is not a choice but a necessity. Beijing has to find ways to stabilize relations with the United States in order to create a favorable regional environment for its periphery policy. Confrontation with the United States will not only complicate China's periphery policy but also render a high cost for its modernization efforts. As one Western reporter pointed out, "the direction of Beijing's relations with the United States could exert a strong influence on China's development plans, forcing funds to be funneled into defense spending instead of economic growth."[56]

Conclusion

To establish and maintain a peaceful security environment in its periphery, China has tried to appear as a benign power that focuses on economic development and has tried to improve relations with its Asian neighbors. However, it has been assertive and even belligerent when dealing with issues relating to what it considers as vital national interests, such as territorial disputes. These two-pronged efforts have often transmitted conflicting signals to its neighbors making it difficult for China to meet the challenges discussed above. The disparity between China's belligerent behavior and "benign" face may bring uncertainties ahead in the Asia-Pacific region, particularly considering the rapid growth of China's economic and military power in the twenty-first century. China could become more assertive in taking back Taiwan and resolving maritime disputes in its own terms. China's regional rivals and weaker neighbors would have to be more concerned about China's might and threats. China's growing power could also rub against long-standing U.S. interests in Asia.

However, this situation has not seemed to bother Beijing's leaders very much. They have become more confident in dealing with other neighbors because of the increase in China's economic and military capacity. In his summary of the security environment for China's rise in the Asia-Pacific region, a Chinese scholar, Yan Xuetong, stated that China and its neighbors have shared the goal of developing a security strategy as all of them have wanted to maintain regional peace and stability. It is not a surprise to any country that China as a big power wants to be placed in an advantageous position in the strategic balance. This is a normal interest and demand in international politics and will not pose a threat to the security of other states. "China has shared certain common security interests with all peripheral countries and has no conflict in the overall interest with any countries although there are some contradictions in national border and reunification issues."[57] But it remains to be seen if this optimistic view can prevail.

Notes

1. Wu Xinbo, "Four Contradictions Constraining China's Foreign Policy Behavior," *Journal of Contemporary China* 10, no. 27: 294.

2. Steven I. Levine, "China in Asia: The PRC as a Regional Power," in Harry Harding, ed., *China's Foreign Relations in the 1980s* (New Haven, CT: Yale University Press, 1982), p. 107.

3. Ibid. Denny Roy, in a book of 1998, still believed that "China has no apparent 'Asian policy.'" Denny Roy, *China's Foreign Relations* (Lanham, MD: Rowman and Littlefield, 1998), p. 8.

4. You Ji and Jia Qingguo, "China's Re-emergence and Its Foreign Policy Strategy," in Joseph Y.S. Cheng, ed., *China Review, 1998* (Hong Kong: Chinese University Press, 1998), p. 128.

5. Wu Baiyi, "The Chinese Security Concept and Its Historical Evolution," *Journal of Contemporary China* 10, no. 27: 278.

6. Liu Huaqiu, "Zhongguo jiang yiongyuan zhixing dudi zhizhu de waijiao zhengce" (China Will Always Pursue Peaceful Foreign Policy of Independence and Self-determination), *Quishi*, no. 23 (December 1997): 3.

7. Liu Tsai-ming, "Zhuanfang xing waijiaobuzhang" (A Special Interview of New Foreign Minister), *Wenhui Bao*, June 27, 1998, p. A3.

8. Wu Baiyi, "The Chinese Security Concept and Its Historical Evolution," pp. 279–280.

9. Yan Xuetong, *Zhongguo de Jueqi, Guoji Huanjing pinggu* (The Rise of China: An Evaluation of the International Environment) (Tianjin, China: Tianjin Renmin Chuban She, 1998), p. 283.

10. Jiang Zemin's Report to the Fifteenth National Congress of the CCP, *Xinhua*, October 16, 1997.

11. You Ji and Jia Qingguo, "China's Re-emergence."

12. Yan Xuetong, *Zhongguo de Jueqi, Guoji Huanjing pinggu*, pp. 234–236.

13. Tian Peizeng, ed., *Gaige Kaifang yilai de Zhongguo Waijiao* (Chinese Diplomacy since the Reform and Opening Up) (Beijing: Shijie Zhishi Chuban She, 1993), pp. 6–7.

14. Xie Yixing, *Zhongguo Dangdai Waijiao Shi* (History of Contemporary Chinese Diplomacy) (Beijing: Zhongguo Qingnian Chuban She, 1996), p. 430.

15. *Beijing Review* 24, no. 40 (October 5, 1981): 11.

16. Suisheng Zhao, "Management of Rival Relations Across the Taiwan Strait: 1979–1991," *Issues and Studies* 29, no. 4 (April 1993): 77–78.

17. Chen Youwei, *Tiananment Shijianhou Zhonggong yu Meiguo Waijiao Niemu* (The Inside Stories of Diplomacy between China and the U.S. after the Tiananmen Incident) (Hong Kong: Zhongzheng Shuju, 1999), pp. 200–211.

18. Yan Xuetong, *Zhongguo de Jueqi, Guoji Huanjing pinggu*, p. 287.

19. The World Bank, *Global Economic Prospects and Developing Countries, 1998/99: Beyond Financial Crisis* (Washington, DC: The World Bank, 1999), p. 34.

20. Wang Yong, "China, ASEAN Stress Peace: Summit Agrees on Approach," *China Daily*, December 17, 1997, p. 1.

21. You Ji, "China and North Korea: A Fragile Relationship of Strategic Convenience," *Journal of Contemporary China* 10, no. 28: 396.

22. Andrew Higgins and Charles Hutzler, "China Sees Key Role for Central Asia in Ensuring Energy Supplies, Stability," *The Wall Street Journal*, June 14, 2001.

23. Ibid.

24. Constantine C. Menges, "Russia, China and What's Really on the Table," *Washington Post*, July 29, 2001, p. B2.

25. John Schauble, "Russia-China Alliance Emerges as a Foil to U.S.," *The Sydney Morning Herald*, June 16, 2001.

26. Robert Marquand, "Central Asians Group to Counterweigh U.S.: Russia, China, and Four Republics Meet to Expand Solidarity and Oppose Separatism," *The Christian Science Monitor*, June 15, 2001.

27. Peter Ferdinand, "China and Russia: A Strategic Partnership?" *China Review*, Autumn/ Winter 1997, p. 19.

28. Henri Eyraud, "From Confrontation to Partnership," *China Strategic Review* 3, no. 1 (Spring 1998): 158.

29. *Renmin Ribao*, December 13, 1999, p. 1.

30. Menges, "Russia, China and What's Really on the Table," p. B2.

31. Reuters, "Russia and China Sign Friendship Agreement," *New York Times*, July 16, 2001.

32. Commentary, "A Productive Summit," *China Daily*, November 12, 1997, p. 4.

33. Tian Peizeng, ed., *Gaige Kaifang yilai de Zhongguo Waijiao*, p. 20.

34. Jiang Zemin's Report to the Fifteenth Party National Congress of the CCP.

35. Harold C. Hinton characterized them into four categories and I regroup them into three. Harold C. Hinton, "China as an Asian Power," in Thomas W. Robinson and David Shambaugh, eds., *Chinese Foreign Policy: Theory and Practice* (Oxford, UK: Clarendon Press, 1995), pp. 352–357.

36. "Sino-Russian Joint Statement on Relations," *China Daily*, November 12, 1997, p. 4.

37. For a study of this crisis, see Suisheng Zhao, ed., *Across the Taiwan Strait: Mainland China, Taiwan, and the 1995–1996 Crisis* (New York: Routledge, 1999).

38. Suisheng Zhao, "Deadlock: Beijing's National Reunification Strategy after Lee Teng-hui," *Problems of Post-Communism* 48, no. 2 (March–April, 2001): 42–51.

39. Mark J. Valencia, Honorable M. Van Dyke, and Noel A. Ludwig, *Sharing the Resources of the South China Sea* (Honolulu: University of Hawaii Press, 1997), pp. 77, 99.

40. UPI news, November 12, 1998.

41. Valencia, Van Dyke, and Ludwig, *Sharing the Resources*, p. 88.

42. Xiaomin Rong, "Explaining the Patterns of Japanese Foreign Direct Investment in China, *Journal of Contemporary China* 8, no. 20 (March 1999): 132.

43. Ren Xiao, "Dongbeiya anquan xingshi de xianzhuang yu weilai" (The Current and Future Security Situation in Northeast Asia), *Guoji Zhanwang* (International Outlook), no. 7 (1996): 11.

44. June Teufel Dryer, "Sino-Japanese Relations," *Journal of Contemporary China* 10 no. 28: 375.

45. China News Digest-Global (electronic news), December 12, 1998.

46. *China Daily*, May 19, 1998, p. 4.

47. Jiang Zemin's Report to the Fifteenth Party National Congress of the CCP.

48. Levine, "China in Asia."

49. Zhang Xizheng, "Zhongguo tong dongmeng de muling huxin huoban guanxi" (The Good Neighboring, Mutual Trust, and Partnership between China and ASEAN), in Zheng Yushou, ed., *Huolengzhan shiqi de zhongguo waijiao* (Chinese Diplomacy in the Post–Cold War Era) (Hong Kong: Tiandi Tushu, 1999), p. 224.

50. Frank Umback, "ASEAN and Major Powers: Japan and China—A Changing Balance of Power?" in Jorn Dosch and Manfred Mols, eds., *International Relations in the Asia-Pacific: New Patterns of Power, Interest, and Cooperation* (New York: St. Martin's Press, 2000), p. 174.

51. For one illumination of the arms race post–Cold War Asia-Pacific, see Desmond Ball, "Arms and Affluence: Military Acquisitions in the Asia-Pacific Region," *International Security* 18, no. 3 (Winter 1993/94): 78–112.

52. Jia Qingguo, "Frustrations and Hopes: Chinese Perceptions of the Engagement Policy Debate in the United States," *Journal of Contemporary China* 10, no. 27: 325.

53. John Pomfret, "China Growing Uneasy about U.S. Relations," *Washington Post*, June 23, 2001, p. A1.

54. "U.S. Policy of Containing China: Result of Biased, Lopsided Information," *Guang-jiaojing* (Wide-Angle Mirror), no. 344, May 16, 2001, p. 33.

55. Shan Min, "What Is the United States Intending to Do with Its China Security Strategy," *Liaowang Zhoukan* (Outlook Weekly), no. 18, April 30, 2001, p. 31.

56. Pomfret, "China Growing Uneasy about U.S. Relations," p. A1.

57. Yan Xuetong, *Zhongguo de Jueqi, Guoji Huanjing pinggu*, p. 238.

15

Constructing the Dragon's Scales

China's Approach to Territorial Sovereignty and Border Relations

Allen Carlson

During the 1980s and 1990s, China's stance on territorial sovereignty and border relations was most strikingly characterized by two strong patterns of continuity. First, Beijing consistently promoted a boundary reinforcing interpretation of the lines between China and her neighbors through the repetition of a set of largely status quo territorial claims. Second, foreign policy elites unfailingly analyzed these boundaries in a manner that sought to legitimize each aspect of the official Chinese position. These continuities have already been featured in the existing literature on China's handling of border relations; however, against this well-publicized backdrop three less obvious, but nonetheless important, patterns of change emerged.

The most notable change was the marked decline in the Chinese use of confrontational territorial claims and threats starting in the early 1980s and continuing through the last decade. Before this period China's official claims mainly consisted of impassioned rhetorical volleys and bitter denunciations of violations of Chinese territory; during the 1980s and 1990s such defensive statements were eclipsed by a set of less hostile declarations. The second modification was an increase in the level of Chinese participation in negotiations concerning both the location and demilitarization of long stretches of China's continental boundary.[1] The third development was a rising wave of attention to China's maritime boundary. Indeed, the vast majority of Chinese commentary on border relations published during this period was directly related to China's ocean territory. In composite these new trends reveal that although Beijing has not relinquished the right to use force to resolve outstanding disputes over territorial sovereignty, reliance on military solutions has been supplemented (if not supplanted) by the utilization of a broader range of diplomatic measures and discursive practices.

These new patterns developed over the course of two distinct phases, the first lasting from the early 1980s through 1988, and the second stretching from the spring of 1989 through the end of the 1990s. During the former period, continuities outweighed new developments, but elites began to tentatively consider novel solutions to China's outstanding boundary disputes. In contrast, over the last decade each of the incremental changes outlined in the preceding paragraph gathered speed and together substantially altered important features of the Chinese position on territorial sovereignty and border relations.

This trend has largely been overlooked within conventional analysis of China's handling of territorial issues, as it does not involve a radical shift in the location of China's boundaries.[2] However, the new emphasis in Beijing on cooperative international legal solutions to outstanding boundary disputes still represents a significant development in the overall Chinese stance on territorial sovereignty. Furthermore, in light of the rising concerns in the region about China's potential role as a destabilizing, revisionist force in the international arena, this development is of real significance to all of those with an interest in China's foreign relations.

Conceptual Issues

In general, China experts have dealt with the issue of territorial sovereignty in an indirect fashion in their analysis and have done so largely within the framework of three broad and, to a certain degree constraining, approaches to border relations. The first of these places an emphasis on collecting and cataloging information on the terrain through which boundaries run and on the treaties that formalize their location.[3] Such political geographic and international legal studies do a remarkable job of tracing the historical evolution of China's border relations and tracking the relationship between various legal claims to specific boundaries and contemporary disputes over the location of such lines. Yet, within such analysis, boundaries, while envisioned as movable, are seen as concrete and relatively objective lines and little consideration is given to the broader issue of territorial sovereignty. Thus, change within the contemporary setting is evaluated almost exclusively with reference to the gains and losses in territorial sovereignty that are caused by shifts in the location of specific segments of the border.

The second conventional frame is defined by an underlying interest in the way in which past empires and the current Chinese state inscribed historical frontiers and modern boundaries into the minds and activities of those residing close to, across, and along them.[4] This type of analysis shows that boundaries are the product of open-ended contests between elites at the state's center and groups at its geographic periphery. Without question such research has breathed new life into the field of Chinese border studies. However, the scholars working within this approach have also tended to neglect the large extent to which territorial sovereignty and its boundaries are the product of interaction between states.

Security studies push the examination of border relations in a notably different direction.[5] This work explores the cause, course, and outcome of military disputes over territorial boundaries. Such research has played a central role in explaining the Chinese stance on contested terrain. However, within this framework, patterns of change and continuity in territorial sovereignty, when considered at all, have largely been measured in terms of the presence or absence of conflict alone.

In sum, those working within each of these main traditions have explored important aspects of the Chinese approach to territorial sovereignty. The geographers and international legal scholars have traced its formal limits, cartographic representations, and international legal bases. The historians and anthropologists have delved into the fascinating issue of the problematic imposition of sovereign boundaries into marginal spaces. The security specialists have examined the conditions surrounding each of China's major boundary disputes. However, all of this work has tended to deemphasize the structural underpinnings of such lines, mainly the principle of territorial sovereignty. Furthermore, it has ignored the broader diplomatic and representational practices that construct sovereign boundaries.

This brief critical overview of the three conventional approaches to China's territorial boundaries is rooted within the new field of critical border studies.[6] Contributors to this turn emphasize the need to develop new conceptual lenses and analytical frameworks to further the study of territorial sovereignty and border relations. In general, such innovation involves conceptualizing sovereign boundaries as institutions constituted by the ongoing enactment of a broad set of social practices between states rather than singular sites of confrontation.[7] In this sense, the disparate boundaries discussed in the literature reviewed above can certainly be dissected solely in terms of the type of relationship that China and her neighbors have developed around them. However, at a more fundamental level they are also distinct manifestations of a singular process: the construction of China's territorial sovereignty. Thus, a comprehensive survey of the most significant of these lines offers not only a collection of data on border issues but also insight into how Chinese elites interpret a foundational aspect of China's sovereign rights.[8]

Building upon such observations, it is possible to identify three main types of diplomatic and representational practices that foreign policy elites within territorially defined states may employ in the process of sovereign boundary construction and maintenance.[9] The first of these is the issuing of official boundary claims. Such claims include pronouncements on general principles for handling territorial issues, specific statements on the location on given segments of the boundary, explanations of why such a vision is correct, and responses to perceived infringements upon a state's territory. As such, official statements represent the attempt to promote a particular interpretation of a state's territorial rights. The second practice involves the analysis of sovereign boundaries by foreign policy elites. While this type of practice does not carry the weight of official territorial claims, it is also less restrained by the protocols of diplomatic discourse. Thus, it tends to reveal a great deal about the range of interpretations of sovereign boundaries that are possible within the context of an individual state's position. The signing, enactment, and observation of international legal agreements constitute the third type of practice. Through making these commitments states formalize their positions on territorial sovereignty. Therefore, by tracing the degree of compatibility between such agreements and other territorial practices it is possible to determine the extent to which a state's territorial aspirations have been realized or denied.

The content of such practices is as diverse as the sovereign boundaries to which they refer. On one level, practices may refer to a single contested territorial point, a stretch of border shared with a sovereign neighbor(s), or be made in a more general sense. After having identified the geographic referent of a state's practices it is possible to code the content of such efforts along three crucial vectors. First, practices forward differing interpretations of the existing location of a state's boundaries. A single practice may promote an expansion of territorial rights, endorse the territorial status quo, or accept a contraction of a state's territory. Second, practices vary according to the degree to which they promote a confrontational or cooperative approach to border relations. Finally, practices may also be categorized according to the manner in which they interpret the meaning of sovereign boundaries themselves. Boundary reinforcing practices promote an interpretation of territorial sovereignty that is absolute and unyielding. In contrast, boundary transgressing practices contain more open, malleable interpretations that acknowledge ambiguity, allow for relatively unrestricted economic and political flows in border regions, deemphasize the sanctity of territorial divisions, and overlook possible infringements or violations of territorial sovereignty. Conceptualizing territorial sovereignty and variation in the state practices that construct its boundaries along these lines creates the analytical space to identify broad patterns of change and continuity within the border relations between states.

Applying this framework to the China case involved determining which segments of China's territorial boundaries would be examined. As I intended to examine China's overall approach to territorial sovereignty, I felt it was essential to include a consideration of both China's continental and maritime border relations within the study. However, in light of the great length of China's sovereign boundaries, it was also necessary to limit the scope of such an examination to the most significant of these lines. Thus, I chose to concentrate on the continental boundaries China shared with the Soviet Union (now Russia, Tajikistan, Kyrgyzstan, and Kazakhstan),[10] India, and Vietnam. As for the ocean boundary, I focused on the segment between China and Vietnam.

Through the late 1970s, the Chinese approach to territorial sovereignty, vis-à-vis each of these segments of China's boundaries, was defined by the waxing and waning of outright military confrontation and a persistent set of official claims and elite analysis designed to legitimize each aspect of Beijing's border policies. Such a position was grounded in concerns about China's ability to defend existing sovereign boundaries and ongoing sensitivity to the loss of territory that took place during the Qing and Republican periods. To what extent did such patterns of confrontation and boundary reinforcement continue to define the Chinese approach to territorial sovereignty during the 1980s and 1990s?

Territorial Practices from 1980 to 1988: Easing Tensions, Bolstering Claims

Over the course of the 1980s, the Chinese approach to territorial sovereignty became more muted. The hypervigilance of the proceeding decades on territorial issues receded and intense sensitivity to perceived encroachments upon Chinese territory was less pronounced. In addition, commendations of neighboring states territorial agendas were toned down, and the more expansive aspects of China's own territorial claims were silenced. Such developments amounted to a slight shift in the manner in which Chinese elites promoted their interpretation of China's territorial sovereignty. For example, a content analysis of the *Beijing Review* from 1982 to 1988 discovered a steady decline in official statements on boundary-related issues.[11] An additional coding of these claims revealed that this drop was largely the product of a reduction in the number of statements issued with regards to China's continental boundaries.[12] Figure 15.1 illustrates these trends.

As significant as these developments were, it is essential to bear in mind that the statements elites issued were quite consistent in their insistence on upholding the territorial status quo. This trend also characterized Chinese elite analysis published during the 1980s. Such commentary placed a singular stress on reinforcing the sanctity of the lines between China and her neighbors. These continuities were underscored by persistent patterns in China's border policies. Although China did not become embroiled in any major military

Figure 15.1   **Territorial Claims Published in** *Beijing Review,* **1982–1988**

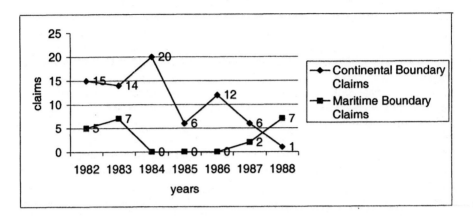

dispute along its boundaries during the 1980s, armed border skirmishes were relatively common through 1987. In addition, Chinese elites took only the most tentative steps toward the negotiating table to resolve outstanding boundary disputes.

The Sino-Soviet Boundary from 1981 to 1988: Deescalating Tensions, Outstanding Issues

During the 1980s the topic of Sino-Soviet border relations was treated with almost total neglect within China's official foreign policy discourse. Indeed, a final country-specific coding of the claims published in the *Beijing Review* revealed not a single official reference to the Soviet border between 1982 and 1988. Elite analysis of the Sino-Soviet boundary was also notably scant. For example, during this period no article dedicated to Sino-Soviet border relations appeared in any of the three major foreign policy journals published in China.[13] In addition, a sampling of articles in these journals that dealt with the broader issue of Sino-Soviet bilateral relations revealed only a smattering of analysis of territorial concerns. I discovered only one article that contained even the briefest mention of the border issue. This was a 1985 piece in *World Economics and Politics* that analyzed Soviet security policies in the Asian region. The article briefly took note of the 1969 to 1972 Soviet troop buildup along the Sino-Soviet and Sino-Mongolian border and added that through the mid-1980s Soviet statements on troop reduction were without basis and in need of further clarification.[14] Two years later the more specialized *Zhongguo bianjiang shidi baogao* (China's Borderland History Report) followed such commentary with an article that protested the legitimacy of Soviet claims to territory south of the "willow wall" in the region of Manchouli in the western section of the eastern segment of the border.[15]

As the Chinese cautiously promoted such territorial claims, it was the Soviets, more specifically Michael Gorbachev, who took the initiative in moving China and the Soviet Union closer to a negotiated settlement of the border issue. Gorbachev made the first move in this direction in 1986 by issuing a series of major foreign policy statements pledging

Soviet acceptance of the main navigational channel, or thalweg, as the principle for locating the river boundary between China and the Soviet Union. Following the extension of such an olive branch, relations between the two countries edged toward a new renaissance. One of the first tangible results of such improved relations was the February 1987 announcement that accompanied the end of the first round of border talks held between the two sides in nine years. The statement read in part that China and the Soviet Union had "agreed to review their entire boundary line, starting with the eastern section."[16] Subsequently, additional talks were held in August of 1987.[17] At this time both sides tentatively agreed to accept the thalweg principle as the basis for negotiations on the location of the river segment of the Sino-Soviet boundary. The following year a group was formed to examine the issue of the eastern border demarcation.[18] However, each of these meetings skirted around the major differences between the two sides. Indeed, they were largely overshadowed by ongoing efforts in both countries to arrange a Sino-Soviet summit.

The Sino-Indian Boundary from 1981 to 1988: Unresolved Differences and the Threat of Renewed Conflict

While meetings between the Indian and Chinese Foreign Ministers in 1979 and 1981 lent an air of normalcy to Sino-Indian relations, tensions between the two countries over their disputed boundary remained quite pronounced throughout most of the 1980s. During this period Chinese elites issued a stream of analysis that justified the official position on the location of China's sovereign boundary vis-à-vis India. The continuity in this discourse can be seen in a series of three articles that were published in the *Journal of International Studies*. The first of these articles appeared in 1982.[19] In it Chen Teqian restated the preexisting Chinese claim that the Sino-Indian boundary had never been delineated and argued that over the course of history both sides had accepted a traditional borderline. Chen then stridently asserted that India had consistently opposed a negotiated settlement of the dispute and claimed, "this is definitely not an issue of an unclear border, but rather one of an ideology of expansion and aggression."[20] The second article was published in 1986.[21] This piece consisted of a polemic against Indian claims. It objected to the Indian promotion of the McMahon line as a basis for settling the boundary dispute and repeated the same list of territorial claims against India that were made in previous articles and foreign ministry statements. The third article appeared in 1988.[22] It focused exclusively on the eastern segment of the boundary; however, it was almost identical to the previous two pieces in strenuously objecting to the McMahon line and Indian occupation of Chinese territory.

Such confrontational analysis set the stage for a short-lived military confrontation between China and India in 1986 and 1987. During this period, China repeatedly charged India with troop incursions into Chinese territory in the region of the eastern sector of the border and railed against the Indian decision to grant statehood to Arunachal Pradesh (a significant portion of which falls within territory that the Chinese claim). Such accusations once again reveal the underlying consistency in the Chinese position on territorial sovereignty. For instance, two *Beijing Review* articles published in 1986 rehashed the argument against India's transgression of the actual line of control (LAC) and attacked the legitimacy of the McMahon line as a basis for border negotiations.[23] Heeding such confrontational claims, during the winter of 1986–1987, the international media was full of reports of an impending reenactment of the 1962 border war.[24] However, the overt tension between the two sides melted away in the spring following the publication of Chinese statements in favor of seeking a peaceful resolution to the conflict.[25] Subsequently, the eighth round of border talks between the two sides were held in November and preparations were made for Rajiv Gandhi's 1988 visit to China.

The Sino-Vietnamese Boundary 1981–1988: From Clashes on the Land to Conflict at Sea

Throughout the 1980s the Chinese stance on the limits of China's territorial sovereignty vis-à-vis Vietnam remained unchanged; however, important shifts in the focus of Chinese practices also took place during this period. During the first half of the decade, Chinese elites concentrated on the land boundary with Vietnam. For example, while neither China nor Vietnam made extensive territorial claims against each other in this segment of their shared boundary, armed incursions across this line remained a common fixture through 1987;[26] however, after 1987, such armed sparring ceased. Instead, the two states became deeply embroiled in a bitter dispute over the location of the maritime boundary in the Beibu Gulf and the South China Sea. Indeed, during this period, Vietnam began to more actively pursue its claims to the disputed waters, and China quickly attacked such moves as unwarranted provocations. This escalating verbal wrangling was part of a broader effort by both sides to cement their claims in the region. Such moves culminated in 1988 when China sent a "survey" team into this maritime territory and an armed clash with Vietnamese forces took place.[27]

In sum, setting aside the prominent exception of the flare-up of the maritime dispute early in 1988, by the end of the 1980s, Chinese elites had moved away from the relatively routine use of force to secure their vision of the scope of China's territorial sovereignty. While territorial differences with each of China's main neighbors remained unresolved, border skirmishes were a less common occurrence than they had been in previous decades, and tentative progress had been made toward dealing with even the most intractable disputes. In each case this turn drew upon the broad shift in Chinese foreign and domestic policy that Deng Xiaoping instituted in the early 1980s and developed in line with the overall evolution of bilateral relations with each neighboring state. However, as the following decade revealed, the degree of Chinese flexibility and their willingness to negotiate was still constrained by deeply held convictions about the legitimacy of China's historically based territorial claims.

Territorial Practices from 1989 to 2001: Stabilizing Land Boundaries, Fighting to Realize Maritime Goals

As is the case with so many facets of Chinese politics, 1989 marked a significant turning point in China's approach to territorial sovereignty. The trend toward deemphasizing territory that began in the early 1980s gathered momentum at this juncture. This development is immediately visible in Figure 15.2, which illustrates the territorial references that appeared in *Beijing Review* from 1989 to 1997.

The most striking aspect of Figure 15.2 is the fact that no claims were issued in 1989,

Figure 15.2 Territorial Claims Published in *Beijing Review*, 1989–1997

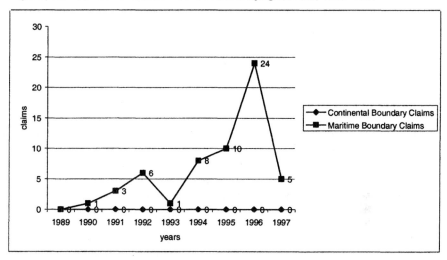

and during the four successive years only eleven territorial claims were reported. Beyond this initial contraction in the Chinese discourse, the rise in claims during the mid-1990s was driven solely by a surge in maritime claims.[28] In other words, during the 1990s, when elites directly referred to territorial sovereignty, they tended to exclusively focus on China's ocean frontier. The specific content of such claims and those made within the broader official arena was unfailingly boundary reinforcing. However, within such a general framework of continuity, over the course of the decade, official statements tended to place an increasing emphasis on stabilizing border relations. As was the case during the 1980s, elite analysis tended to mirror the official discourse. Most such analysis resonated with the officially sanctioned position on territorial sovereignty. However, interview data collected in 1997–1998 and 2001 revealed a higher level of combativeness and defensiveness than was found in open and official sources about China's territorial rights and dissatisfaction with the Chinese leadership for making too many compromises in negotiations with neighboring states. As significant as such opinions are, they should still be viewed against the backdrop of China's expanding participation in bilateral and multilateral territorial negotiations during the 1990s. Such developments amount to a significant shift in the Chinese stance on territorial sovereignty, albeit one that was consistently framed by an underlying attachment to status quo interpretations of China's sovereign boundaries. In other words, while the lines between China and her neighbors were still being reinforced, elites were utilizing a different mix of territorial practices to accomplish this goal.

*The Sino-Soviet (Russian and Central Asian) Boundary after 1988:
Cooperation through Bilateral to Multilateral Negotiations*

Throughout the 1990s, the official Chinese discourse on territorial issues with Russia and the Central Asian Republics was minimal and largely consisted of a series of endorse-ments of negotiating successes between China and each of these states. Elite analysis was also limited, but it contained a persistent trace of dissatisfaction in the elite community with the flexibility in Beijing's policies on the northern boundary. As intriguing as such muted voices of dissent are, their significance pales against the plethora of bilateral and multilateral negotiations that were established during the late 1980s and 1990s and the series of international agreements that emerged out of such talks.

The first step in this direction took place in May of 1989 when the Chinese and Soviet leadership utilized the Deng-Gorbachev summit to formally announce an agreement for resolving their outstanding differences over territorial issues. Both sides pledged that the dispute should be settled "on the basis of the treaties concerning the present Sino-Soviet boundary."[29] Out of this agreement an expert group on the demilitarization of the border was formed and met for the first time in November 1989. In the spring of the following year a formal Sino-Soviet agreement on border troop reduction was signed during Li Peng's trip to Moscow. The agreement stated that, "both sides will reduce military forces to the lowest levels suited to normal good-neighborly relations between the two countries on a equal basis of mutual security."[30]

The next major international legal agreement reached between the two sides came in 1991 during the lead up to Jiang Zemin's May summit meeting with Gorbachev. At this time a new treaty was signed that established the exact location of the eastern segment of the border. The treaty in no way touched upon the enormous tracts of land China had claimed in the past. Instead, it essentially formalized the existing de facto location of the boundary between the two states. However, reflecting Chinese demands, it also ensured that in principle a number of islands in the Wusuli River segment of this sector would be returned to Chinese control. Indeed, the treaty took particular note of the fact that the location of the border in each of the river segments of the eastern boundary was to be based on the main navigation line.[31]

It was widely reported in the Chinese media that this agreement resulted in the location of the vast majority of the disputed sections of the Sino-Soviet border.[32] However, the ratification of the treaty was not greeted with universal satisfaction in China. Evidence of such unease can be found in a *South China Morning Post* report on a *neibu* (internal publication) article written by a Chinese scholar. The author reportedly criticized the treaty for giving away too much historically Chinese territory to the Soviets at a time when the Soviet Union was on its last leg.[33] Regardless of how authoritative such a report was, the airing of territorial grievances at such a delicate stage in Sino-Soviet relations is indicative of two important trends. First, the persistence in elite circles of historically grounded understandings of the legitimate scope of China's territorial sovereignty vis-à-vis the Soviet Union. Second, the initial uncertainty that existed in Beijing over how to handle the new border dynamic that emerged out of the collapse of the Soviet Union.

Despite such underlying tensions, in the aftermath of the breakup of the Soviet Union at the end of 1991, Chinese elites opted to quickly stabilize relations and the territorial status quo with each of China's new neighbors. This policy was first forwarded through a flurry of reassurances between Beijing and Moscow that emphasized the maintenance of normal relations. These statements were soon complemented by a series of joint communiqués announcing the establishment of diplomatic relations between China and Kazakhstan, Tajikistan, and Kyrgyzstan.[34] China also opened negotiations on the location of the boundary and troop reduction in the border region to each of the Central Asian states.

In spite of this new focus on central Asia, Chinese leaders did not neglect the Sino-Russian relationship. Indeed, in 1992, six top-level meetings took place between China and Russia. The most significant of these meetings was Boris Yeltsin's mid-December visit to China. During this trip both sides again pledged to continue negotiations on the location of the border and agreed to extend the border troop reduction and confidence-building measures (CBMs) they had signed in 1990.[35]

After dealing with the challenges and opportunities that the breakup of the Soviet Union presented China in 1992, the year 1993 was a period of relative quiet in Sino-Russian relations. However, during the following year a formal agreement on the location of the western sector of the border was signed. This treaty noted that the short 55 kilometer western border was to be located by two existing markers and would be jointly demarcated by a red line on the official maps of both countries. In addition, both sides pledged to conduct joint surveys designed to specifically delineate the border.[36]

Chinese elites added to this stabilization of the northern frontier with the signing of a new boundary treaty with Kazakhstan. This treaty reiterated the importance of using existing agreements and norms of international law to determine the location of the border. In short, the agreement amounted to an acceptance by both China and Kazakhstan of the existing boundary line.[37]

Building upon such developments, China, the Central Asian Republics, and Russia signed a historic CBM border agreement in April 1996. This CBM was given prominent coverage in the Chinese media and was ratified by the National People's Congress in August. Its most celebrated component was a pledge by all five signatories to establish a 100-kilometer buffer zone in the border region.[38] Complementing this commitment, a border agreement between Kyrgyzstan and China was signed in July during Jiang's visit to the central Asian state. With the signing of this last agreement, only the Tajikistan segment of the former western segment of the Sino-Soviet boundary remained undefined. However, in 1999, China and Tajikistan reached a tentative agreement on their outstanding territorial differences.[39]

In sum, at the end of 1990s, the disputes that had been the focal point of Chinese territorial practices in preceding decades had each been resolved through negotiations and border agreements. Furthermore, via the 1996 multilateral security agreement, and the formal CBM mechanisms it established, such stabilization extended to a partial demilitarization of China's northern frontier. The relative ease with which these tasks were accomplished and the virtual absence of confrontational territorial claims against the Russians and central Asians throughout this period indicates that Chinese elites viewed territorial issues through the prism of rational power concerns. In other words, territorial sovereignty was simply one issue to be dealt with within the context of a careful consideration of broader national interests.

However, as previously mentioned, the memory of historical transgressions against Chinese territory remained quite strong. For example, in 1992, a *World Economics and Politics* article commented on the persistence of an unresolved boundary dispute between Russia, Kazakhstan, and China.[40] The article also took note of the extensive amount of territory China lost as a result of the unequal treaties Russia forced upon the Qing dynasty.

Similar claims appeared in 1996, well after all existing Sino-Russian territorial differences had been formally settled. In his review of the history of the PRC's foreign relations, Xie Yixian included the familiar litany of historical grievances against Russia.[41] In addition, a pair of internal publications reiterated the core historical claims that China had as opposed to what was now Russian and central Asian territory.[42] Furthermore, during interviews conducted in 1997 and 1998, three junior scholars voiced their dissatisfaction with the border agreement China had reached with Kazakhstan in 1994.[43] Each of these interviewees noted that some elites had criticized this agreement for compromising too much on territory the Qing empire lost to Russia in the 1800s. They also reported that there was concern among some elites over the likelihood that China would fail to aggressively pursue historical territorial claims in its border relations with Tajikistan.

The Sino-Indian Boundary after 1988: Nagging Differences, Finessing Cooperation

Whereas the Chinese were able to resolve each of the outstanding territorial issues they faced to the North during the 1990s, a negotiated settlement to the Sino-Indian boundary dispute proved to be quite elusive. Indeed, at the end of the decade, Beijing and New Delhi were a little closer to compromising on issues of territorial sovereignty than they were at its start. However, against this backdrop, two significant changes took place in the Chinese approach to Sino-Indian border relations. First, throughout this period, the official and elite Chinese discourse on the Sino-Indian boundary literally dried up. Quite simply, the accusations against Indian transgressions of Chinese sovereign boundaries that had been the main source of official claims on such lines through the mid-1980s ceased to appear in the *Beijing Review* after 1988. Second, negotiations and talks about handling (if not solving) border relations gradually replaced the military standoffs and rhetorical confrontations that had defined the approach of both states to the boundary through 1988. In this sense, the southern frontier also became a site for securing relatively status quo interpretations of the scope of China's territorial sovereignty.

Just as Gorbachev's 1989 trip to Beijing heralded the normalization of Sino-Soviet relations, the visit of the Indian Prime Minister Rajiv Gandhi to China in December of 1988 ushered in a new era in the Sino-Indian relationship. The visit was marked by an effort by both states to establish a more normal relationship and create a less hostile security environment. Within this context both sides attempted to downplay the importance of the outstanding boundary dispute that divided the two states. Thus, the joint press communiqué issued during the summit emphasized the need "for a fair and reasonable settlement of the boundary question while seeking a mutually acceptable solution to this question."[44]

One concrete result of the communiqué was the reestablishment of Sino-Indian border talks in the form of a new joint working group. This group convened for the first time in July 1989 and subsequently met on a regular basis throughout the following decade.[45] While no substantive agreement was produced in this initial round of talks, Chinese reporting on the meetings feature the solid prospects for developing more cooperative border relations. Such commentary was an extension of the positive developments that had taken place during the 1988 summit. Moreover, it is quite clear that the display of normal diplomacy that the talks represented was particularly valuable for Beijing at a time when China was facing international isolation in reprisal for the suppression of the Tiananmen demonstrations.

Despite these signs of progress and meetings of the joint working group in the fall of 1990 and spring of 1991, no new major agreement between the two sides was reached until December 1991. The context for such a development was a state visit by Li Peng to India. In the lengthy joint communiqué that marked the occasion, both China and India agreed to make efforts to "arrive at an early and mutually acceptable solution to the boundary question through friendly relations."[46] More specifically, it was agreed that "periodic meetings between the military personnel in the border areas should be held on a regular basis."[47] High level meetings and negotiations took place again in 1992, yet it was not until 1993 that a new CBM pertaining to the boundary dispute was reached. The agreement emphasized that the two sides would create a tranquil and peaceful environment along the border through friendly negotiations and the reduction of military forces in the region of the LAC. In addition, both sides pledged not to engage in military exercises in disputed regions, negotiate any unexpected problems, not violate airspace, do no harm to the other side's principled position, hold mutual inspections, and establish an experts group on demilitarization of the border. More specifically, the agreement included "measures about interaction between forward posts and making information available to both sides about military activities along the line of actual control."[48] However, it did not contain any details on the level of troop reductions nor did it touch on the underlying territorial differences between the two sides.

One of the main substantive results of this agreement was the formation of an experts group that met multiple times in the following years. The initial meetings of this forum took place against the backdrop of Chinese concerns related to Indian support for Tibetan independence, activities within India, and subsequent reports of military buildups in the border region.[49] Not surprisingly, such tensions meant that few concrete accomplishments were made in the first rounds of talks in terms of addressing basic territorial concerns. However, the institutionalization of contact between the two sides in the experts group and the joint working group did lay the groundwork for a second, more detailed CBM measure between the two states.

The precursor to this second agreement was announced at the eighth round of joint working-group talks in August of 1995 when China and India reiterated their commitment to implement aspects of the 1993 CBM agreement. Such an assertion laid out the terms for the dismantling of border posts in the Wangdong area along the eastern sector of the border, while also calling for a meeting of border security authorities.[50] Subsequently, in late November 1996, Jiang Zemin paid an official state visit to India, and while no official joint statement or communiqué was issued during his trip, the event was marked by the announcement of the second Sino-Indian CBM.[51]

This agreement, in conjunction with the 1993 CBM, led to a significant reduction in tensions in the border region. As Sidhu and Yuan have observed, while the two CBMs are limited in scope and contain few verification measures, they "are often regarded as crowning achievements in the long process of normalizing bilateral relations between the two countries."[52] However, they did not move the two states appreciably closer to a settlement of the boundary dispute. Furthermore, the growing rapprochement between Beijing and

Delhi was disrupted in the spring of 1998, when India held a series of nuclear tests. John Garver has attributed the timing of the Indian decision to test to the combination of unease in India over China and an underlying commitment of the newly elected Bharatiya Janata Party (BJP) to develop India's nuclear program. Regardless of what the real Indian motives for the action were, some Indian officials attempted to justify testing with references to the "China threat."[53]

At a time when China was feverishly working to counter charges within the region (and from the United States) about the destabilizing nature of China's economic and military rise, such accusations struck a particularly raw note in Beijing. They led to an angry Chinese response and the postponement of the previously scheduled 1998 meeting of the joint working group. However, Beijing could not simply condemn India for testing as such a move would reinforce existing Indian concerns about China and contribute to the very image of a combative Beijing that the Chinese were seeking to diffuse. Thus, Chinese elites pushed for a normalization of bilateral relations at the start of 1999, and an eleventh meeting of the joint working group was held in April of the same year.[54] While no new major CBMs have been signed during the ensuing period, there have been limited indications that the scope of negotiations incrementally widened in 2000.[55]

In sum, during the 1990s, the Chinese position on securing the mountainous southern limits of China's territorial sovereignty shifted from that of military posturing to participation in frequent negotiating sessions and an incremental expansion in limited confidence building measures. While the substance of both is open to debate, by the end of the decade the border region was more stable than it had been since well before the 1962 war between the two states. This development was largely a product of a relative convergence of interests between the two sides to cement a more predictable relationship. Bracketing territorial differences via the mechanism of confidence-building measures and institutionalized negotiating sessions allowed such a process to move forward. Chinese elites also utilized these developments to promote an image of China that was nonaggressive and cooperative to offset charges of an emerging "China threat" in the region.

However, neither Chinese elites nor their Indian counterparts were willing or able to address the underlying territorial differences between them and make the compromises necessary to resolve the boundary dispute. The intractability of such differences is located in the static strategic importance of the territory at stake.[56] The disputed section of the western sector of the boundary includes an expanse of territory that contains the only main land route between Xinjiang and Tibet and, as such, constitutes a significant logistical link for the Chinese military in western China. The fate of this territory also has obvious ties to the Indian struggle with Pakistan over the fate of Kashmir and the location of the boundary between the two states along the Siachen glacier. At the same time, control of the southern slope in the eastern sector is considered to be of central importance to India's security vis-à-vis China. In other words, compromise by either side would have significant costs that neither Beijing or New Delhi is willing to bear. Yet, beyond such rational calculations, the intractability of the dispute is also tied to the degree to which Chinese elites have internalized historically based interpretations of the extent of Chinese territory in the region.[57] This factor was clearly evident in an interview I conducted in 1997 with a retired Ministry of Foreign Affairs official who was personally involved with the abortive border negotiations between China and India in the 1950s. According to the interviewee, it was only possible to understand the contemporary Chinese stance on territorial issues with reference to the historical losses China had suffered prior to the establishment of the PRC. He observed that from the time of the Opium Wars "China's sovereignty and territorial control was abused and oppressed, therefore China developed a very sensitive attitude toward territorial issues, because China was forced to give up land and the integrity of its territory."[58] This general observation was followed by a recitation of the historical legitimacy of each of the territorial claims China has made in relation to the Sino-Indian boundary dispute and the aggressiveness of India posturing in the border region. While the interviewee acknowledged that the negotiations between India and China had gained some momentum in the 1990s, he did so in the most reluctant fashion.[59]

The Sino-Vietnamese Boundary after 1988: Containing Potentially Explosive Differences at Sea

New patterns and points of emphasis also characterized the Chinese approach to the boundary with Vietnam. The land boundary that had been the site of such extensive combat through the mid-1980s was quietly transformed into a zone of trade and commerce in which the major threat to both sides came more from illicit drug trafficking and smuggling than from each other. Indeed, China and Vietnam promoted a series of successful negotiations on delineating and demarcating this previously contested frontier and by the end of the decade were working on cooperative policing arrangements to control trade in the newly flourishing border region. As the territorial differences between the two sides in this area were never an object of serious contention, this change largely amounted to one of cooperative institutionalization of the existing boundary. In contrast, Chinese elites continued to sharply contest Vietnamese claims to the Beibu Gulf and especially the South China Sea (including the Xisha and Nansha island chains). However, such confrontational discursive practices were offset by an emphasis by both sides on the need to negotiate.

This shift in Chinese discursive practices from land to maritime boundaries was quite apparent in elite journals. For example, throughout the 1990s, no articles in the major foreign policy journals were dedicated solely to the former boundary. In contrast, between 1988 and 1991, a series of three articles on the island chains in the Beibu Gulf and South China Sea appeared in their pages. The first of these articles began with an assertion of Chinese control over the Xisha and Nansha islands since ancient times; it then protested that China's legitimate territorial rights had been distorted in Vietnamese claims to the islands.[60] The second article asserted that there was no international legal basis for the territorial claims that Vietnam had made over the Nansha Islands.[61] The article concluded, "Regardless of whether the issue is examined from the perspective of discovery, private activities, government administration or individual government's recognition, China has effective sovereignty over the Nansha archipelago, and this is entirely in accord with the demands of international law."[62] The third article, while not focused on Sino-Vietnamese border relations, protested Vietnamese territorial claims to both the Xisha and Nansha islands and Vietnamese use of military force in the Nansha archipelago.[63]

Despite such defensive commentary, bilateral relations between China and Vietnam began to improve in 1991 when the General Secretary of the Vietnamese Communist party visited Beijing.[64] During this trip, a general agreement on border affairs was reached. This agreement, while refraining from delving into specific territorial issues, contained assurances that both sides would work together to "preserve stability and tranquility of the border region between the two countries to make the border between the two countries one of peace and friendship."[65]

The following year, in a move that infuriated the Vietnamese, the National People's Congress approved a resolution designed to formalize China's authority over the South China Sea.[66] However, despite such controversy, the following year the normalization of relations between China and Vietnam continued, highlighted by trips by Qian Qichen and Li Peng to Hanoi. At the end of the Li Peng visit, a joint communiqué established government-level negotiations on both the land and ocean boundary disputes.[67] The first such government-level talks were held in August 1993, and in October both sides recommitted to a resolution of territorial issues according the general principles agreed to in 1991. Furthermore, new working groups on the land border and the Beibu Gulf were established at this time.[68]

Building upon this 1993 agreement, both sets of joint working groups began meeting in 1994. However, such talks were followed by a round of accusations over the granting of oil contracts by both Vietnam and China in disputed territory. Indeed, during this period, the Chinese repeatedly charged Vietnam with seizing a group of Chinese fishermen.[69] Nonetheless, the conflict was eventually quelled through meetings between Qian Qichen and the Vietnamese foreign minister later in the month. In August, both sides agreed to hold additional meetings of the Beibu Gulf experts group.

At this juncture, the focus of the South China Sea dispute decisively shifted toward the Nansha Islands and the contrasting claims between China and the Philippines to the maritime territory in which they are located. Indeed, contestation between Beijing and Manila over Mischief Reef came to a head in 1995 and continued to be the source of tension in the region throughout the rest of the decade. The emergence of this aspect of China's territorial practices in the region quickly overshadowed the outstanding differences between Beijing and Hanoi.[70] However, as significant as this turn of events was, China and Vietnam continued to map out a cautiously cooperative approach to their disputed maritime possessions. For example, the first meeting of the maritime experts group took place in mid-November 1995. In the official commentary on these meetings, the Chinese attempted to balance their claim to contested territory with an indication of a general willingness to negotiate. Thus, the lead Xinhua report on the meeting began with an assertion of Chinese sovereignty "over the Nansha Islands and the sea waters around them," but added that the dispute over the islands should "be settled by peaceful means rather than by force or the threat of using force."[71] It concluded that China was interested in resolving the dispute with Vietnam through negotiation and the use of international law and encouraged the two sides to temporarily set aside differences and "seek joint development or cooperation in various forms."[72]

During the following years, the two sides placed an emphasis on the cooperative economic relationship that was maturing along the land boundary; however, Beijing and Hanoi continued to trade barbs over the granting of oil claims in disputed maritime territory.[73] Yet, against this backdrop of rhetorical sniping, China and Vietnam pledged to accelerate the negotiating process.[74] More important, border talks continued to be held and both sides placed a strong emphasis on the progress that was being made toward resolving outstanding issues. These developments were highlighted in an October 1998 *Xinhua* report, which referred to the "consensus" Jiang Zemin had reached the previous year with the Secretary of the Vietnamese Communist party to "sign the land border treaty before the year 2000."[75]

This high level endorsement of the negotiating process led to an early 1999 joint statement that announced the institutionalization of a very limited maritime CBM measure. The statement committed both sides not to "take any action that may complicate and enlarge disputes, nor resort to the use of force or threat by force," and to "hold consultations whenever divergence occurs."[76] Later that fall the two sides signed a new formal agreement on the land border.[77] By the end of the following year, a formal agreement was also reached on the Beibu Gulf.[78]

Despite such remarkable progress, the two countries remained sharply divided over the location of the maritime boundary that extends through the South China Sea. As with the boundary dispute with India, it is well known that this contested Sino-Vietnamese ocean territory has important strategic value to both sides. Its security value to Vietnam is obvious, while for China the South China Sea gives it a broad security buffer off the southwestern seaboard and provides Beijing with a major presence in Southeast Asia. In other words, unlike the other disputed segments of the boundary between China and Vietnam, any compromise on the South China Sea and the Nansha Islands would have real costs for both sides. Such a price includes the potential loss of fishing rights in the region and, more important, to the reported oil reserves located under this maritime territory. Furthermore, resolution of the South China Sea dispute between Vietnam and China is not simply a bilateral matter, as has been strikingly demonstrated by the competition that has emerged between China and the Philippines to assert sovereign authority over this region. Not only are multiple players involved, but with the inclusion of Vietnam in the Association of Southeast Asian Nations (ASEAN) in 1995, the question of a multilateral resolution of the dispute became more prominent. Chinese elites have reacted to this prospect with utmost caution and consistently attempted to contain negotiations within the existing bilateral framework.

Such efforts have grown out of a rising concern in Beijing with portraying China as a stabilizing force in the region, a strong conviction in the legitimacy of Chinese claims, and frustration over the lack of international recognition of what is seen as Chinese restraint on territorial issues. For example, the interest in cooperating was given full voice by an influential Sino-U.S. relations scholar at Peking University who noted that China had been quite pragmatic in dealing with the South China Sea issue. The interviewee observed that in this region China had agreed to put aside questions of sovereignty and "even in territories where there are disputes over whether it is ours or yours, we are still willing to go ahead and mutually develop such territory."[79] A Chinese Institute of International Studies scholar, who had been involved in regional track to meetings on maritime issues, echoed these sentiments in 2001. She noted that while historical influences limited Chinese flexibility on the South China Sea, Chinese leaders were now more confident and therefore willing to compromise on territorial issues. Indeed, she argued that many Chinese elites were "more interested in accumulating international prestige and being accepted as part of international society and as a result more relaxed and not so concerned about territorial sovereignty and security."[80] In contrast, other interviewees emphasized the mounting frustration in elite circles over the failure to resolve the South China Sea dispute. A retired government official, who had been part of China's delegation to the Law of the Sea talks at the United Nations, most succinctly expressed this sentiment. He noted, "Look at how much constraint China has shown in the South China Sea. . . . They say we are the aggressors, on the contrary, we have suffered great indignities as a result of our restraint in the region, and I don't know how much longer we can do this if provocations continue."[81]

Conclusion: The Ongoing Construction of Chinese Territorial Sovereignty

This chapter has shown that, during the last two decades, Chinese foreign policy elites have consistently worked on constructing the dragon scales of the modern Chinese state, the boundaries of China's territorial sovereignty. They have done so in a manner that persistently reinforces the line between China and the rest of the region. However, against this static backdrop, elites have utilized a changing set of discursive and diplomatic practices to realize such territorial aspirations. The combativeness of earlier official commentary was replaced by broad silences on territorial issues and measured analysis of contemporary progress toward resolving outstanding disputes. In addition, very few expansive claims were issued (especially in regards to continental boundaries). More significantly, negotiations, border treaties, and CBMs became the most prominent mechanism in China's approach to territorial issues.

In each of the cases considered in this article, the Chinese used such practices to secure highly status quo interpretations of the scope of China's territorial sovereignty. This conservative approach to sovereign boundaries stands in stark contrast to the dire warnings issued by many of those who tend to emphasize China's apparently aggressive and revisionist agenda within Asia. However, the insistence upon the necessity of maintaining relatively static interpretations of Chinese territorial sovereignty via these practices is also reflective of the strict limits that still exist with regards to the Chinese willingness to cooperate and compromise with China's continental and maritime neighbors. In this sense, the intractability of the South China Sea dispute with Vietnam (and other ASEAN states) and the elusiveness of substantive progress in negotiations on the location of the Sino-Indian boundary is indicative of the limits of Chinese flexibility.

Notes

1. To a certain extent this move marked a return to the "good neighbor" policies promoted during the early years of the People's Republic of China (PRC). However, the contemporary period is distinguished by the fact that cooperative border measures extended beyond the minor border states to include China's main territorial neighbors. I would like to thank an anonymous reviewer of this chapter and M. Taylor Fravel, for pointing out the similarities between the two time periods

2. For recent partial exceptions to this oversight, please see Brantley Womack, "International Relationships at the Border of China and Vietnam: An Introduction," *Asian Survey*, no. 6 (2000): 981–986; Gu Xiasong and Brantley Womack, "Border Cooperation between China and Vietnam in the 1990s," *Asian Survey*, no. 6 (2000): 1042–1058; and Waheguru Pal Singh Sidhu and Jin-dong Yuan, "Resolving the Sino-Indian Border Dispute: Building Confidence through Cooperative Monitoring," *Asian Survey*, no. 2 (2001): 351–376.

3. For examples of such analysis, see Luke T. Chang, *China's Boundary Treaties and Frontier Disputes* (London: Oceana Publications, 1982); Ying Cheng Kiang, *China's Bound-*

aries (Berkeley: The Institute of China Studies, 1984); Chih H. Lu, *The Sino-Indian Border Dispute: A Legal Study* (New York: Greenwood Press, 1986); Byron Tsou, *China and International Law: The Boundary Disputes* (New York: Praeger, 1990); Victor Prescott, *The South China Sea: Limits of National Claims* (Kuala Lumpur: Maritime Institute of Malaysia, 1998); and Greg Austin, *China's Ocean Frontier: International Law, Military Force and National Development* (St. Leonards, Australia: Allen and Unwin, 1998).

4. For examples of such analysis, please see Linda Benson and Ingvar Svanberg, *China's Last Nomads: The History and Culture of China's Kazaks* (Armonk, NY: M.E. Sharpe, 1998); Jonathan Lipman, *Familiar Strangers: A History of Muslims in Northwest China* (Seattle: University of Washington Press, 1997); Ralph A. Litzinger, *Other Chinas: The Yao and the Politics of National Belonging* (Durham, NC: Duke University Press, 2001); and Katherine Palmer Kaup, *Creating the Zhuang: Ethnic Politics in China* (Boulder, CO: Lynne Rienner, 2000).

5. The most influential early example of work conducted within this vein is Allen Whiting's *The Chinese Calculus of Deterrence: India and Indochina* (Ann Arbor: University of Michigan Press, 1975). Please also see Sung An Tai, *The Sino-Soviet Territorial Dispute* (Philadelphia: Westminster Press, 1973); George Ginsberg and Carl F. Pinkle, *The Sino-Soviet Territorial Dispute, 1949–64* (New York: Praeger, 1978); Tsien-hua Tsui, *The Sino-Soviet Border Dispute in the 1970s* (Ontario: Mosaic Press, 1983); Pao-min Chang, *The Sino-Vietnamese Territorial Dispute* (New York: Praeger, 1986); Eric Hyer, "The South China Sea Disputes: Implications of China's Earlier Territorial Settlements," *Pacific Affairs*, no. 1 (1995): 34–55; and Bob Catley and Makmur Keliat, *Spratlys: The Dispute in the South China Sea* (Brookfield, UK: Aldershot, 1997).

6. For examples of this literature, please see Anssi Paasi, *Territories, Boundaries and Consciousness: The Changing Geographies of the Finnish-Russian Border* (New York: John Wiley, 1996); Simon Dalby, and Geroid Tuathail, *Rethinking Geopolitics* (New York: Routledge, 1998); and Heikki Eskelinen, Ilkka Liikanen, and Jukka Oksa, eds., *Curtains of Iron and Gold* (Brookfield, VT: Ashgate, 1999).

7. This emphasis on the institutional aspect of territorial boundaries draws on Anssi Paasi's "Boundaries as Social Processes: Territoriality in the World of Flows," in David Newman, ed., *Boundaries, Territory and Postmodernity* (London: Frank Cass, 1999), p. 73.

8. Therefore, throughout the chapter analysis is focused on China's international boundaries and the efforts Chinese elites have made to secure such lines. The chapter leaves aside the question of disputed regions such as Taiwan, Tibet, and Xinjiang. These issues, while having obvious implications for the ultimate size and shape of China's territorial sovereignty, involve broader concerns relating to sovereign jurisdiction and the right to self-determination. For a more detailed explanation of the difference between territorial and jurisdictional sovereignty, please see Allen Carlson, "Constructing a New Great Wall: Chinese Foreign Policy and the Norm of State Sovereignty," Ph.D. diss., Yale University, Political Science Department, 2000.

9. Such a move is not designed to provide an exhaustive list of all such practices, but rather to highlight the most significant measures that elites use. This effort is part of a larger research project that examines the role of diplomatic and representational practices in defining sovereignty's meaning in international politics. Please see Allen Carlson, "Unifying China, Integrating with the World: Securing Chinese Sovereignty in the Reform Era," unpublished manuscript, 2002.

10. In the early 1990s, there was some confusion over the appropriate romanization of each of these Central Asian Republic's names. Throughout this chapter, except when original sources differed, I use these now commonly accepted spellings.

11. All content analysis was conducted by the author. The recording unit used was the sentence. Each sentence that included a specific reference to sovereignty, interference, territory, boundary, and any segment of China's borders, was counted as a single official claim. Only statements attributed to government officials, official spokespeople, or quoted from offi-

cial publications such as the *Xinhua General News Service (Xinhua)* and *People's Daily*, were designated as official claims. The entire contents of each issue were coded. However, since coding was conducted of the mainland edition of *Beijing Review*, it did not include the section that was added to the North American edition in the late 1980s.

12. Claims were coded as containing a land referent if specific mention was made of any of the following key phrases: any segment of China's continental borders, territorial sovereignty (not specified as sea), and air space. Claims coded in the sea category contained reference to any of the following: territorial waters, ocean/maritime borders, Diaoyu Islands, South China Sea, Beibu Gulf, Xisha/Nansha islands. In this chapter, the Chinese terms for each of these disputed territories are used in analyzing the Chinese discourse on sovereign boundaries.

13. These journals include *Shijie jingji yu zhengzhi* (World Economics and Politics), *Xiandai guoji guanxi* (Journal of Contemporary International Relations), and *Guoji wenti yanjiu* (Journal of International Studies).

14. Liu Jiangshui, "Sulian de Dongya zhanlue ji Woguo duice" (The Soviet Union's Far Eastern Strategy and China's Policies), *World Economics and Politics*, no. 2 (1985).

15. Lu Yiran, "Buo Liutiaobian 'Guoji' shuo" (Refuting the "Willow Wall" Borderline), *China's Borderland History Report*, no. 1 (1987).

16. "Sino-Soviet Border Talks End in Moscow," *Xinhua*, February 23, 1987, Lexis-Nexis Universe: World News, Online.

17. "Chinese Foreign Ministry Spokesman on Sino-Soviet Boundary Talks," *Xinhua*, August 5, 1987, Lexis-Nexis Universe: World News, Online.

18. "Chinese, Soviet Working Groups of Experts End First Meeting on Boundary Issue," *Xinhua*, February 1, 1988, Lexis-Nexis Universe: World News, Online.

19. Chen Tiqiang, "Zhong-Yin bianjie wenti de falu fangmian" (The Legal Aspects of the Sino-Indian Border Issue), *Journal of International Studies*, no. 1 (1982).

20. Ibid., p. 13.

21. Jing Hui, "You guan Zhong-Yin bianjie zhengduan de yixie qingkuang he beijing" (Some Facts about the Situation and Background of the Sino-Indian Boundary Dispute), *Journal of International Studies*, no. 2 (1986).

22. Jing Hui, "Zhong-Yin dongbianjie zhenxiang" (The Real Facts of the Eastern Sector of the Chinese-Indian Border), *Journal of International Studies*, no. 1 (1988).

23. "Spokesman Rejects Indian Border Claim," *Beijing Review*, July 26, 1986; and "On the True Situation in Sumdorong Chu Valley Area on the East Sector of the Sino-Indian Boundary," *Beijing Review*, September 1, 1986.

24. For an example, please see, "A Fresh Chill at the Border," *Asiaweek*, January 4, 1987.

25. Please see, "China Urges Tranquility on Border," *China Daily*, May 8, 1987; and "Sino-Indian Cooperation Urged," *China Daily*, June 17, 1987.

26. For an authoritative Chinese overview of Vietnamese provocations in the border region at this time, please see the 1987 and 1988 volumes of *Zhonghua Renmin Gongheguo waijiaobu, waijiao bianji shi*, ed. *Zhongguo waijiao gailan* (Outline of China's Foreign Relations) (Beijing: Shijie zhishi chubanshe).

27. For examples of Chinese commentary on these events, please see "China Reiterates Indisputable Sovereignty over Xisha, Nansha Islands," *Xinhua*, March 24, 1988, Lexis-Nexis Universe: World News, Online; "People's Daily Commentator Condemns Disgusting Conduct of Hanoi Authorities," *Xinhua*, March 29, 1988, Lexis-Nexis Universe: World News, Online.

28. The 1996 spike in claims was largely related to the sudden escalation of the Sino-Japanese dispute over the Diaoyu Islands. This facet of China's maritime claims lies beyond the scope of this article. However, the general level of Chinese restraint in dealing with the issue tends to support the arguments made on these pages.

29. "Sino-Soviet Joint Communiqué," *Xinhua*, May 18, 1989, Lexis-Nexis Universe: World News, Online.

30. "Sino-Soviet Agreements Signed to Boost Ties," *China Daily*, April 26, 1990.

31. *Guowuyuan gongbao* (State Council Gazetteer), no. 4 (1992): 105–107.

32. For example, a monograph published in 1996 stated that the treaty settled 97 percent of the eastern border and left only two small regions in dispute. Please see Liu Dexi, Sun Yan, and Liu Songbin, *Sulian jietihou de Zhong-E guanxi* (Sino-Russian Relations after the Disintegration of the Soviet Union) (Heilongjiang: Heilongjiang jiaoyu chubanshe,1996), p. 197.

33. "China: Russian Border Deal Criticized," *South China Morning Post*, February 21, 1992, Lexis-Nexis Universe: World News, Online.

34. Please see "China, Tadzhikistan Establish Diplomatic Relations," *Xinhua*, January 5, 1992, Lexis-Nexis Universe: World News, Online; "China, Kazakhstan Establish Diplomatic Relations," *Xinhua*, January 4, 1992, Lexis-Nexis Universe: World News, Online; "China Establishes Relations with Kirghizstan," *Agence France Presse*, January 6, 1992, Lexis-Nexis Universe: World News, Online.

35. "China, Russia Issue Joint Declaration," *Xinhua*, December 18, 1992, Lexis-Nexis Universe: World News, Online.

36. *State Council Gazetteer*, no. 31 (1994): 1205–1208.

37. Ibid., pp. 1178–1194.

38. Yu Lei, "Five-Way Border Pledge Signed," *China Daily* (April 27, 1996). Subsequently, each of the five signatory states (the "Shanghai Five") participated in a series of multilateral summits. These meetings produced additional CBMs at the end of the decade, and in 2001 Uzbekistan joined the "Shanghai Five."

39. "China, Tajikistan Sign Four Documents," *Xinhua*, August 13, 1999, Lexis-Nexis Universe: World News, Online.

40. Yin Weiguo, "Dulianti neibu ji yu zhoubian lingguo de lingtu zhan wenti" (Commonwealth of Independent State's International Affairs) and the Issue of Neighboring States Territory), *World Economics and Politics*, no. 8 (1992).

41. Xie Yixian, *Zhongguo dangdai waijiao shi* (The History of Contemporary China's Diplomacy) (Beijing: Zhongguo qingnian chubanshe, 1996), pp. 254–256.

42. Mao Zhenfa, *Bian fanglun* (A Discussion of Border Defense), (Internal Publication: Junshi kexue chubanshe, 1996); and Yang Gongsu, *Zhonghua Renmin Gongheguo he waijiao lilun yu shixian* (The Theory and Practice of the Diplomacy of the PRC) (Internal Circulation, 1997).

43. Interview, Foreign Affairs College, March 29,1998; interview, Peking University, April 8,1998; and interview, Institute of Contemporary International Relations, July 4, 1997.

44. "China, India Issue Joint Communiqué," *Xinhua*, December 23, 1988, Lexis-Nexis Universe: World News, Online.

45. "India, China Favor Early Solution to Border Issue," *Xinhua*, July 9, 1989, Lexis-Nexis Universe: World News, Online. Please also see Sidhu and Yuan, "Resolving," pp. 354–358.

46. "Full Text of Sino-India Joint Communiqué," *Xinhua*, December 16, 1991, Lexis-Nexis Universe: World News, Online.

47. "Sino-Indian Joint Communiqué," December 16, 1991.

48. "China, India Agree on More Confidence-Building Measures," *Xinhua*, June 28, 1993.

49. For example, please see Robert Barnett, "Army Doubles Troops in Tibetan Border Area," *South China Morning Post*, April 25, 1994, Lexis-Nexis Academic Universe: World News, Online.

50. "Sino-Indian Boundary Talks Make Good Progress," *Xinhua*, August 20, 1995.

51. *Renda gongbao* (Gazetteer of the National People's Congress), no. 3 (1997): 479–485.

52. Sidhu and Yuan, "Resolving," p. 360.

53. John Garver, *Protracted Contest: Sino-Indian Rivalry in the Twentieth Century* (Seattle: University of Washington Press, 2001), pp. 336–337.

54. "China Hopes for Proper Settlement of Sino-Indian Border Issue," *Xinhua*, April 27, 1999, Lexis-Nexis Universe: World News, Online.

55. Sidhu and Yuan, "Resolving," p. 358.

56. For a more extensive discussion of these issues, please see Garver, *Protracted*, pp. 80–96, 98–100.

57. While the same process may have taken place in India, I have not examined Indian approaches to territorial sovereignty in enough detail to offer even a tentative observation on this aspect of the Indian stance.

58. Interview, retired foreign ministry official, May 20, 1997.

59. Ibid.

60. Dai Kelai and Yu Xiangdong, "'Fubian Zalu' yu suowei 'Huangsha,' 'Changsha' wenti'" (The Ancient "Fubian" Text and the So-Called Issue of "Huangsha" and "Changsha"), *Journal of International Studies*, no. 3 (1989).

61. Cheng Xiaoxu and Zhang Wenbin, "Zong guojifa kan Woguo dui Nanshaqundao de zhuquan" (Looking at Chinese Sovereignty over the Nansha Islands from the Perspective of International Law), *World Economics and Politics*, no. 1 (1990).

62. Ibid., p. 66.

63. Chen Ning, "Nanhai zhuquan fenzheng xingshi yu Woguo de duice" (Trends in the Dispute over the Sovereignty of the Southern Sea and China's Countermeasures), *World Economics and Politics*, no. 11 (1991).

64. For a detailed overview of this period in Sino-Vietnamese border relations, please see Ramses Amer, "The Territorial Dispute between China and Vietnam and Regional Stability," *Contemporary South East Asia*, no. 1 (1997): 86–114.

65. *Guoji tiaoyue ji* (Collection of International Treaties), (1991), p. 85.

66. Text of "The Law of the People's Republic of China on Its Territorial Waters and Their Contiguous Areas," *Xinhua*, February 25, 1992.

67. *State Council Gazetteer*, no. 32 (1992): 1439–1441.

68. "China, Vietnam Sign Accord on Territorial, Border Issues," *Xinhua*, October 19, 1993.

69. For example, please see "China Asks Viet Nam to Release Detained Chinese Fishing Boats," *Xinhua*, (July 4, 1994), Lexis-Nexis Universe: World News, Online.

70. This aspect of the South China Sea dispute lies beyond the reach of this chapter, but is an important indicator of the Chinese commitment to secure lines that are viewed as consistent with their interpretations of the legitimate scope of China's territorial sovereignty.

71. "China Has Indisputable Sovereignty over Nansha: Official," *Xinhua*, November 14, 1995, Lexis-Nexis Universe: World News, Online.

72. Ibid.

73. "Vietnam Demands Halt to Chinese Offshore Drilling," *Reuters World Service*, March 15, 1997; and Jane Macartney, "China Sees No Quick Fix for Vietnam Oil Dispute," Reuters North American Wire, April 10, 1997, Lexis-Nexis Universe: World News, Online.

74. "Vietnam, China Agree to Speed Up Border Talks," *Reuters World Service*, April 21, 1997, Lexis-Nexis Universe: World News, Online.

75. "Jiang Zemin Meets Vietnamese Prime Minister," *Xinhua*, October 20, 1998, Lexis-Nexis Universe: World News, Online.

76. "China and Vietnam Issue Joint Statement," *Xinhua*, February 27, 1999, Lexis-Nexis Universe: World News, Online.

77. "China, Vietnam Sign Land Border Treaty," *Xinhua*, December 30, 1999, Lexis-Nexis Universe: World News, Online.

78. "China-Vietnam Agreement," *New York Times*, December 26, 2000, Lexis-Nexis Universe: World News, Online.

79. Interview, Peking University, June 26, 1997.

80. Interview, Chinese Institute of International Studies, December 21, 2001.

81. Interview, retired State Oceanic Administration official, April 23, 1998.

16

Rituals, Risks, and Rivalries

China and ASEAN

Ho Khai Leong

The People's Republic of China (PRC) and the Association of Southeast Asian Nations (ASEAN) are two gigantic entities in the Asia-Pacific region, and their interaction will to a great extent affect the future and prospects of the entire region.[1] After the establishment of diplomatic ties with Singapore and Brunei in 1991, China now enjoys official relations with all of the ASEAN states. Sino-ASEAN relations have been the subject of many studies since that time.[2] One of the most common arguments encountered in the literature is that the success or failure of the relations will to a large extent depend on the political will of both sides and the perception of each other as allies or rivals. At present, both sides are working strenuously at confidence-building measures to reassure the other party of their sincere desire to maintain a mutually beneficial, cooperative relationship.

This chapter argues that, despite reassurances and determination by both sides to increase interaction and cooperation, there remain areas of contention and contestation in Sino-ASEAN relations. This argument is supported by the fact that in the last two decades, the PRC's influence has seeped into more and more of Southeast Asia and the PRC itself has climbed inexorably toward regional prominence. After its long isolation, self-imposed or otherwise, the PRC has been able to integrate itself successfully into the political life, economy, and security interests of the region; however, the PRC's market potential, military capability, and its enormous size have both excited and threatened the Southeast Asian states. While the Southeast Asian states have been willing to engage this emerging regional power, they are also wary of the potential risks when dealing with China. Increased trade, investment, and cultural exchanges between the two regions have made both sides increasingly aware of the opportunities and challenges involved in further developing relations. The relationship is not without its problems; although there is a general consensus that it is on a much better footing at present than in recent decades. The development of Sino-ASEAN relations now and in the next decade can be best described under three subheadings: rituals, risks, and rivalries.

Rituals

As in all diplomatic and official endeavors, Sino-ASEAN relations have been coined positively in official terms. The public image is one of cooperation, friendliness, and common interests. There is little question that both sides would like to improve political relations, especially after the end of the Cold War. Indeed, the trend toward normalization of Sino-ASEAN relations in large part was a result of intensified Sino-Soviet hostility and the partial rapprochement between Washington and Peking.[3] Such geopolitical considerations have a major impact on present as well as future diplomatic developments.

Institutionally, PRC-ASEAN relations operate at three levels: the "unofficial" level, the "semi-official" level, and the "official" level.[4] These established links enable both China and ASEAN to build bridges and exchange views on prospective issues and problems. While these institutional links are established based on trust and goodwill, there is an element of ritualism that runs through the interactions.

The PRC government has been making goodwill gestures to Southeast Asian states for some time, and in the past couple of years it has intensified its diplomatic efforts. For example, speaking at the annual ASEAN and China Dialogue meeting (ASEAN+1) in Bangkok in July 2000, Chinese Foreign Minister Tang Jiaxuan said the frequent exchanges of visits between leaders from both sides have vigorously promoted the Sino-ASEAN commitment to good-neighborliness and friendship. The Sino-ASEAN dialogue at all levels, he continued, has further deepened a mutual understanding and confidence between the peoples of China and ASEAN countries. By issuing or signing framework documents governing the bilateral cooperation, the two sides have laid out a blueprint for cooperation in all areas for the new century. On economic issues, the PRC was pleased to see increasing dynamic Sino-ASEAN cooperation with momentum particularly strong in such areas as investment, finance, science and technology, agriculture, and industry. The third set of cooperative projects under the Sino-ASEAN Joint Cooperation Committee are progressing smoothly, with the personnel-exchange program, workshops on understanding modern China, and remote sensing and transgenetic plants successively launched.

On security issues, China has always supported the Treaty of Amity and Cooperation in Southeast Asia and the Southeast Asia Nuclear Weapon Free Zone Treaty. Tang noted that China and the ASEAN countries concerned have pledged to settle their differences peacefully and have made notable progress exemplified by the signing of the China-Vietnamese land boundary treaty[5] and China's active participation with ASEAN in discussing the formulation of a code of conduct on the South China Sea.[6]

A similar version of these diplomatic rituals is reflected in the visit by Chinese Vice-President Hu Jintao to the region in July 2000. Hu repeated many of the official statements and assured the region that the PRC government will adhere to its policy of maintaining good neighborly and friendly relations with countries in the region. In a typical diplomatic fashion, he affirmed, "China cannot achieve development in isolation of Asia, and Asia cannot realize prosperity without China. It is a consistent policy of the Chinese government to strengthen the good neighborly and friendly relations with its surrounding countries." Citing that as early as the 1950s, China and some Asian countries jointly initiated the Five Principles of Peaceful Coexistence, Hu said, "It is on that basis that China has established friendly and cooperative relations with all Asian countries successively."

Difficult areas in the diplomatic relations of China and ASEAN were ritualistically brushed aside during Hu Jintao's visit to the region. While PRC government officials attempted to address these disputes, the open dialogue inevitably took on a diplomatic tone. Speaking before the Indonesian Council on World Affairs, Hu reassured the ASEAN states that, as for the historical differences between China and its neighboring countries, the PRC has always proceeded from the overall interest of all parties concerned and stood for a peaceful solution of these differences through consultations. The proposal of joint development has often been heard, especially in regard to the South China Seas dispute. "Facts have proven that China is an important driving force for Asia's development as well as an important force for safeguarding stability in Asia. China's development will pose no threat to any country,"[7] Hu reassured the listeners.

Overall, on the diplomatic and conference tables at least, China sought to expand multi-level exchanges and cooperation with ASEAN in the areas of trade, economics, science, and technology. It is true that China and ASEAN are highly complementary to each other in terms of agriculture, mechanical and electrical equipment manufacturing, medicine, transportation, and other fields, where both sides have the potential for extensive cooperation.

ASEAN is quick to respond to China's goodwill and friendliness. It has publicly acknowledged that it recognizes the influential role that the PRC has been playing in the promotion of peace, security, and prosperity in the region and the world at large. While it is apprehensive of the growing military influence of the PRC, it has nevertheless maintained a fairly rational and diplomatic approach in engaging the PRC in regional affairs. For example, Malaysia's foreign minister, Datuk Seri Syed Hamid Albar, has reiterated that relations between Beijing and ASEAN have grown steadily at various levels in recent years and that ASEAN looks forward to continuing to work closely with China in regional affairs and in addressing international issues of mutual interest and concern. In order to consolidate ASEAN-China relations, it was important for ASEAN to make full use of the various mechanisms that have been established such as the ASEAN-China Senior Officials Consultations, ASEAN PMC plus China, and the Asean+3 framework.

The ASEAN Regional Forum (ARF), of which China is a member, is also another link where ARF members share views on the security outlook for the Asia Pacific region. ASEAN by and large considers the promotion of an inclusive type of cooperative security arrangement involving ASEAN and China necessary for the region as this could encourage and facilitate constructive engagement within the region as opposed to exclusive arrangements. In a short span of time, both sides have made tremendous progress in various areas of cooperation, such as China's commitment to cooperative development manifested by the number of projects implemented under the dialogue. More projects such as the second ASEAN-China Economic and Trade Seminar, ASEAN-China Workshop on Remote Sensing Cooperation, and ASEAN China Workshop on Transgenic Plants are scheduled to be implemented in 2000.[8]

There are two observations to be made here. First, a network of multilevel dialogues—official, semi-official, and unofficial—between the PRC and ASEAN has been well established. The PRC obviously wants to reassure ASEAN countries that its intention in the region is purely nonmilitary in nature and that its principles of peaceful coexistence are alive and well. ASEAN members, on the other hand, are reciprocating and responding in kind. While apprehensive about China's intentions in the region, ASEAN maintains an approach of constructive engagement with their giant neighbor. Second, these diplomatic and ritualistic efforts, even at the rhetorical level, have contributed significantly to the stability and peaceful development of the region. These can be viewed as confidence-building measures that both sides would like to further enhance. The real contribution in terms of policy initiatives is doubtful. For example, the ARF is seen primarily as a means of engaging China in a multilateral security dialogue but without any expectation of solving disputes or building a comprehensive regional security structure.[9] Sino-ASEAN relations seem to be cordial considering the cooperation taking place at various levels; however, there are also potential risks that both sides cannot ignore.

Risks

The greatest potential threat to the diplomatic efforts between China and ASEAN states is the dispute in the South China Sea. The Paracel and Spratly island groups, believed to be rich in oil and gas, are claimed in whole or part by Vietnam, China, Brunei, Malaysia, the Philippines, and Taiwan. Indeed, the Spratly Islands dispute remains an important source of contention between China and Southeast Asian states. The potential, large scale, suboceanic oil deposits have made the Spratlys and other uninhabited islands of the South China Sea the subject of conflicting territorial claims by PRC, Taiwan, Vietnam, the Philippines, Malaysia, and Brunei. Bilateral and multilateral talks over these South China Sea islands have begun, but there remain major issues to be resolved by all parties.[10]

Beijing claims the Paracel and Spratly islands, as well as the surrounding waters and their natural resources. The PRC claims to have taken a reasonable approach to deal with the dispute, but the Southeast Asian states are still wary of China's aims in the South China Sea as there were discoveries of a new Chinese facility on territory claimed by the Philippines. There have also been occasional conflicts, such as the detention of sixty-two Chinese fishermen by Philippine authorities for alleged illegal fishing. And in an incident where an oilrig was constructed on disputed territory, the Chinese agreed to dismantle the structure instead of confronting Hanoi. By contrast, the Sino-Vietnamese row over the Spratlys in 1988, when Vietnam was still the grouping's enemy in Cambodia, ended with two of its ships sunk and seventy-seven navy men killed. It is clear that parties involved want to be cautious in dealing with such disputes as they do not want to jeopardize the goodwill that both sides have tried hard to forge.

The issue of disputed territory has been raised in the ASEAN Regional Forum (ARF). Although the ARF is not set up to arbitrate territorial disputes, it has increasingly been used for that purpose. Indeed, for ARF participants, the Spratlys continue to loom large on their agenda. From the outset, Beijing has been reluctant to have the matter discussed and settled in such multinational settings, presumably for fear of an anti-China alliance being formed among claimants. In April 1998, however, China softened its position and said it was willing to discuss the Spratlys dispute multilaterally, as ASEAN has long demanded.

Recent developments suggest that there has been "limited" progress in this area. A consultation group meeting that discussed the disputed territories was held in Hanoi, Vietnam, on October 11, 2000. According to the Vietnamese foreign minister, Nguyen Dy Nien, ASEAN and China agreed to finalize a code of conduct governing competing claims in the South China Sea by the end of 2000. Apparently, ASEAN and China have agreed in principle to develop and adopt the Regional Code of Conduct for the South China Sea.[11] PRC Foreign Minister Tang also issued similar statements after his visit to Hanoi,[12] but there were still unresolved problems. First, both sides have yet to agree on the code's geographic scope. China wants the code to cover only the Spratlys, while Vietnam wants the code to also apply to the Paracels, an island chain claimed by both Beijing and Hanoi. Second, the terms of operations are still unclear. ASEAN sources said another major point in the code that is still being sorted out is the phraseology of a ban on new occupation of uninhabited islands in the disputed area. Diplomats said Beijing wants a much weaker formulation of the clause under which the claimants to the islands will undertake only to "refrain from any action that will complicate the situation," a wording they say is already contained in existing agreements. The ASEAN draft explicitly stipulates a "halt to any new occupation of reefs, shoals and islets in the disputed area."[13]

The South China Seas dispute will continue to be on top of the agenda for Sino-ASEAN relations in the foreseeable future. It is unlikely that a significant breakthrough will be made without major concessions from either side. The Regional Code of Conduct may ease tensions for a while, but major disagreements will continue. As one observer put it, "It would be a mistake to believe that a regional code of conduct will do much to alter the volatile mix of history, sovereignty and resources that has made the South China Sea one of the potential flashpoints of East Asia. For a start, Asean [sic] and China have an understanding that a code won't be used as a legal instrument to settle overlapping territorial and jurisdictional claims."[14]

Rivalries

Military Rivalries

Before the release of the Cox Commission report in the United States, many politicians and observers were already pointing to China as the next great threat to world security (U.S. national security). Indeed a number of members of Congress in the United States have been hyping the "Chinese threat" theory since the Soviet Union broke apart and the "Soviet menace" disappeared. For the Southeast Asian states, the concern that China poses a threat to the region has also been a regular and familiar tune, sung by many.[15] In order to placate such fears, the Chinese government constantly assures its Asian neighbors that China poses no threat of any kind.

From a political and security perspective, whether China will constitute a threat to the Southeast Asian region is really a matter of perception.[16] The ASEAN states are not in complete agreement as far as this issue is concerned. There are those who believe that China is a benign power that is incapable of domination. This school of thought suggests that the regime in Beijing is far more preoccupied with managing the monumental forces unleashed at home by its economic reforms than with creating instability within the region. The Malaysians, for example, certainly take a more balanced view. The Malaysian governing elite regards China as a key factor in regional security. Prime Minister Mahathir was quoted as saying, "The U.S. is saying we are threatened by China, but I don't see the threat from China as being any worse than the threat from the U.S."[17] However, as far as China's defense buildup is concerned, Malaysia's attitude is less certain. Former Defense Minister Najib remarked, "Malaysia is not worried about China's defence modernization so long as she does not develop long-range offensive capability such as strategic nuclear-capable bombers and carrier battle groups."[18]

ASEAN, however, is taking measured steps at the same time to strengthen itself. The recent admission of Burma (Myanmar) into ASEAN was perceived by many as a strategy to reinforce the bargaining power of the organization and thus a way to contain China's influence in the region. Although there is the lingering concern of political incompatibility, ASEAN now encompasses the whole Southeast Asian region. "For ASEAN, China's growing presence in bordering ASEAN countries is a major concern."[19]

In a recent forum, China's Foreign Minister Tang expressed concern over joint military exercises that some of the ASEAN countries conduct with U.S. armed forces. His concern apparently was raised privately to ASEAN's foreign ministers and was not an open remark. China howled in protest earlier this year when the Philippine military and the U.S. armed forces conducted a series of joint military exercises in Philippine waters near the Spratlys and charged that the exercises were directed at China.[20] Malaysian Foreign Minister Syed Hamid Albar, who acted as the coordinator for ASEAN, explained to Tang that the issue of multilateral military exercises is an internal matter and is an issue best left to the concerns of the countries involved.

In sum, both China and ASEAN member states have expressed concern for the other's military intentions or military alliances. While ASEAN members' governments may not have a common position about China's rising power, they have all expressed concern about its rising military spending and hence its intentions.[21] After the Cold War, ASEAN still looks to the United States not only to provide security guarantees but also for access to its market and technology. China, on the one hand, does not want too much American military involvement in the region, as it would challenge China's dominance and influence. Beijing and Washington's rivalry in the region will probably intensify as a result of China's military ascendance and Washington's determination to play a major security role in the Asia-Pacific region;[22] therefore, ASEAN is eager to promote interaction and dialogue between the PRC and the United States in the hopes that they will successfully manage their relations within the ARF framework.[23]

Economic Rivalries

Trade is arguably the most significant aspect of Sino-ASEAN relations. Although there is undeniably the possibility of economic competition threatening the cooperation of the PRC and ASEAN states, it is important for both sides to realize the complementary aspects of trade and the potential for mutual benefit. In the 1950s and 1960s, direct trade between China and Southeast Asia was limited, and bilateral trade depended heavily on indirect trade via the entrepots of Hong Kong and Singapore. In the 1970s, the momentum picked up, and economic as well as political relations between China and Southeast Asian states improved significantly as Malaysia, the Philippines, and Thailand established diplomatic relations with the PRC. In the 1980s, Sino-ASEAN relations entered a new phase as China embarked on its four modernizations program, established an open-door policy, and began actively seeking foreign capital investments, which included investments from Southeast Asian states. The 1990s saw further expansion of economic relations between China and ASEAN states, as the Southeast Asian countries sought to expand their foreign markets in China to help counteract the debilitating effects of the world economic recession. (See Table 16.1.)

Table 16.1

China's Trade with ASEAN States, 1990 and 1995 (in US$ million)

Country	1995 Exports	1995 Imports	1990 Exports	1990 Imports
Brunei	34	0.3	5.5	—
Indonesia	1,440	2,052	220	326
Malaysia	1,280	2,065	302	298
Philippines	1,030	276	121	270
Singapore	3,500	3,398	1,863	461
Thailand	1,752	1,611	610	147
Vietnam	720	332	—	—
Total	9,756	9,734.3	3,121.5	1,502

Source: China's Foreign Trade Yearbook, 1996–97 (Beijing: Chinese Ministry of Foreign Trade), and Qingxin Wang, "In Search of Stability and Multipolarity: China's Changing Foreign Policy towards Southeast Asia after the Cold War," *Asian Journal of Political Science* 6, no. 2 (December 1998).

The competition with Taiwan in the region not withstanding, China's trade with ASEAN has continued to grow in the last few decades. In 1985, the value of China's trade with ASEAN was only US$3.2 billion. It had increased to US$4.6 billion in 1990 and to US$20 billion in 1995. Foreign direct investment by ASEAN states in China has also increased rapidly (approximately US$6 billion in 1995).[24]

Despite the increased level of economic interdependence, frictions and difficulties with individual countries remain. For example, Singapore's investments in the Singapore Suzhou Industrial Park were plagued with problems, which were only recently resolved. It is suggested that the economies of ASEAN and China have some similarities that result in a competitive rather than a complementary relationship. For this reason, it is important to encourage frequent consultations as a means to avoid unnecessary competition and friction in the future.[25]

All said, the potential for further development of China-ASEAN economic relations is great. China's entry into the World Trade Organization (WTO) will certainly stimulate the growth of bilateral trade between the two regions. The opportunities for trade and investment as well as other areas of cooperation will certainly be enormous; however, the PRC's relationship with Southeast Asia has been complicated by Taiwan's economic presence in the region. Taiwan's economic growth in the postwar period has often been dubbed a "miracle."[26] It is argued that this success was due to a combination of causal factors, such as the export-oriented industrialization policy of the Nationalist government and the high-saving function of the Taiwanese society.[27] From the mid-1970s, foreign trade grew exponentially, and it now plays an important role in Taiwan's economy. By 1990, exports had soared to US$67,214 million compared to US$19,810 in 1980, and imports were valued at US$54,716 million compared to US$19,733 million ten years earlier.[28]

In the 1990s, Taiwan's treasury accumulated huge amounts of capital from both foreign earnings and domestic savings, and domestic businessmen began to turn their attention else-

Table 16.2

Taiwanese Investment toward Southeast Asia (in US$ million)

Country	Year	Approval by local government		Approval by Taiwanese government	
		Value	Number of cases	Value	Number of cases
Thailand	1990	782.69	144	149.39	39
	1992	289.92	44	83.29	23
	1994	477.49	88	57.53	12
	1996	2688.20	45	71.41	9
Malaysia	1990	2383.00	270	184.88	36
	1992	602.00	237	155.73	17
	1994	1149.60	100	101.13	17
	1996	310.40	79	93.53	12
Philippines	1990	140.65	158	123.60	16
	1992	9.27	27	1,21	3
	1994	292.37	42	9.60	10
	1996	52.70	22	74.25	20
Indonesia	1990	618.20	94	61.87	18
	1992	563.30	23	39.93	20
	1994	2487.50	48	20.57	12
	1996	534.60	111	82.61	13
Vietnam	1990	104.43	18	0.00	0
	1992	531.41	23	20.17	9
	1994	524.01	76	108.38	33
	1996	844.61	61	100.48	25

Source: Chen Tain-jy, ed., *Taiwanese Firms in Southeast Asia* (Cheltenham, UK: Edward Elgar, 1998), pp. 12–13.

where for investment. Overseas investments were also encouraged by the policies implemented by the Nationalist government around 1985. In the beginning of the 1990s, major revisions continued to be made with regard to banking and foreign exchange regulations, which saw a relaxing of controls over both incoming and outgoing capital, thus further increasing capital investment mobility outside the island state.

Taiwan's use of economics as a functional instrument of its "flexible diplomacy" in the 1980s and 1990s is hardly surprising. "Economic diplomacy" and "flexible diplomacy" are really two sides of the same coin. There is continuity in the pragmatic approach of combining trade and investment in pursuit of a viable foreign policy and exerting its statehood with the ultimate objective of establishing diplomatic relations with the countries concerned.

Taiwan's official announcement that it was pursuing a "southward policy" can only be understood with the rationale of flexible diplomacy in mind. The southward policy had several objectives. First, it sought to divert local companies' investment funds from China to Southeast Asia. Second, it aimed to make Southeast Asia its intermediary for investing in mainland China when Hong Kong reverted back to the PRC in 1997. Third, the policy looked to strengthen Taiwan's diplomatic relations with Vietnam and the ASEAN states. It is quite clear that these objectives have significant political underpinnings, integrating the economic with the political. The former foreign minister, Frederick Chien, was quoted as saying, "The 'Southward Policy' is mainly intended for economic cooperation and is not directed against any third power. No one should read anything more into it."[29] Despite repeated denials, there is practically no one in the region who believes that this is in fact the case.[30]

The involvement of Taiwan in Southeast Asia should not be overstated, but Taiwan clearly has a competitive edge in the region. The situation in Southeast Asia has changed enough to make it more and more flexible and convenient for Taiwan to attain a balance between its economic and political-military interests. In many ways, Taiwan's presence in the region "has the potential of undercutting China's negotiating leverage with Southeast Asia."[31] In the long run, Taiwan is certain to be seen as a rival for commercial and political influence in the region. Normalization of relations between Taiwan and the PRC would help put the economic relationship between the two countries in the public light for all of Southeast Asia to see and would serve to strip away the illusions many still have about Chinese trade. In a broader sense, more normal ties with the PRC might test the strength and confidence of Taiwan in their own political, economic, and social modernization more fully than can be tested by relations with the smaller states of Southeast Asia.

Another recent development that adds to the economic rivalry of China and ASEAN is the increased attraction of China as a high-tech investment area. From the mid-1990s, China has become popular for foreign high-tech investment as a result of its strategies of attracting multinational corporations (MNCs) to establish large as well as medium high-tech industries in various parts of China. The rush to China has produced a "hollowing out" effect for ASEAN, which was initially a favorite investment region for many American, European, and Japanese MNCs. Indeed, direct foreign investment in ASEAN countries has slowed markedly from the mid-1990s, especially in the last three years. By 2000, direct foreign investment in the region had dropped some 41 percent from 1996. This was largely a result of China's liberalization of its economic policy, making it far easier on foreign firms to invest in high-tech industries there. For example, it is reported that, in 2000, Motorola unveiled its intention to build an integrated semiconductor plant in China, bypassing ASEAN.[32]

Conclusion

China's greatest concern in the coming decades will continue to be economic modernization. While many observers still believe that China poses a threat to the Asia-Pacific region, ASEAN states, in general, maintain a fairly pragmatic position toward China. Southeast Asian states' concerns regarding the Chinese military stem primarily from the Spratlys disputes and China's hostile posturing toward Taiwan. This critical dispute serves as a reminder that Sino-ASEAN relations cannot be taken for granted. However, if both sides can push security concerns to the background and concentrate on their economic relations, the future development of Sino-ASEAN relations holds a lot of promise. By increasing economic interdependence and deepening political ties, policy-makers on both sides will realize the benefits of cooperation, and as a result, Sino-ASEAN relations will become less ideological and more geared to the economic needs of both sides.

It is reasonable to expect Sino-ASEAN relations in the coming decades to maintain interaction based on these three variables: rituals, risks, and rivalries. A network of multi-level dialogues and frequent exchange of visits by senior leaders from both sides will continue to increase mutual trust, support, and understanding. There are, however, problematic areas, such as the South China Sea dispute, which both sides cannot ignore. While limited progress has been made, potential risks still exist that threaten to weaken the relationships. As the globalization trend surges ahead with great momentum, China and ASEAN will be confronted with a more competitive situation in terms of direct foreign investment and high-tech capital investments. The increasing presence of Taiwan in Southeast Asia adds to the complexity of this economic rivalry.

Since its establishment of diplomatic ties with the Southeast Asian states, China has shown more flexibility in its approach to modernization and foreign policy, and this is encouraging to the ASEAN leaders. Undoubtedly, China will become an important political, security, and trading partner of the ASEAN states in the years to come. However, the prospects for greater economic and political interaction will depend largely on how China's economic reforms continue to unfold in the post-Deng and post-Jiang periods. Equally important is how the ASEAN states meet the challenges of democratization and how successfully they sustain their economic growth following the currency crisis.

Notes

1. The country members of ASEAN are Thailand, Myanmar, Malaysia, Singapore, Brunei, Indonesia, Laos, Vietnam, and the Philippines.
2. See, for example, Leszek Buszynski, "China and the ASEAN Region," in Stuart Harris and Gary Klintworth, eds., *China as a Great Power: Myths, Realities, and Challenges in the Asia-Pacific Region* (New York: St. Martin's Press, 1995); Lee Lai To, "ASEAN-PRC Political and Security Cooperation: Problems, Proposals and Prospects," *Asian Survey*, no. 11 (November 1993): 1095–1104; Wang, Qingxin, "In Search of Stability and Multipolarity: China's Changing Foreign Policy towards Southeast Asia after the Cold War," *Asian Journal of Political Science* 6, no. 2 (December 1998); Ross H. Munro, "China's Changing Relations with Southeast, South, and Central Asia," in Hafeez Malik, ed., *The Roles of the United States, Russia, and China in the New World Order* (New York: St. Martin's Press, 1997); and Allen Whiting, "ASEAN Eyes China," *Asian Survey* (April 1997).
3. Edwin W. Martin, *Southeast Asian and China: The End of Containment* (Boulder, CO: Westview Press, 1977), p. 87.
4. Lee Lai To, "Some Thoughts on ASEAN and China: Institutional Linkages," in Richard Grant, ed., *China and Southeast Asia* (Washington, DC: Center for Strategic and International Studies, 1993).
5. For an analysis of Vietnam-China relations, see Ang Cheng Guan, "Vietnam–China Relations since the End of the Cold War," *Asian Survey* 33, no. 12 (December 1998).
6. "Chinese Foreign Minister Hails Cooperation with ASEAN," *BBC Monitoring*, July 28, 2000.
7. "Hu Jintao on PRC's Policy on Asia," *World News Connection*, July 24, 2000.
8. "ASEAN Recognises China's Role in Promoting Peace," Bernama, The Malaysian National News Agency, July 28, 2000.

9. Allen Whiting, "ASEAN Eyes China," *Asian Survey* (April 1997): 300.

10. John W. Garver, "China's Push through the South China Sea. The Interaction of Bureaucratic and National Interest," *The China Quarterly*, no. 132 (December 1992); Mark J. Valencia, China and the South China Sea Disputes, Adelpi Paper No. 298, London, Royal Institute for Strategic Studies, 1995.

11. "Vietnam to Keep ASEAN's Focus on Economy, Regional Security," Dow Jones International News, July 30, 2000.

12. "In the interest of maintaining peace and stability in the region, China has actively participated in the discussion with ASEAN countries on the formulation of a code of conduct on the South China Sea and notable progress has been made in this regard. All in all, we are moving forward step by step toward the objective of the good-neighborly partnership of mutual trust, a goal jointly set out by China and ASEAN countries." Ibid.

13. "China, ASEAN Fail to Reach Agreement on Spratlys," October 12, 2000, Associated Press.

14. Barry Wain, "At Loggerheads with Beijing," *Asian Wall Street Journal*, October 6, 2000.

15. See, for example, Shee Poon Kim, "Is China a Threat to the Asia-Pacific Region?" in Wang Gungwu and John Wong, eds., *China's Political Economy* (Singapore: World Scientific, 1998).

16. Chang Pao-min, "China and Southeast Asia: The Problem of a Perception Gap," *Contemporary Southeast Asia* 9, no. 3 (December 1987): 181–193.

17. *Asiaweek*, August 11, 1993, p. 21.

18. *Straits Times*, July 13, 1993.

19. "Widening Gap of Political Incompatibility among ASEAN Members Reviewed," World News Connection, July 17, 2000.

20. "Code of Conduct in S. China Sea to Be Concluded by Nov.," Japan Economic Newswire, July 28, 2000. The Philippines is among those Southeast Asian countries that has a joint military exercises agreement with the U.S. military. Some other ASEAN countries also have similar agreements, including Thailand and Singapore.

21. David B.H. Denoon and Wendy Friedman, "China's Security Strategy: The View from Beijing, ASEAN and Washington," *Asian Survey* (April 1996).

22. See Ezra F. Vogel, *Living with China, U.S.-China Relations in the Twenty-first Century* (New York: W. W. Norton, 1997).

23. Jurgen Haacke, "The ASEANization of Regional Order in East Asia: A Failed Endeavor?" *Asian Perspective* 22, no. 3 (1998): 7–47.

24. Wang, Qingxin, "In Search of Stability and Multipolarity: China's Changing Foreign Policy towards Southeast Asia after the Cold War," *Asian Journal of Political Science* 6, no. 2 (December 1998); Fred Herschede, "Trade between China and ASEAN: The Impact of the Pacific Rim Era," *Pacific Affairs* 64, no. 2 (1991).

25. Djisman S. Simandjuntak, "Recent Developments in and Prospects for Pacific Cooperation and Possibilities Therein for ASEAN-China Economic Relations," in Chia Siow Yue and Cheng Bifan, eds., *ASEAN-China Economic Relations: In the Context of Pacific Economic Development and Cooperation* (Singapore: Institute of Southeast Asian Studies, 1992).

26. There is an abundance of literature on this topic. See, for example, Thomas B. Gold, *State and Society in the Taiwan Miracle* (Armonk, NY: M. E. Sharpe, 1986); Tsiang S. C., "Taiwan's Economic Miracle: Lessons in Economic Development," in Arnold C. Harberger, ed., *World Economic Growth*, pp. 301–326 (San Francisco: Institute of Contemporary Studies, 1984); and Lim Chong Yah, "Taiwan's Economic Miracle: A Singaporean Perspective," in Seiji Naya and Akira Takayama, eds., *Economic Development in East and Southeast Asia* (Honolulu: ISAS and East West Center, 1990). See also Denis Fred Simon, ed., *Beyond the Economic Miracle* (Armonk, NY: M. E. Sharpe, 1991).

27. Lim Chong Yah, "Taiwan's Economic Miracle," p. 53.

28. Monthly Statistics of Exports and Imports, Department of Statistics, Ministry of Finance, January 20, 1992, p. 10.

29. Julian Baum, "Looking South," *Far Eastern Economic Review*, March 10, 1994, p. 18.

30. For discussions on Taiwan's investment in Southeast Asia, see the following: Shan Yee Lee, "Trade between Taiwan, R.O.C. and ASEAN Countries," in *Proceedings of the Eight Asian Pacific Cultural Scholars' Convention*, Asian-Pacific Cultural Center, July 21–27, 1986, Taipei; Ho Khai Leong, "The Changing Political Economy of Taiwan-Southeast Asia Relations," *The Pacific Review* 6, no. 1 (1993): 31–40; Samuel C.Y. Ku, "The Political Economy of Taiwan's Relations with Southeast Asia," *Contemporary Southeast Asia* 17, no. 3 (December 1995): 282–297; Chen Tain-jy, ed., *Taiwanese Firms in Southeast Asia* (Cheltenham, UK: Edward Elgar, 1998); Gerald Chan, "Sudpolitik: The Political Economy of Taiwan's Trade with Southeast Asia," *The Pacific Review* 9, no. 1 (1996); Chen, Hurng-Yu, "Taiwan's Economic Relations with Southeast Asia," in Gary Klintworth, ed., *Taiwan in the Asia-Pacific in the 1990s* (St. Leonards, New South Wales, Australia: Allen & Unwin in association with Department of International Relations, Australian National University, 1994); Gary Klintworth, *New Taiwan, New China: Taiwan's Changing Role in the Asia-Pacific Region* (Melbourne, Australia: Longman, 1995).

31. Wang, Qingxin, "In Search of Stability and Multipolarity: China's Changing Foreign Policy towards Southeast Asia after the Cold War," *Asian Journal of Political Science* 6, no. 2 (December 1998).

32. G. Pierre Goad, "Anaemic Asean," *Far Eastern Economic Review*, September 7, 2000.

About the Editor and Contributors

The Editor

Suisheng Zhao is an associate professor and Executive Director of the Center for China-U.S. Cooperation at the Graduate School of International Studies, University of Denver. He was the recipient of the 1999–2000 Campbell National Fellowship at the Hoover Institution of Stanford University, founder and editor of the *Journal of Contemporary China*, and a member of the Board of Directors at the U.S. Committee of the Council for Security Cooperation in the Asia Pacific (USCSCAP). An author and editor of four books, his most recent are *Across the Taiwan Strait: Mainland China, Taiwan, and the Crisis of 1995–96* and *China and Democracy: Reconsidering the Prospects for a Democratic China*. His next book, *Chinese Nationalism and the Construction of a Nation-State*, will be forthcoming from Stanford University Press in 2004. His articles have appeared in *Political Science Quarterly, The China Quarterly, World Affairs, Asian Survey, Journal of Northeast Asian Studies, Asian Affairs, Journal of Democracy, Pacific Affairs, Communism and Post-Communism Studies, Problems of Post-Communism, The Journal of East Asian Affairs, Issues and Studies, Journal of Contemporary China*, and elsewhere.

The Contributors

Allen Carlson is an assistant professor in the government department at Cornell University, New York. In 2000 he completed his dissertation entitled, "Constructing a New Great Wall: Chinese Foreign Policy and the Norm of State Sovereignty" in the political science department at Yale University in New Haven, Connecticut, which he is revising for publication. Professor Carlson has also recently contributed a piece to the China Policy Series published by the National Committee on United States-China Relations, and is coediting (with J. J. Suh and Peter Katzenstein) a volume on Asian security. His current research interests include international relations theory, Asian security, and Chinese foreign relations.

Joseph Y. S. Cheng is chair and professor of political science at the City University of Hong Kong and the founding editor of the *Hong Kong Journal of Social Sciences* as well as the *Journal of Comparative Asian Development*.

Lowell Dittmer is professor of political science at the University of California at Berkeley and editor of *Asian Survey*. He has authored four books, coauthored two books, coedited three books, and written many chapters on various aspects of Chinese foreign and domestic policy. His most recent works include *Informal Politics in East Asia* (coeditor, 2000), *Liu Shaoqi and the Chinese Cultural Revolution* (revised edition, 1997), *Chinese Politics under Reform* (1993), *China's Quest for National Identity* (coeditor, 1993), and *Sino-Soviet Normalization and Its International Implications* (1992).

Ho Khai Leong is senior lecturer at the department of political science, National University of Singapore (NUS). He has also taught in the Master of Public Policy (MPP) program at NUS. He specializes in public policy theories, comparative governance, and political economy in the Asia-Pacific region. He has published in journals such as *The Asian Journal of Political Science*, *The Pacific Review*, *The Asian Journal of Public Administration*, *Pacific Focus*, *Southeast Asian Affairs*, *Asian Survey*, and *Asian Perspective*. His latest publication is *The Politics of Policy-making in Singapore* (Singapore: Oxford University Press, 2000). He was a visiting fellow at the Japan Institute of International Affairs, Japan, in 1992. He received his Ph.D. from Ohio State University in 1988.

Lau Siu-kai is professor of sociology at the Chinese University of Hong Kong. Currently on leave from the university, he is head of the Central Policy Unit of the Hong Kong Special Administrative Region government. Professor Lau's research interests are Hong Kong's social and political development as well as comparative politics. He is the author of *Society and Politics in Hong Kong* (Hong Kong: Chinese University Press, 1982) and coauthor (with Kuan Hsin-chi) of *The Ethos of the Hong Kong Chinese* (Hong Kong: Chinese University Press, 1988), and many articles published in international journals.

Rex Li is senior lecturer in international relations at Liverpool John Moores University and an associate editor of *Security Dialogue* (International Peace Research Institutes, Oslo/ Sage, London). In recent years, he has been a visiting lecturer at the Joint Services Command and Staff College, UK Defence Academy. He is regularly interviewed by the BBC World Service commenting on East Asian and international security affairs. From 1993 to 1995 he was coorganizer of the British Pacific Rim Research Group's seminar series funded by the UK Economic and Social Research Council. He has research interests in international relations theory, Asia-Pacific security, Chinese security perceptions and policy, U.S.-China relations, and China-Taiwan relations. He is coeditor of *Fragmented Asia: Regional Integration and National Disintegration in Pacific Asia* (1996) and *Dynamic Asia: Business, Trade and Economic Development in Pacific Asia* (1998). His recent work has appeared in *The Journal of Strategic Studies*, *Security Dialogue*, *The World Today*, *World Defence Systems*, *Contemporary Politics*, *Journal of Contemporary China*, *Pacifica Review*, *Asia Pacific Business Review*, and several edited books published by Routledge, Palgrave, and Lynne Rienner.

Liu Ji was vice-president of the Chinese Academy of Social Sciences in Beijing.

Jonathan D. Pollack is chairman of the Strategic Research Department and chair of the Asia-Pacific Studies Group at the Naval War College, Newport, Rhode Island. He has published extensively on China's political and strategic roles, the international politics of Asia, U.S. Pacific strategy, and Chinese technological and military development. His current research includes Chinese security policy debate since September 11, alternative U.S.-Chinese strategic futures, future policy options toward North Korea, and leadership of a joint U.S.-Japanese strategic assessment of longer-term trends in regional and global security.

Andrew Scobell is a specialist on Asian politics and security in the Strategic Studies Institute at the U.S. Army War College in Carlisle, Pennsylvania, and adjunct professor of political science at Dickinson College. Prior to these current positions, he was associate professor of political science at the University of Louisville. He is the author of *China's Use of Military Force: Beyond the Great Wall and the Long March* (2003).

Wu Xinbo is a professor at the Center for American Studies, Fudan University, and the vice-president, Shanghai Institute of American Studies. He teaches China-U.S. relations and writes widely about China's foreign policy, Sino-American relations, and Asia-Pacific issues. He is the author of *Dollar Diplomacy and Major Powers in China, 1909–1913* (Shanghai, 1997) and has published numerous articles and book chapters in China, the United States, Japan, Germany, South Korea, and India. He is also a frequent contributor to China's and international newspapers. He earned his Ph.D. in international relations from Fudan in 1992. He spent one year at George Washington University as a visiting scholar in 1994, was a visiting fellow at the Asia-Pacific Research Center, Stanford University, and the Henry Stimson Center in Washington, DC, in the fall of 1997, and was a visiting fellow at the Brookings Institution from January to August 2000.

Yao Yunzhu is a colonel at the Academy of Military Science of the People's Liberation Army, in Beijing, China.

Jing-dong Yuan is a senior research associate at the Center for Nonproliferation Studies, and teaches Chinese Politics and Northeast Asian security and arms control at the Monterey Institute of International Studies in California. He received his Ph.D. in political science from Queen's University in 1995 and has had research and teaching appointments at York University, the University of Toronto, and the University of British Columbia. His research focuses on Asia-Pacific security, global and regional arms control and nonproliferation issues, and China's defense and foreign policy. He is the coauthor of *China and India: Cooperation or Conflict?* (2003) and has published articles in *Asian Survey, Contemporary Security Policy, International Herald Tribune, Jane's Intelligence Review, Los Angeles Times*, and *Nonproliferation Review*, among others.

Zhang Junbo is a major general and president of the Political Academy of the People's Liberation Army in Beijing, China.

Zhang Wankun is associate head, Department of Public Administration, School of Management, Shengzhen University, China. His research areas include Chinese foreign policy, international relations theories, and strategic culture.

Zi Zhongyun is a senior fellow and former director of the Institute of American Studies at the Chinese Academy of Social Sciences, and editor-in-chief of *Meiguo Yanjiu* (Journal of American Studies in China). She has held the positions of guest professor at Peking University and Nanjing University–Johns Hopkins University Center for Chinese and American Studies. She is the president of the Society for Chinese Scholars of Sino-U.S. Relations, member of the Executive Committee of The Chinese Association of American Studies, and member of the Standing Committee of Society for Pacific Affairs. Professor Zi has been extensively published and is the author of several books and a series of articles about international relations, foreign policy, and East-West culture. She has won the title of Scholar of Outstanding Contribution of the Nation in China.

Index